The Animal Kingdom

VOLUME 13:
THE CLASSES ANNELIDA,
CRUSTACEA, AND ARACHNIDA

GEORGES CUVIER
EDITED AND TRANSLATED BY
EDWARD GRIFFITH
WITH EDWARD PIDGEON

CAMBRIDGE
UNIVERSITY PRESS

CAMBRIDGE UNIVERSITY PRESS

Cambridge, New York, Melbourne, Madrid, Cape Town,
Singapore, São Paolo, Delhi, Mexico City

Published in the United States of America by Cambridge University Press, New York

www.cambridge.org
Information on this title: www.cambridge.org/9781108049665

This edition first published 1833
This digitally printed version 2012

ISBN 978-1-108-04966-5 Paperback

CAMBRIDGE LIBRARY COLLECTION

Books of enduring scholarly value

Life Sciences

Until the nineteenth century, the various subjects now known as the life sciences were regarded either as arcane studies which had little impact on ordinary daily life, or as a genteel hobby for the leisured classes. The increasing academic rigour and systematisation brought to the study of botany, zoology and other disciplines, and their adoption in university curricula, are reflected in the books reissued in this series.

The Animal Kingdom

Georges Cuvier (1769–1832), made a peer of France in 1819 in recognition of his work, was perhaps the most important European scientist of his day. His most famous work, Le Règne Animal, was published in French in 1817; Edward Griffith (1790–1858), a solicitor and amateur naturalist, embarked in 1824, with a team of colleagues, on an English version which resulted in this illustrated sixteen-volume edition with additional material, published between 1827 and 1835. Cuvier was the first biologist to compare the anatomy of fossil animals with living species, and he named the now familiar 'mastodon' and 'megatherium'. However, his studies convinced him that the evolutionary theories of Lamarck and St Hilaire were wrong, and his influence on the scientific world was such that the possibility of evolution was widely discounted by many scholars both before and after Darwin. Volume 13 covers annelids, crustaceans and arachnids.

THE

ANIMAL KINGDOM

ARRANGED IN CONFORMITY WITH ITS ORGANIZATION,

BY THE BARON CUVIER,

MEMBER OF THE INSTITUTE OF FRANCE, &c. &c. &c.

WITH

SUPPLEMENTARY ADDITIONS TO EACH ORDER,

BY

EDWARD GRIFFITH, F.L.S., A.S.,

CORRESPONDING MEMBER OF THE ACADEMY OF NATURAL SCIENCES OF PHILADELPHIA, &c.

AND OTHERS.

VOLUME THE THIRTEENTH.

LONDON:

PRINTED FOR WHITTAKER, TREACHER, AND CO.

AVE-MARIA-LANE.

MDCCCXXXIII.

THE

CLASSES

ANNELIDA, CRUSTACEA,

AND

ARACHNIDA,

ARRANGED BY THE

BARON CUVIER,

WITH

SUPPLEMENTARY ADDITIONS TO EACH ORDER

BY

EDWARD GRIFFITH, F.L.S., A.S., &c.

AND

EDWARD PIDGEON, Esq.

LONDON:
PRINTED FOR WHITTAKER TREACHER, AND CO.
AVE-MARIA-LANE.

MDCCCXXXIII.

LIST OF PLATES.

ANNELIDA.

CRUSTACEA.

ARACHNIDA.

THIRD GRAND DIVISION

OF THE

ANIMAL KINGDOM.

THE ARTICULATED ANIMALS.

THIS third general form is altogether as characteristic as that of the vertebrated animals; the skeleton is not interior, as in the latter, but it is not reduced to a nullity, as in the mollusca. The articulated rings which surround the body, and frequently the limbs too, supply its place; and as they are almost always sufficiently hard, they can afford all the necessary points of support for locomotion; so that we find in this division, as in the vertebrated, the different movements of walking, running, leaping, swimming, and flying. It is only the families destitute of feet, or whose feet have only soft and membranaceous articulations, which are confined to the movement of reptation. This external position of the hard parts, and that of the muscles in their interior, reduce each articulation to the form of a case, and allow it only two kinds of motion. When it is attached to the neighbouring articulation by a firm juncture, which happens in the limbs, it is fixed there by two points, and can never move, only by ginglymus, that is to say, on a single plane, which requires more numerous articulations to produce the same variety of motion. From this also results a greater loss of force in the muscles, and consequently more general weakness in each animal, in proportion to its size.

But the joints which compose the body have not always this sort of articulation: they are most usually united merely by flexible membranes, or emboxed one within the other; and then their motions are more varied, but do not possess the same force.

The system of organs on which the articulated animals are most similar to each other, is that of the nerves.

Their brain, situated on the œsophagus, and furnishing nerves to the parts which adhere to the head, is very small. Two cords, which embrace the œsophagus, are continued along the belly, uniting, from space to space, into double knots or ganglia, from which proceed the nerves of the body and of the limbs. Each of these ganglia performs the functions of a brain, for the surrounding parts, and suffices to preserve their sensibility for a certain time, when the animal has been divided. If we add to this, that the jaws of these animals, when they have any, are always lateral, and move from without to within, and not from top to bottom, and that there has not yet been discovered in any of them a distinct organ of smell, we shall have pretty nearly expressed all that may be said of them in general. But the existence of the organs of hearing; the existence, number, and form of those of sight; the product, and the mode of generation *; the nature of respiration; the existence of the organs of circulation; and even the colour of the blood, present great variations, which it will be necessary to study in the different sub-divisions.

* A remarkable discovery on this subject is that of M. Herold, that in the egg of the crustacea and arachnida, the vitellus communicates with the interior by the back. See his dissertation on the egg of spiders, Marburg, 1824, and that of M. Rathke on the egg of the Astaci, Leipzig, 1829.

Distribution of the Articulated Animals into Four Classes.

THE articulated animals, whose inter-relations are equally numerous and varied, nevertheless present themselves under four principal forms, whether internal or external.

The ANNELIDA, or RED-BLOODED WORMS, constitute the first. Their blood, generally of a red colour, like that of vertebrated animals, circulates in a double and closed system of arteries and veins, which has sometimes one or many hearts or fleshy ventricles, sufficiently marked. They respire by organs which are sometimes developed externally, and sometimes remain at the surface of the skin, or sink into its interior. Their body, more or less elongated, is always divided into numerous rings, the first of which, named the head, is scarcely different from the others, except in the presence of the mouth, and the principal organs of sensation. Many have their gills uniformly distributed along their body or its middle; others, which generally inhabit tubes, have them all at the anterior part. Those animals never have articulated feet, but the greater number, instead of feet, have *setæ*, or bundles of stiff and moveable hairs. They are generally hermaphrodites, and some of them have need of a reciprocal intercourse. The organs of their mouth consist sometimes of jaws, more or less strong, sometimes of a simple tube; those of the external senses, of fleshy tentacula, and sometimes articulated; and of some blackish points, which have been regarded as eyes, but which do not exist in all the species.

The CRUSTACEA constitute the second form or class of articulated animals. They have articulated, and more or less complicated limbs, attached to the sides of the body; their blood is white: it circulates by means of a fleshy ventricle, placed in the back, which receives it from the gills, situated on the sides of the body, or under its posterior part, and whither it returns by a ventral canal, which is sometimes double. In the final species, the heart, or dorsal ventricle, is itself elongated, like a canal. All these animals have antennæ, or articulated filaments, attached to the fore-part of the head, generally four in number, many transverse jaws, and two compound eyes. It is only in some few species that a distinct ear is to be found.

The third class of articulated animals is that of the ARACHNIDA, which, like a great number of crustacea, have the head and thorax united in a single piece, with articulated limbs on each side, but the principal viscera are enclosed in an abdomen, attached to the hinder part of this thorax. Their mouth is armed with jaws, and the head provided with simple eyes, varying in number; but they never have any antennæ. Their circulation is performed by a dorsal vessel, which sends forth arterial branches, and receives venous ones from them; but their mode of respiration varies, some of them having true pulmonary organs at the sides of the abdomen, and others receiving the air through tracheæ, like the insects. Both, however, have lateral apertures—true stigmata.

The INSECTS are the fourth class of articulated animals, and at the same time the most numerous of the whole animal kingdom. Some genera excepted, such as the Myriapoda, where the body is divided into a great number of nearly equal articulations, the insects have it divided into three parts; the head provided with antennæ, eyes, and mouth; the thorax, or corslet, to which the feet are attached, and wings, where there are any; and the abdomen, which is suspended behind the thorax, and encloses the principal viscera. The insects which have wings do not receive them until a certain age, and often pass through two forms, more or less different, before arriving at the perfect state of the winged insect. In all their states they breathe by tracheæ, that is, by elastic vessels which receive the air through stigmata, pierced on the sides, and distribute it, by infinite ramifications, into all parts of the body. There is but a vestige of a heart, which is a vessel attached along the back; it undergoes alternate contraction, but no branches to it have been discovered, so that we must believe that the nutrition of the parts takes place by imbibition. It is probably this sort of nutrition which has necessitated the sort of respiration proper to insects, because the nutritive fluid not being contained in vessels *, and not capable of being directed to circumscribed pulmonary organs, in search of air, it was necessary that the air should be spread through the whole body, to reach the fluid. This is also the reason why insects have no secretory glands, but only long spongy vessels, which appear to absorb, by their great surface, in the mass of nutritive fluid, the proper juices which they should produce †.

The insects vary infinitely in the forms of their organs of the mouth and of digestion, as well as in their industry and mode of living. Their sexes are always separate.

The crustacea and arachnida were for a long time united with the insects under one common name, and they resemble them in many respects, both in external form, and in the disposition of the organs of motion, of sensation, and even of manducation.

* M. Carus has recognized some regular movements in the fluid which fills the body of the larvæ of certain insects; but these movements do not take place in a close system of vessels, as in the superior animals. See his treatise, intitled, "Discovery of a Simple Circulation of the Blood," &c., in German, Leipzig, 1827, 4to.

† See on this subject, my Memoir on the nutrition of insects, printed in 1799, in the M. de la Soc. d'Hist. Nat., &c. Paris, Baudouin. An. vii. 4to. p. 32.

FIRST CLASS OF ARTICULATED ANIMALS.

THE ANNELIDA*,

ARE the only invertebrated animals with red blood; it circulates in a double system of complicated vessels †.

Their nervous system consists in a double knotted cord, like that of the insects.

Their body is soft, more or less elongated, and often formed into a very considerable number of segments, or at least of transverse folds.

Almost all of them live in the water (the earth-worms or lumbrici excepted); many bury themselves in holes in the bottom, or form tubes there with the ooze, or other substances, or even exude a calcareous matter, which produces a sort of tubular shell.

* I established this class, distinguishing it by the colour of its blood and by other attributes, in a memoir read at the Institute in 1802. See Bullet. des Sc. Messidor. An. X. where I have principally described the organs of circulation. M. Lamarck adopted, and named it *Annelida,* in the extract from his course of Zoology, 1812. Bruguières had previously joined it to the order of intestinal worms; and Linnæus, more anciently still, had placed a part of it among the mollusca, and another with the intestinal worms.

† It has been asserted, that the Aphroditæ had not red blood. I think that I have observed the contrary in the *Aphrodita Squamata.*

Division of the Annelida into Three Orders.

THIS not very numerous class, presents in its respiratory organs, the bases of satisfactory divisions.

Some have gills in the form of plumes, or arbusculæ, attached to the head, or to the anterior part of the body. Almost all of them inhabit tubes. These we call TUBICOLÆ.

Others have on the middle part of the body, or along the sides, gills in the form of trees, tufts, laminæ, or tubercles, in which the vessels ramify. Most of them live in the mud or ooze, or swim in the sea. The smaller number have tubes. We name them DORSIBRANCHIA.

Others, in fine, have no apparent gills, and respire, either by the surface of the skin, or as some believe, through internal cavities. Most of them live freely in water or mud; some only in humid earth. We call them ABRANCHIA.

The genera of the first two orders have all stiff hairs, or bristles, and of a metallic colour, issuing from their sides, sometimes simple, sometimes in bundles, and supplying the place of feet. But in the third order there are some genera destitute of such support *.

The special study which M. Savigny has devoted to those feet, or organs of locomotion, has caused him to distinguish, 1. The foot itself, or the tubercle which supports the bristles: sometimes there is but one to each ring; sometimes there are two, one above the other; this is what is named simple or double oar. 2. The bristles which compose a bundle for

* M. Savigny has proposed a division of Annelida, according to their having bristles for locomotion or not; these last being reduced to the leeches. M. de Blainville, who has adopted this idea, makes of the *Annelida* which have bristles his class of *Entomozoa chetopoda,* and of those which have none, that of *Entomozoa apoda;* but (what M. Savigny did not do) he intermingles, among the apoda, several of the intestinal worms.

each oar, and vary much in form and consistence, sometimes forming true spines, sometimes fine and flexible hairs, often denticulated, barbed, &c.* 3. The cirri, or fleshy filaments adherent either above or beneath the feet.

As to their organs of the senses, the annelida of the first two orders, have generally, at the head, tentacula, or filaments, to which, notwithstanding their fleshy consistence, some moderns give the name of antennæ, and many genera of the second and third have black and shining points, which, with some reason, have been regarded as eyes. The organization of their mouth varies greatly.

* See, on this subject, the memoirs of M. Savigny, on the invertebrated animals; and those of MM. Audouin, and Milne Edwards, on the *Annelida*.

FIRST ORDER OF ANNELIDA.

THE TUBICOLÆ *.

SOME form a calcareous homogeneous tube, resulting probably from their transudation, like the shell of the mollusca, to which, however, they do not adhere by muscles. Others construct it by agglutinating grains of sand, fragments of shells, and particles of mud, by means of a membrane, which is doubtless also transuded. There are some, in fine, whose tube is almost entirely membranous, or horny. To the first of these belong,

SERPULA, *Lin.*,

Whose calcareous tubes cover, by twisting round them, stones, shells, and all submarines bodies. The section of these tubes is sometimes round, and sometimes angular, according to the species.

The body of the animal is composed of a great number of segments. Its anterior part is spread into a disc, armed on each side with several parcels of coarse hairs, and on each

* M. Savigny, joining the *Arenicolæ* to this order, changes the name into *Serpulaceæ*. M. de Lamarck, adopting the same arrangement, changes the name *Serpulacea* into *Sedentaria*. My genera of *Tubicolæ* are M. Savigny's family of *Amphitrite*. With M. de Lamarck, they compose those of *Amphitritea* and *Serpulacea*. With them, M. de Blainville forms his order of *Entomozoa chetopoda heterocrisina*; but contrary to his own definition he has introduced there *Spio* and *Polydorus*.

side of its mouth is a plume of gills in the form of a fan, usually tinted with lively colours. At the base of each plume is a fleshy filament, one of which, either on the right side, or the left, indifferently, is always elongated, or dilated at its extremity, into a disc, variously configurated, which serves as an operculum, and closes the aperture of the tube, when the animal has retired into it *.

The common species *Serpula contortuplicata* †, Ell. Corall. xxxviii. 2., has round tubes twisted, and three lines in diameter. Its operculum is funnel-shaped, and its gills often of a fine red, or varied with yellow, or violet. It speedily covers with its tubes, vases, and other objects, thrown into the sea.

We have, on our coasts, a smaller species, with a claviform operculum, armed with two or three small points. (*Serp. vermicularis*, Gm.) Müll. Zool. Dan. lxxxvi. 7. 9. &c. The gills are sometimes blue. Nothing is more agreeable to see than a group of these serpulæ, when they are well expanded.

In others, the operculum is flat, and bristled with more numerous points ‡.

There is one in the Antilles (*Serpula gigantea*, Pall. Miscell. x. 2. 10.) which sojourns among the madrepores, and whose tube is often surrounded by their masses. Its gills are spirally rolled when they re-enter the tube, and its operculum is armed with two small, branching horns, like the antlers of a deer §.

* The most common Serpula having this disc in the form of a funnel, naturalists have taken it for a proboscis, but it is not pierced, and the other species have it more or less claviform.

† It is the same animal as the *Amphitrite penicillus*, Gm. or *Proboscidea*, Brug. *Probosci-plectanas*, Fab. Columm. Aquat., c. xi. p. 22.

‡ These are the *Galeolaria*, Lam. An operculum is visible on them, Berl., Schr. IX. iii. 6.

§ The same as *Terebella bicornis*, Abildg., Berl., Schr., IX. iii. 4. Seb.

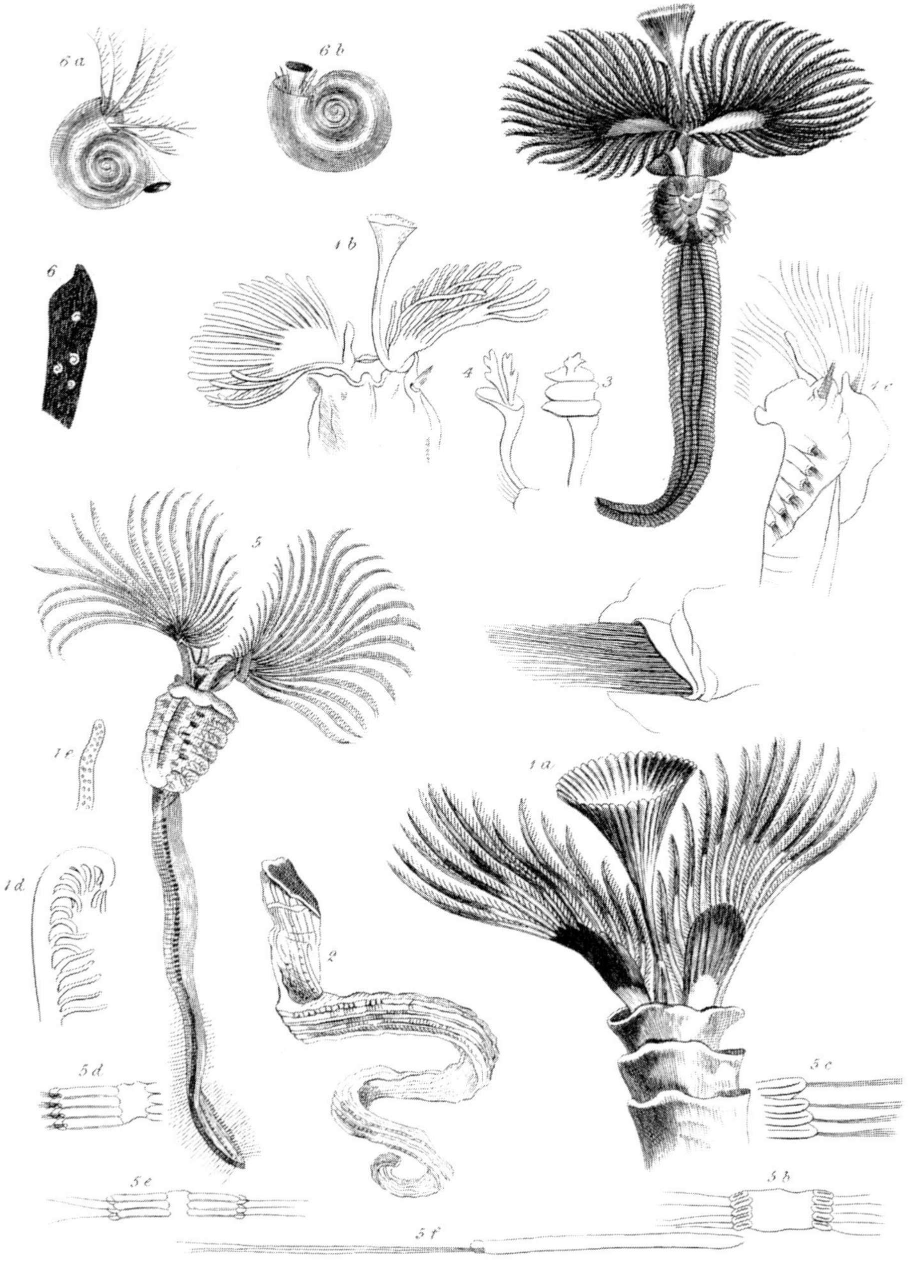

1 *Serpula contortuplicata.*
2 *Tube of Serpula costalis.*
3 *Operculum of Serp. stellata.*
4 *Operculum of Serp. bicornis.*
5 *Sabella protula.*
6 *Spirorbis nautiloides.*

London, Published by Whittaker & C°. Ave Maria Lane. 1832.

M. de Lamarck distinguishes:

SPIRORBIS, *Lam.*, in which the branchial threads are much less numerous, three or four on each side; their tube is a tolerably regular spiral, and they are usually very small *

SABELLA, *Cuv.*, †

Have the same kind of body, and the same fan-like gills as the Serpulæ. But the two fleshy filaments adherent to the gills both terminate in a point, and form no operculum; they are sometimes even entirely wanting. Their tube most frequently appears composed of grains of clay, or very fine mud, and is rarely calcareous.

The known species are pretty large, and their branchial plumes are delicate and brilliant in the extreme.

Some of them, like the Serpulæ, have a membranous disc on the anterior portion of the back, through which pass the first pairs of bundles of hairs. Their branchial combs are spirally convoluted, and their tentacula are reduced to slight folds ‡.

III. xvi. 7, and the *Actinia* or *Animal-flower*, Hom. Lect. on Comp. Anat. II. pl. I. On this spiral convolution of the gills, M. Savigny establishes his subdivision of *Serpulæ Cymospiræ*, of which M. de Blainville has since made a genus. Add. *T. Stellata*, Gm. Abildg., loc. cit., f. 5. remarkable for an operculum formed of three plates strung together.

* *Serpula Spirillum*. Pall. nov. act. Petrop. V. pl.V. f. 21. *Serp. Spirorbis*. Mull. Zool. Dan. III. lxxxvi. 1—6.

† This name of *Sabella* designates in Linnæus and Gmelin various animals with factitious, and not transuded tubes. We confine it to those which resemble each other in their proper characters. M. Savigny has employed it as we have, with the exception of our first division, which he places among his Serpulæ. M. de Lamarck calls our Sabellæ, Amphitritæ.

‡ M. Savigny leaves this division among the Serpulæ, and makes of it his *Serpulæ Spiramellæ*, of which M. de Blainville has since made his genus *Spiramella*.

The Mediterranean possesses a fine and large species, with a calcareous tube, like that of the Serpulæ, with orange gills, &c. *Sabella protula*, Nob., or *Protula Rudolphii* Risso *.

In others, there is no membranous disc in front; their branchial combs form two equal spirals †.

There are sometimes two ranges of filaments on each comb ‡.

In others again, one of the two combs only is thus formed; the other, which is smaller, envelopes the base of the first, *Sabella unispira*, Cuv.; *Spirographis Spallanzanii*. Viviani. Phosph. Mar. pl. IV. and V. §

There are some whose gills only form round the mouth a simple funnel, but with numerous filaments, crowded, and strongly ciliated on the internal surface. Their hairy feet are almost imperceptible; *Sabella villosa*, Cuv. New species.

Some, in fine, have been described, which have but six filaments, arranged like a star. *Tubularia Fabricia*, Gm. Fabr. Grœul. p. 450; it is the genus *Fabricia*, of Blainville.

TEREBELLA, *Cuv.*,

Inhabit, like most of the Sabellæ, a factitious tube; but it is composed of grains of sand and fragments of shells. More-

* The existence of this magnificent species, and the calcareous nature of its tube, are incontestible, notwithstanding the doubt expressed. Dict. des Sc. nat. lvii. p. 432, note. The *Sabella bispiralis* (*Amphritrite volutacornis*, Trans. Linn. VII. vii.) differs very little from it. I dare not affirm that it is the same as Seb. I. xxix. 1. erroneously cited by Gmelin and Pallas, under *Serpula gigantea*, for this figure exhibits no disc.

† These are the simple Sabellæ of Savigny. *Amphitrite reniformis*, Mull. Ven. xiv. or *Tubularia penicillus*, id. Zool. lxxxix. 1. 2. or *Terebella reniformis*, Gm.; *Amphitrite infundibulum*, Montag., Lin. Trans., IX. viii.; *Amph. vesiculosa*, id. ib. XI. v.

‡ These are the *Sabellæ Astartæ*, Sav., such as *Sabella grandis*, Cuv., or *Indica*, Sav.; *Tubularia magnifica*, Shaw, Linn., Trans. V. ix.

§ These are the *Sabellæ Spirographicæ* of Savigny.

over, their body has much fewer rings, and their head is differently ornamented. Numerous filiform tentacula, capable of considerable extension, surround the mouth; and on the neck are gills in the form of arbuscula, and not fan-shaped.

Linnæus, in his twelfth edition, had given this name to an animal described by Kaehler, and which might belong to this genus, because it was supposed that it pierces stones. M. de Lamarck has employed this name, (An. sans Vert. p. 324.) for a *Nereis*, and for a *Spio*. The *Terebellæ* of Gmelin, comprehend *Amphinomæ*, *Nereides*, *Serpulæ*, &c. At the present day, MM. Savigny, Montagu, Lamarck, and Blainville, employ this name as I do, and as I had proposed, Dict. des Sc. Nat. ii. p. 79.

We have several of them on our coasts, a long time comprehended under the name of *Terebella Conchilega*, Gm. Pall. Miscell. ix. 14—22, and for the most part remarkable for tubes formed of thick fragments of shells, and the edges of their aperture elongated into several small branches formed of the same fragments, and serving to lodge the tentacula.

The greater number have three pairs of gills, which in those whose tube has branches, issue through a hole destined for that purpose.

These are the simple terebellæ of M. Savigny, such as *Tereb. medusæ*. Sav., Eg., Annal., I. f. 3; *Ter. cirrhata*, Gm. Mull., Ver. xv.; *Ter. gigantea*, Montagu., Linn., Trans. XII. ii.; *T. Nebulosa*, Id. Ibid. 12. 2.; *T. constrictor*, Id. Ibid. 13. 1.; *T. venusta.*, ibid. 2.; he also names a *T. cirrhata*, ibid. xii. 1., but which does not appear to be the same as that of Muller. Add *T. variabilis*, Risso, &c.

N.B. M. Savigny has two other divisions of Terebellæ, his *T. Phyzellæ*, which have but two pairs of gills; and his *T. Idaliæ*, which have but one. Among these last would come *Amphitrite cristata*, Mull., Zool., Dan., LXX. i. 4.; *Amph. ventricosa*, Bosc. Vers. I. vi. 4—6.

AMPHITRITE, *Cuv.*,

Are easily recognized by bristles of a golden colour, arranged like combs, or a crown, in one or more rows, on the anterior part of the head, where they probably serve as a defence, or, perhaps, as a means of crawling, or gathering the materials for their tube. Around the mouth are very numerous tentacula, and on the commencement of the back, on each side, gills in the form of combs.

This genus, such as it is in Muller, Bruguières, Gmelin, and Lamarck, also comprehends the *Terebellæ* and *Sabellæ*. I reduced it, in 1804, to its present limits. (Dict. des Sc. Nat. ii. p. 78.) Since then, M. de Lamarck has changed my divisions into genera, his *Pectinariæ* and *Sabellariæ*, which M. Savigny calls *Amphictenæ* and *Hermellæ*. The name *Amphitrite* has been transferred by M. de Lamarck, to my *Sabellæ*. M. Savigny, on the contrary, makes it the name of a family.

Some of them compose light tubes, in the form of regular cones, which they carry along with them. Their gold-coloured bristles form two combs, whose teeth are directed downwards. Their very ample and multifold intestine is usually filled with sand.

These are the *Pectinariæ*, Lam.; the *Amphictenæ* of Savigny: the *Chrysodontes* of Aken; and the *Cistenæ* of Leach. This perpetual changing of names—and in the present case there was not even the pretext of a change of limits in the group—will end by rendering the study of nomenclature much more difficult than that of facts.

On our coasts, we have, of this division, *Amphitrite auricoma Belgica*, Gm., Pall., Miscell. ix. 3—5, whose tube, two inches long, is formed of small round grains of divers colours. It is the same as the *Sabella Belgica*, Gm., Klein., tab. i. 5. Echinod. xxxiii. A. B., and as *Amph. auricoma*, Mull.,

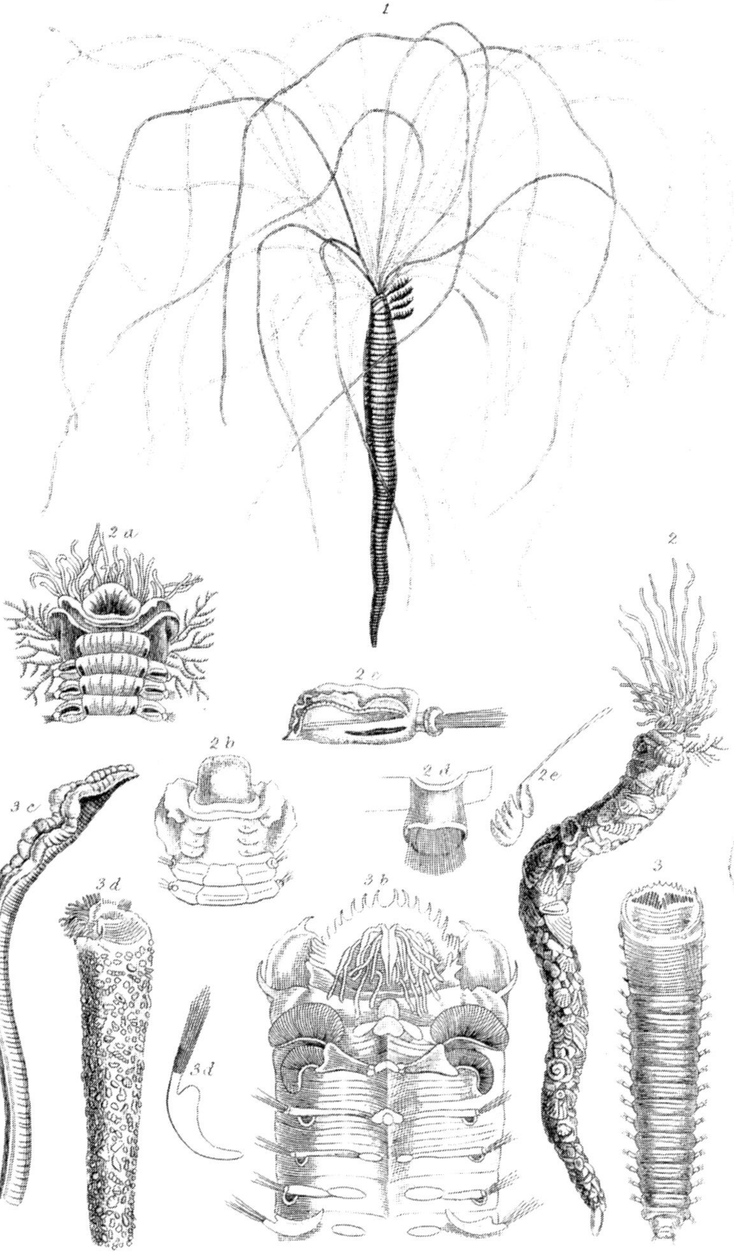

1 *Terebella variabilis.*

2 *Ter. medusa.*

3 *Amphitrite Ægyptia.*

London. Published by Whittaker & C° Ave Maria Lane 1832.

Zool., Dan. xxvi., of which Bruguières has made his *Amphitrite dorée.*

The South Sea produces a larger species, *Amphitrite auricoma Capensis,* Pall., Miscel. ix. 1, 2, whose tube, thin and polished, has the appearance of being transversely fibrous, and formed of some dried, soft and stringy substance. It is the same as *Sabella chrysodon,* Gm., Bergius Mem. de Stockh. 1765, ix. 1, 3.; as *Sabella capensis,* Id. Stat, Mull., Nat. Syst. VI. xix. 67, which is only a copy of Bergius; as *Sabella Indica,* Abildgaart., Berl., Schr. IX. iv. See also Mart., Slabber., Mem. de Flessing. I. ii. 1—3.

Other Amphitritæ inhabit factitious tubes, fixed to submarine bodies. Their gilded bristles form on their heads many concentric crowns, from which results an operculum, which closes their tube when they contract themselves in it, and the two parts of which may be separated. They have a cirrus to each foot. Their body is terminated behind by a tube bent towards the head, without doubt to emit the excrements. I have found a muscular gizzard in them. They are the *Sabellariæ* of Lamarck, and the *Hermellæ* of Savigny.

Such is the species found along our coasts, *Sabella alveolata,* Gm., *Tubipora arenosa,* Linn. Ed. xii. Ellis. Corall. xxxvi., in which the tubes, united to each other in a compact mass, exhibit their orifices, pretty regularly arranged, somewhat like the cells of bees.

N. B. It is here, perhaps, that the *Amphitrite plumosa* of Fab., Faun., Grœn. p. 288; and Mull., Zool., Dan. xc., ought to come. But the descriptions are so obscure, and agree so little together, that I dare not place it. M. de Blainville makes of it his genus *Pherusa.*

Another, *Amph. ostrearia,* Cuv., establishes its tubes on the shells of oysters, and is said to injure their propagation very much.

I suspect that it is to this order that we must refer,—

SYPHOSTOMA, *Otto*,

Which have at each articulation, superiorly, a bundle of fine hairs, inferiorly, a simple *seta*, or bristle, and at the anterior extremity two parcels of strong and gilded bristles. Under these bristles is the mouth, preceded by a sucker, surrounded with many soft filaments, which may perhaps be gills, and accompanied by two fleshy tentacula. The knotted medullary cord is observable through the skin of the belly. They live embedded in mud.—*Siphostoma diplochaites,* Otto; *Siph. uncinata,* Aud. et Edw. Littoral. de la France, Annél. pl. ix. fig. 1.

Hitherto has always been placed in this vicinity—

DENTALIUM,

Which have a shell like an elongated cone, arched, open at the two ends, and which has been compared to an elephant's tusk in miniature. But the recent observations of M. Savigny, and especially of M. Deshayes, render this classification very doubtful. See "Monographie du Genre *Dentalė*, Mem. de la Soc. d'Hist. Nat. de Paris," t. ii. p. 321.

This animal does not appear to have any sensible articulation, nor lateral *setæ*; but it has a membranous tube in front, in the interior of which is a sort of foot, or fleshy and conical operculum, which closes its orifice. On the base of this foot is a small and flatted head, and gills in the form of plumes, and visible in the nape. If the operculum remind us of the foot of the *vermetæ* and *siliquariæ*, which have been placed among the mollusca, the gills strongly remind us of those of the Amphitritæ and Terebellæ. Ulterior observations on their anatomy, and principally on their nervous and vascular system, will resolve this problem.

Some of these have the shell angular, or longitudinally striated.—*Dent. Elephantinum,* Martini; I. i. 5. A.; *D.*

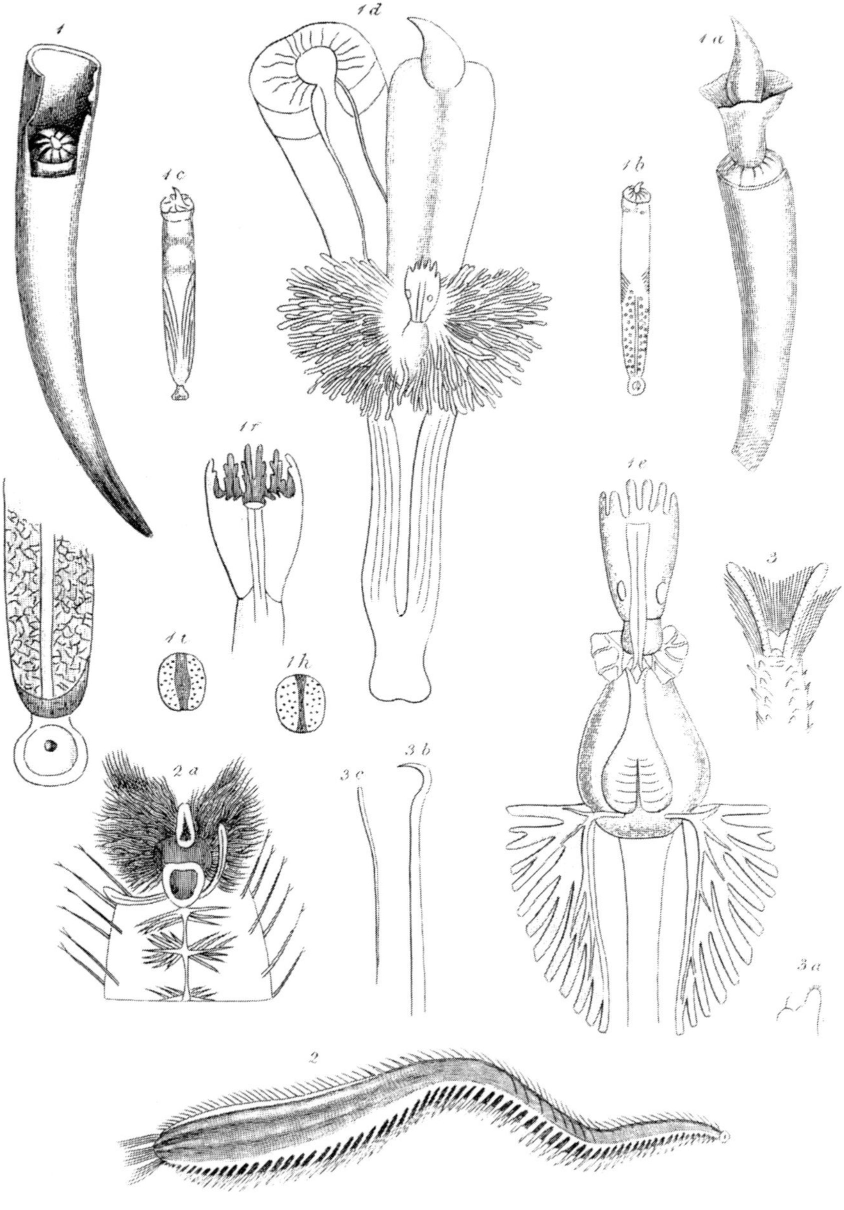

1 *Dentalium entalis*

2 *Siphostoma diplochaitos*

3 *Details of Siph. uncinata.*

London, Published by Whittaker & C.º Ave Maria Lane 1832.

aprinum, ib. 4. A.; *D. striatulum*, ib. 5. B.; *D. arcuatum*, Gualt. x. G.; *D. sexangulum. Dent. dentalis*, Rumph. Mus. xli. 6; *D. Fasciatum Martini*, Conch. i. 1. 3. B.; *D. rectum*, Gualt. x. H., &c.

Others have round shells: *Dent. entalis*, Martini, I., i. 1, 2, &c.

SECOND ORDER OF ANNELIDA.

THE DORSIBRANCHIA,

HAVE their organs, especially their gills, distributed pretty equally along their whole body, or at least along its middle portion.

We shall place at the head of the order, the genera whose gills are the most developed.

ARENICOLA, *Lam.*,

Have gills in the form of arbuscula, on the rings of the middle part of their body only; their mouth is a fleshy proboscis, more or less dilatable, and neither teeth nor tentacula, nor eyes are visible. The posterior extremity is not only destitute of gills, but also of the parcels of silky hairs, which furnish the rest of the body. No cirrhus exists to any ring of the body.

M. Savigny has made of this genus, a family which he names *Thelethusa*, and which has been adopted by his successors.

The known species, *Aren. piscatorum*, Lam., *Lumbricus marinus*, L. Pall. Nov. Act. Petrap. II. i., is very common on the sand on the sea-shore, where the fishermen go in search of it with spades, to use it as a bait. It is about a foot long, of a reddish colour, and diffuses, when it is touched, an abundant yellowish fluid. It has thirteen pairs of gills. Add.

Arenicol. clavata. Ranzani. dec. i. p. 6. pl. i. f. 1.; if indeed it be a distinct species.

AMPHINOME, *Brug.*,

Have on each of the rings of the body a pair of gills, in the form of a tuft or plume, more or less complicated, and to each of their feet two packets of separate bristles, and two cirrhi. Their proboscis has no jaws.

This genus has been justly withdrawn by Bruguières from the *Aphroditæ* of Pallas, and the *Terebellæ* of Gm. With M. de Savigny it is the type of a family, which he names *Amphinome*, also adopted by his successors.

M. Savigny divides them into

CHLOEIA, *Sav.*, which have five tentacula to the head, and the gills in the form of a tripinnate leaf.

The Indian Ocean produces one of them, the *Amphinome chevelue*, Brug.; *Terebella flava*, Gm., Pall., Miscel., viii. 7—11., extremely remarkable for its long bundles of lemon-coloured bristles, and the beautiful purple plumes of its gills. Its form is broad and depressed; it has a vertical crest on the snout.

And into

PLEIONE, *Sav.* AMPHINOME, *Blainv.*, which, with the same tentacula, have gills in the form of tufts. They are also of the Indian seas, and there are some very large; *Terebella caruncolata*, Gm.; *Aphr. car.* Pall., Miscell. viii. 12, 13; *Ter. rostrata*, ib. 14—18; *Ter. complanata*, ibid. 19—26; *Pleione alcyonia*, Sav., Eg. Annel. ii. f. 3.

He adds here, EUPHROSYNE, *Sav.*, which have but a single tentaculum in the head, and whose tree-like gills are very much developed and complicated, *Euphrosyne laureata*, Id., ibid. f. 1; *E. mirtosa*, id. ib. 2.

N.B. It is also near Amphinome that the genus ARISTENIA should come, *Sav.*, Eg. Annel. pl. ii. f. 4; but it is only established on a mutilated individual.

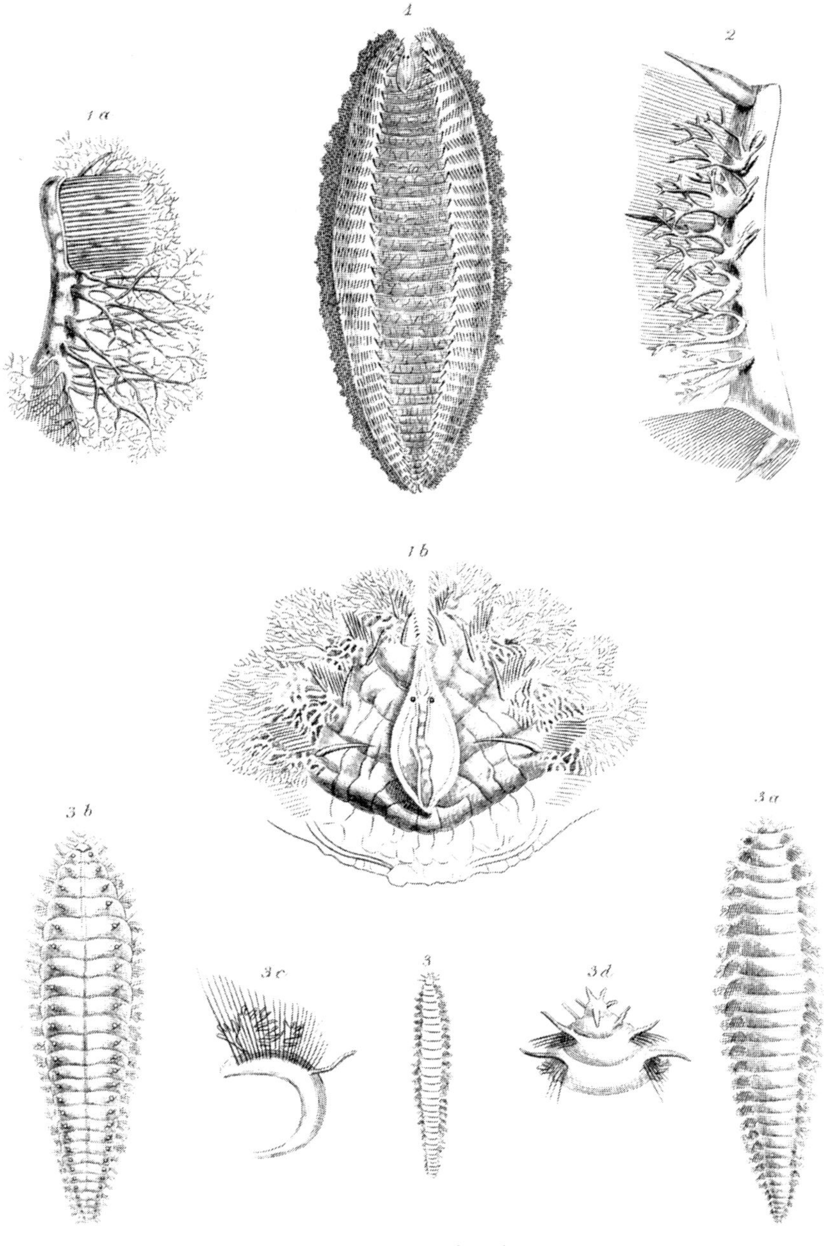

1 *Euphrosine laureata.*

2 *Gills of Euph. mirtosa.*

3 *Hipponoe Gaudichodii.*

London, Published by Whittaker & C° Ave Maria Lane, 1832.

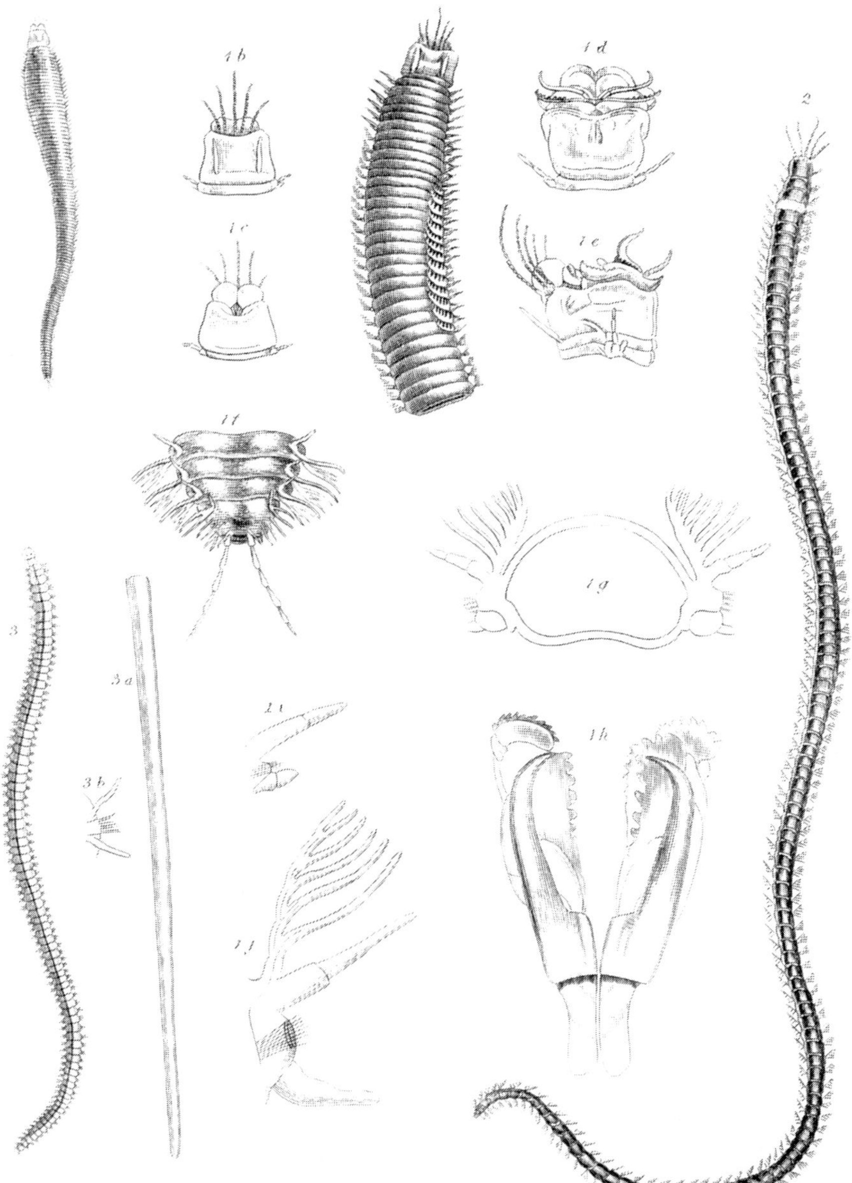

1 Eunice (Leodice Sav) antennata.

2 Eun. sanguinea.

3 Eun. tubicola.

London, Published by Whittaker & C.º Ave Maria Lane, 1832.

MM. Audouin and Edwards approximate to the amphinomæ, HIPPONÖE, which have no caruncle, and but a single bundle of bristles, and a single cirrhus to each foot.

There is a species belonging to Port Jackson, *Hipponöe Gaudichaudii*, Ann. des Sc. Nat. t. xviii. pl. vi.

EUNICE, *Cuv.*,

Have also gills in the form of plumes, but their proboscis is powerfully armed with three pairs of corneous jaws differently formed. Each of their feet has two cirrhi, and a bundle of bristles. Their head has five tentacula above the mouth, and two at the nape. In some species only there are two small eyes.

Eunice is the name of a *nereis* in Apollodorus. M. Savigny makes it the name of a family, and gives to the genus the name of LEODICE; M. de Blainville has changed these names, first into *Branchionereis*, and then into *Nereidon*.

The sea of the Antilles possesses one species of more than four feet long, *Eun. gigantea*, Cuv., which is the largest of the known annelida.

There are several on our coasts less considerable in size; such as *Nereis Norwegica*, Gm., Mull., Zool. Dan. I. xxix. 1; *N. pinnata*, ib. 2; *N. Cuprea*, Bosc., vers. I. v. 1.; *Leodice Gallica*, and *L. Hispanica*, Sav; *Leod. antennata*, Sav. Annel. pl. iii. fig. 1—4; *Eun. harassii*. ib. fig. 5—11.

M. Savigny distinguishes from them, under the name of MARPHISÆ, species otherwise very similar, but which want the two tentacula in the nape. Their upper cirrhus is very short; *Ner. sanguinea*, Montag., Linn. Trans. xl. pl. iii.

A species at least closely allied, *N. Tubicola*, Mull., Zool. Dan. I. xviii. 1—5, inhabits a corneous tube.

It is probably near Eunice that should come the *Nereis crassa*, Mull., Vers., pl. xii., which M. de Blainville, without having seen it, proposes to refer to the genus ETEONE of M.

Savigny, which nevertheless would appear to have gills altogether different.

After these genera with complicated gills, we may place those in which they are reduced to simple laminæ, or even to slight tubercles, or, in fine, in which the cirrhi alone supply their place.

There are some which still are related to Eunice by the strong armature of their proboscis, and by their antennæ being of odd numbers.

Such are,

LYSIDICE, *Sav.*, which with jaws similar to those of Eunice, or even more numerous, and often of odd numbers, have but three tentacula, and some cirrhi for gills, *Lysidice Valentina,* Sav.; *L. Olympia,* id.; *L. galatina,* id., Eg., Annel. p. 53.

AGLAURA, *Sav.*, have also numerous jaws, and of odd numbers, as seven, nine, &c.; but they want tentacula, or have them altogether concealed. Their gills are also reduced to the cirrhi.

I unite the AGLAURÆ and the ŒNONÆ of Savigny; and even certain species without tentacula, which MM. Audouin and Milne Edwards leave in the Lysidicæ, *Ag. fulgida,* Eg. Annel. v. 2; *Œnone lucida,* ib. f. 3.

NEREIS, (proper) *Cuv.*, LYCORIS, *Sav.*, have tentacula of even numbers, attached to the sides of the base of the head, and a little further forwards, two others, biarticulate, between which are two simple ones. They have but one pair of jaws in their proboscis; their gills only form small laminæ, over which is spread a net-work of vessels. There are, besides, to each foot two tubercles, two bundles of bristles, a cirrhus above, and another below.

We have a tolerable number of them on our coasts, *Nereis versicolor,* Gm., Mull., Wurm. vi.; *N. Fimbriata,* id. viii. 1—3; *N. Pelagica,* id. vii. 1—3; *Terebella rubra,* Gm.,

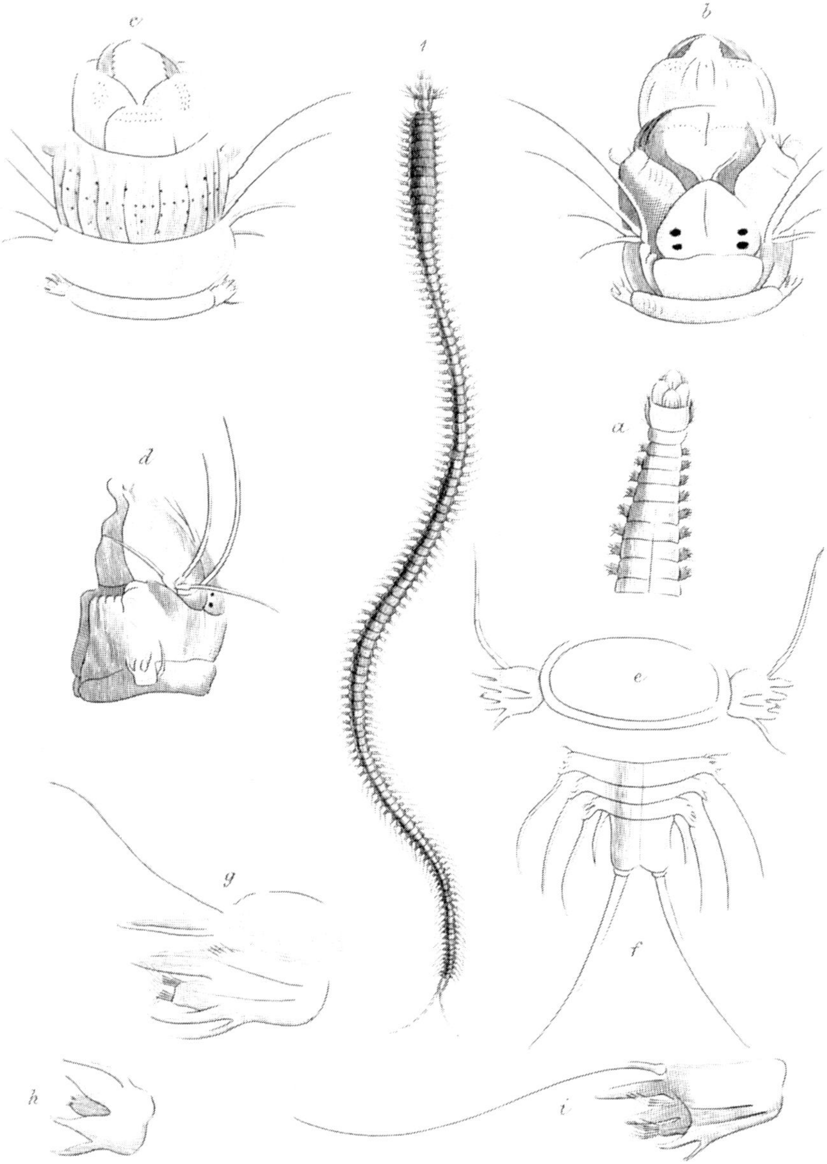

Nereis nuncia — Cuv.

London, Published by Whittaker & C.° Ave Maria Lane, 1832.

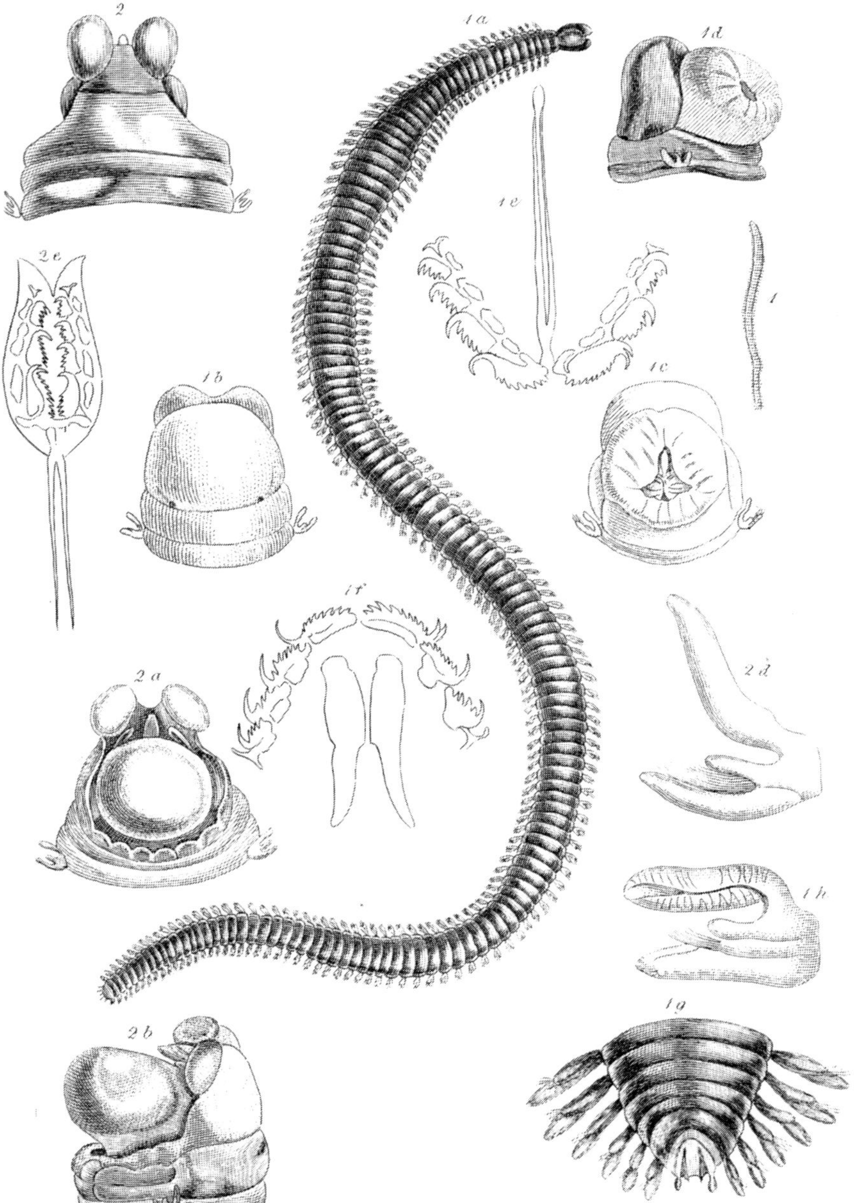

1 *Œnone lucida.*

2 *Aglaura fulgida.*

London, Published by Whittaker & C.º Ave Maria Lane. 1832.

Bommé, Mem. de Flessing. vi. 357. fig. 4, A. B.; *Lycoris Egyptia*, Eg. Annel. pl. iv. f. 1; *Lycoris nuntia*, id. ib. f. 2; *N. beaucoudraisii*. Aud. and Edw. Littor. de la France. Annel. pl. iv. fig. 1—7; *N. pulsatoria*, ib. fig. 8—13.

N. B. The *Nereis verrucosa*, Mull., vers. pl. vii. and *incisa*, Ott., Fabr. Soc. d'Hist. Nat. de Copenh. v. 1st part, pl. iv. f. 1—3, appear to have the head of lycoris, but with long filaments in place of gills. They require a fresh examination.

Near these Nereides are grouped many genera, likewise with slender body, and gills reduced to simple laminæ, or even to simple threads or tubercles. Many are destitute of jaws or of tentacula.

PHYLLODOCE, *Sav.*, NEREIPHYLLA, *Blainv.*, have, like the true nereides, an even number of tentacula, at the sides of the head, and four or five small ones besides, in front. They have eyes; their large proboscis, furnished with a circle of very short fleshy tubercles, shows no jaws, and they are especially distinguished by having gills in the form of rather broad leaves over-lapping each other in a single row on each side of the body, and over them are spread some vessels extremely ramified.

Nereis lamellifera Atlantica, Pall., Nov. act. Petrop. ii. pl. v. fig. 11—18; perhaps the same as the *Nereiphylle de Pareto*, Blainv., Dict. des Sc. Nat.; *N. flava*, Ott., Fab., Soc. d'Hist. Nat. de Copenh. v. part i. pl. iv. f. 8—10.

N. B. *N. viridis*, Mull., vers. pl. xl. of which M. Savigny, without having seen it, proposes to form the genus EULALIA and the two EUNOMIA, *Risso*, Europ. Merid. iv. p. 420., appear to me to be also *phyllodocæ*; perhaps we should even refer to them the *Nereis pinnigera*, Montag., Linn. Trans. IX. vi. 3; and the *Nereis stellifera*. Mull., Zool. Dan. pl. lxii. f. 1., of which M. Savigny, without having seen it, proposes to make a genus, under the name of LEPIDIA; and the

N. longa, Ott. Fab., which M. Savigny places with *N. flava*, in his genus ETEONE: all these annelida should be carefully examined afresh, according to the detailed method of M. Savigny.

We must not confound these phyllodocæ of M. Savigny with those of M. Ranzani, which are allied to the aphroditæ, and particularly to the polynoæ.

ALCIOPA, *Aud.* and *M. Edwards.*

They have pretty nearly the mouth and tentacula of phyllodoce; but their feet, independently of the tubercle which supports the bristles, and the two foliaceous cirrhi, or gills, present two branchial tubercles, which occupy their superior and inferior edges, *Alciopa Reynaudii*, Aud. and Edw. Of the Atlantic Ocean. The pretended *Nais Rathke*, Soc. d' Hist. Nat. de Copenh. v. part i. pl. iii. f. 15, may very probably be an Alciopa.

SPIO, *Fab.* and *Gmel.*

The body is slender; there are two very long tentacula, which have the appearance of antennæ, eyes in the head, and on each segment of the body a gill on each side, in the form of a simple filament. These are the small worms of the Arctic Ocean, which inhabit membranous tubes, *Spio seticornis*, Ott., Fab., Birl., Schr. VI. v. 1—7; *Spio filicornis*, ib. 8—12. The POLYDORÆ, Bosc. vers. I. v. 7., appear to me to belong to this genus.

SYLLIS, *Sav.*, have tentacula of an uneven number, articulated like the beads of a rosary, as well as the superior cirrhi of the feet, which are very simple, and have but a single parcel of bristles. It appears that there are some varieties relative to the existence of their jaws, *Syllis monilaris*, Sav., Eg. Annel. iv. f. 3. copied. Dict. des Sc. Nat.—N. B. The *Nereis armillaris*, Mull. vers. pl. ix., of which, without

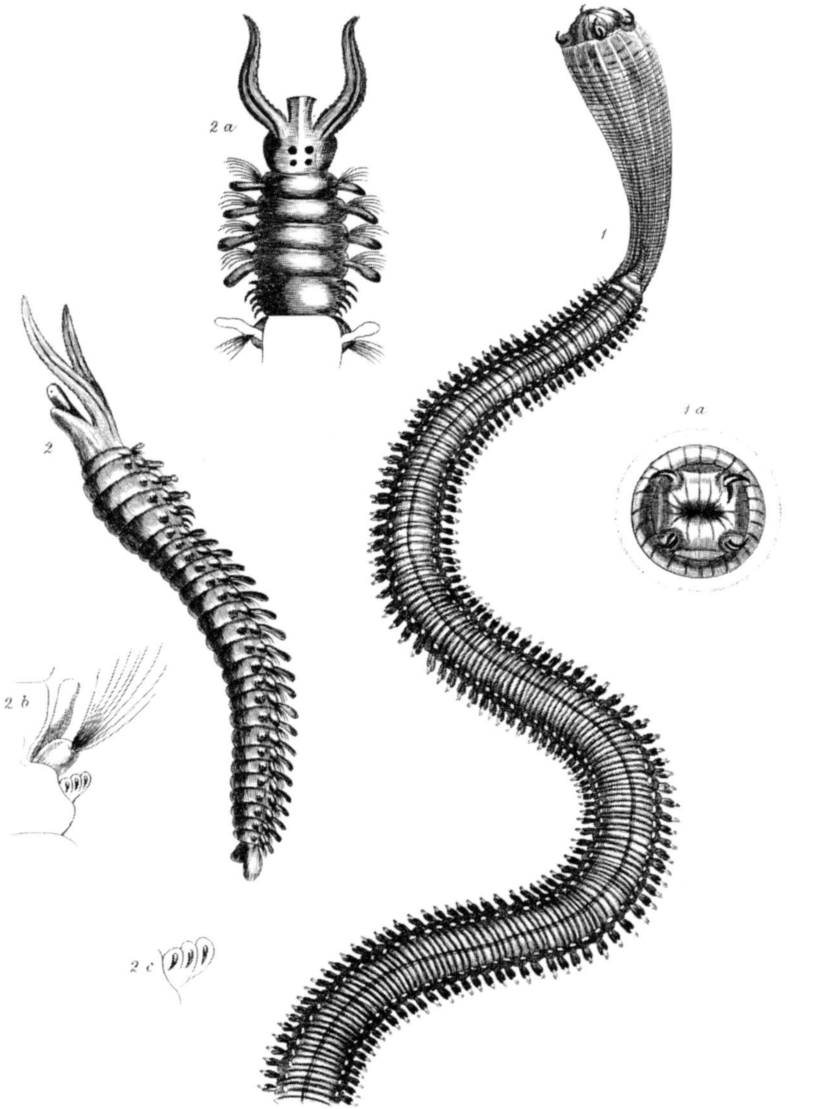

1 *Glyceres dubitosa.* 2 a. *Upper part of the Head.*

1 a *Front view of the Head.* 2 b. *An appendix.*

2 *Spio seticornis.* 2 c. *One of the setaceous crooks.*

London, Published by Whittaker & C.º Ave Maria Lane, 1832.

having seen it, M. Savigny proposes to form a genus, which he names LYCASTIS, has tentacula and cirrhi, bead-like, as in Syllis; but its tentacula are represented of an even number. It has also need of a new examination.

GLYCERA, *Sav.*, are recognized by their head being in the form of a fleshy and conical point, which has the appearance of a little horn, and the summit of which is divided into four very small tentacula, scarcely visible. The proboscis of some still presents jaws. It is said that in others there is no appearance of this kind, *Nereis alba*, Mull. Zool., Dan. lxxii. 6, 7.; *Glyc. Meckelii*, Aud. and Edw. Littor. de la France. Annel. pl. vi. fig. 1.

NEPHTHYS, *Cuv.* The proboscis of phyllodoce, but no tentacula, and on each foot two bundles of bristles, very much separated, between which is a cirrhus, *Nephthys Hombergii*, Cuv. represented in the Dict. des Sc. Nat.

LUMBRINERA, *Blainv.*,

Want tentacula; their body, very much elongated, has at each articulation only a very small forked tubercle, from which issues a small bundle of bristles. If there be an external organ of respiration, it can only be an upper lobe of this tubercle, *nereis ebranchiata*, Pall., Nov., Act. Petrop. ii. pl. vi. f. 2; *Lombrinère brillant*, Blainville, pl. of the Dict. des Sc. Nat.; *Lumbricus fragilis*, Mull., Zool. Dan. pl. xxii., of which M. Blainville makes, but with some hesitation, his genus SCOLETOMA.

N. B. The SCOLOLOPES, Blainville, which are known only by the figure of Abildgaardt, *Lumbricus squamatus*, Zool. Dan. IV. clv. 1—5., have a very slender body, with numerous rings, to each of which is a cirrhus serving as a gill, and two bundles of bristles; the inferior of which seems to issue from a fold of the skin, compressed like a scale; their head has neither jaws nor tentacula.

ARICIA, *Sav.*, want both teeth and tentacula. Their body, which is elongated, supports upon the back two ranges of lamellated cirrhi, and their anterior feet are furnished with denticulated crests, which are not found on the other feet. *Ar. Cuvierii*, Aud. and Edw. Litt. de la France, Annel. pl. vii. p. 1—13.

The *Lumbricus Armiger*, Mull., Zool., Dan. pl. xxii. f. 4 and 5, of which, without having seen it, M. de Blainville proposes to form a genus under the name of SCOLOPLE, appears to want both teeth and tentacula, and to have simple small bundles of short bristles on its first segments, and a bifid wart, a small bristle, and a branchial lamina, long and pointed, on the others.

Our coasts of the Atlantic possess some species of many of these genera.

HESIONE have the body short, tolerably thick, and composed of few and indistinct rings; a very long cirrhus, which probably performs the function of gills, occupies the upper part of each foot, which also has another inferior one, and a bundle of bristles. Their proboscis is large, and without jaws or tentacula. We have some from the Mediterranean, *Hesione Splendida*, Sav. Egg. Annel. pl. iii. f. 3; *H. festiva*, id. ib. p. 41; *Hes. pantherina*, Risso., Eur., Mir. iv. p. 418.

OPHELINA, *Sav.*

The body thick and short, with feebly-marked rings, and scarcely perceptible bristles; some long cirrhi, serving as gills on the two thirds of its length. The mouth contains, at the palate, a denticulated crest; its lips are surrounded with tentacula, of which the upper two are larger than the others.

N. B. It is probably in this neighbourhood that should come the *Nereis prismatica* and *bifrons*, Fabric. Soc. d'Hist. Nat. de Copenh. v. prem. part pl. iv. p. 17—23.

CIRRHATULUS, *Lam.*,

Have a very long filament, serving as gills, and two small bundles of bristles to each of the articulations of the body, which are very numerous and very crowded; there is, moreover, a line of long filaments around the nape. Its head, but little marked, has neither tentacula nor jaws, *Lumbricus cirrhatus*, Ott., Fabr., Faun., Grœnl. f. 5., from which the *Terebella tentaculata*, Mont., Linn., Trans. IX. vi. and the *cirrhinère filigère*, Blainv. pl. Dict. des Soc. Nat., do not appear to me to differ in genus; *Cirrh. Lamarckii.*, Aud. and Edw. Littoral. de la France, Annelid., pl. vii. f. 1—4.

PALMYRA, *Sav.*,

Are recognised by their upper fasciculi, the bristles of which are large, flatted, disposed like a fan, and shine like the most finely polished gold. Their inferior fasciculi are small; their cirrhi and gills but little marked. They have the body elongated, two tentacula tolerably long, and three very small.

But one is known, from the Isle of France, one or two inches long, *Palmyra aurifera*, Sav.

APHRODITA, *Lin.*,

Are easily recognized in this order by the two longitudinal ranges of broad membranous scales which cover their back, to which, through a very groundless assimilation, the name of elytra has been given, and under which their gills are concealed, in the form of small fleshy crests.

Their body is generally of a flatted form, and shorter and broader than in the other Annelida. We observe in their interior a very thick and muscular œsophagus, capable of being protruded externally like a proboscis. The intestine is unequal, and furnished on each side with a great number of branching cœca, the extremities of which are fixed between the bases of the bundles of bristles, which serve as feet.

M. Savigny distinguishes HALITHEÆ, which have three tentacula, and between two of them one very small crest. They are destitute of jaws.

We have one of these upon our coasts, which is one of the most beautiful of animals, in point of colour, *Aphrodita aculeata*, L. Pall. Misc. v. ii. 1—13. It is oval, six or eight inches long, and two or three broad. The scales of its back are covered and concealed by a sort of stuff resembling tow, which arises from the sides. From these same sides spring groups of strong spines, which partly pierce the tow, and bundles of flexuous bristles, shining with all the brilliancy of gold, and changing into all the tints of the rainbow. They do not yield in beauty to the plumage of the humming-bird, nor to the most lively lustre of precious stones. Lower down is a tubercle, from which spines in three groups issue forth, and of different dimensions, and, finally, a fleshy cone. Forty of these tubercles may be counted on each side, and between the first two are two small fleshy tentacula. There are fifteen pairs of broad scales, and sometimes inflated, on the back, and fifteen small branchial crests on each side.

Some of these Halitheæ have none of this tow-like stuff on the back, and our seas produce one species of them—*Aphrod. hystrix*, Sav.

Another subdivision of the Aphrodita is that of POLYNOË, Sav., EUMOLPE, Oken, which have no tow on the back; their tentacula are five in number, and their proboscis contains strong and corneous jaws.

We have several species on our coasts.

Aphrod. squamata, Pall. Misc. Zool. vii. 14. Littor. de la France, Annel. pl. i. f. 10—6; *Polyn. lævis*. Aud. and Edw. ib. pl. ii. f. 11—18; *Aphrod. punctata*, Mull. Vers. xiii; *Aphrod. cirrhosa*, Pall. Misc. Zool. viii. 3—6; *A. lepidota*, id. ib. 1—2; *Aphr. Clava,* Montg. Lin. Trans. IX. vii., which is at least greatly allied to *Aphr. plana,* Mull. Vers. xix; *Polynoë impatiens,* Sav. Egg. Annel. pl. iii. f. 2; *Polynoë muricata,* id. ib. f. 1.

SIGALION, *Aud. and Milne Edw.*, are of a much more elongated form than the other aphroditæ; they have cirrhi to all the feet. *Sigalion Mathiedæ,* Aud. and Edw. Littor. de la France, Annel.

The ACOETES, of the same, have cirrhi which alternate with the elytra for a considerable length; their jaws are stronger, and better denticulated, *Acoetes Pleei,* Aud. and Milne Edw. Collect. of the Museum. The Antilles possess a large species, which inhabits a tube of the consistence of leather.

N. B. The *Phyllodoce maxillosa,* of M. Ranzani, called POLYDONTE by Renieri, and *Eumolpe maxima* by Oken, appears very much allied to ACOETES. Its proboscis and jaws are the same, and neither of the genera has perhaps been described from complete individuals.

There remain still many of the Annelida too imperfectly described to be well characterized: such as *Nereis cœca,* Fabr. Soc. d'Hist. Nat. de Copenh. part i. pl. iv. f. 24—28; *N. longa,* id. ib. f. 11—13; *N. Aphroditoides,* ib. 4—7; ib. f. 11—13; *Branchiarius quadrangulatus,* Montag. Linn. Trans. xii. pl. xiv. f. 5; *Diplotes hyalina,* id. ib. f. 6 and 7; and the pretended *Hirudo branchiata,* of Archib. Menzies, Lin. Trans. i. pl. xvii. f. 3. Nor have I placed the MYRIANA, and two or three other genera of M. Savigny, for want of an opportunity of re-examining them.

CHŒTOPTERUS, *Cuv.*,

A new and singular genus that can only come here; the mouth without jaws, and proboscis furnished above with a lip, to which are attached two very small tentacula. Then comes a disk, with nine pairs of feet; then a pair of long silky bundles, like two wings. The gills, in the form of laminæ, are attached, rather below than above, and predominate along the middle of the body.

There is one species, *Chætopterus Pergamentaceus,* eight or ten inches long, which inhabits a tube of the substance of parchment. It belongs to the sea of the Antilles.

THIRD ORDER OF THE ANNELIDA.

THE ABRANCHIA,

HAVE no organ of respiration externally apparent, and seem to respire, some like the lumbrici, by the entire surface of their skin; others, like the Hirudines, by interior cavities. They have a closed circulating system, most generally filled with red blood, and a knotted nervous cord, like all the Annelida. There are some which still have setæ, or bristles serving for locomotion, and others which are destitute of them, which gives rise to the establishment of two families.

The first family, that of

SETIGEROUS ABRANCHIA, or PROVIDED WITH SETÆ, comprehends the *Lumbrici* and the *Naides* of Linnæus.

LUMBRICUS, L.,—vulgo, *earth-worms*—characterized by a long cylindrical body, divided by wrinkles into a great number of rings, and by a mouth without teeth, necessarily required subdivision.

LUMBRICUS, *Cuv.* (proper),

Are destitute of eyes, of tentacula, of gills, and of cirrhi; a tubercle, or sensible enlargement, particularly at the season of reproduction, seems to attach them to each other in intercourse. In their interior there is a strait rugose intestine, and some whitish glands towards the middle of the body, which appear to serve the purposes of generation. It is certain that they are hermaphrodites; but it may be possible that their approximation serves only to excite them mutually to self-fecundation. According to Mr. Montague, the eggs descend between the intestine and the external envelope, as far as around the rectum, where they disclose. The young ones come out alive through the anus. M. L. Dufour says, on the contrary, that they form eggs analogous to those of the Hirudines. The nervous cord is only a series of an infinity of small ganglia, crowded one against the other.

M. Savigny subdivides them again: his Enterion have under each ring four pairs of little bristles, eight in all.

Every body knows the *common earth-worm, lumbricus terrestris*, L., with a reddish body, attaining nearly a foot in length, with one hundred and twenty rings, and more. The enlargement is towards the anterior third of the body; under the sixteenth ring are two pores, the use of which is unknown.

This animal pierces the earth in all directions, and swallows much of it; it also eats roots, ligneous fibres, animal substances, &c. In the month of June it issues from the earth at night, for the purposes of generation.

What I have here mentioned is common to many species, which M. Savigny was the first to distinguish : he has characterized twenty of them. See my analysis of the labours of the Academy of Sciences, 1821. M. Dugès distinguishes six, but he does not refer them exactly to those of M. Savigny.

N. B. Muller and Fabricius speak of lumbrici, with two setæ to each ring, of which M. Savigny proposes to form his genus Clitellio, *Lumb. minutus*, Fab., Faun. Grœnl. f. 4., and of lumbrici with four or six setæ; but their descriptions, which are old, have need of being confirmed and completed, before we can classify their species.

HYPOGÆON, *Sav.*, have an additional odd pair of setæ on the back of each ring: they belong to America. *Hypogæon hirtum*, Sav., Egg. Annel. p. 104.

MM. Audouin and Milne Edwards also distinguish TROPHONIA, which have on each ring four bundles of shorter setæ, and a great number of long and brilliant ones, which surround the mouth.—*Trophonia barbata*, Aud. and Edw. Littor. de la France, Annel. pl. x. f. 13—15.

NAIS, *Linn.*,

Have the body elongated, and the rings less marked than the lumbrici. They live in holes, which they excavate in mud, at the bottom of the water, and from which they protrude the anterior part of their body, which they keep incessantly moving. There are visible on the head of many some small black points, which may be taken for eyes. They are small worms, and their power of reproduction is as astonishing as that of the hydræ. Many of them exist in our fresh waters.

Some have setæ tolerably long, *Nais Elingius*, Mull. Wurm, ii; *N. littoralis*, id. Zool. Dan. lxxx.

And sometimes a long proboscis in front, *Nais proboscidea*, id. Wurm, i. 1—4., of which M. de Lamarck makes his genus STYLARIA.

Or several small tentacula at the posterior extremity, *Nais digitata*, Gm.; *cœca*, Mull. ib. v.; of which M. Oken makes his genus PROTO.

Others have very short setæ, *Nais vermicularis*, Gm., Rœs. III. xciii. 1—7; *N. Serpentina*, id. xcii., and Mull. iv. 2—4; *Lumbricus tubifex*, Gm., Bonnet. vers d'eau douce, iii. 9—10. Mull. Zool. Dan lxxxiv.; *Lumb. lineatus*, Mull. Wurm. iii. 4—5.

We may approximate to this genus certain annelida hitherto referred to the lumbrici, which fabricate for themselves tubes of clay, debris, &c., in which they live, *Lumbricus tubicola*,

Mull. Zool. Dan. lxxv.; *Lumb. sabellaris,* ib. civ. 5. M. de Lamarck unites them with *Nais tubifex,* and makes of them his genus TUBIFEX; but a new examination of them is necessary.

CLYMENA, *Sav.,*

Also appear to belong to this family; their body, tolerably thick, has but few rings, which for the most part have a range of strong setæ; and a little higher, on the dorsal side, is a bundle of finer setæ. Their head has neither tentacula nor appendages; their posterior extremity is truncated and radiated. They also inhabit tubes, *Clymena amphistoma,* Sav., Egg. Annel. pl. i. f. 1; *Cl. lumbricalis,* Ot., Fabr., Aud. and Edw. Littor. de la France, *Annel.* pl. x. f. 1—6; *Cl. Ebiensis,* Aud. and Edw. Littor. de la France, f. 8—12.

The second family, or that of

ABRANCHIA, WITHOUT SETÆ, comprehends two great genera, both aquatic.

HIRUDO, *Lin., the* LEECH.

Have the body oblong, sometimes depressed, and wrinkled transversely. The mouth is surrounded with a lip, and the posterior extremity provided with a flatted disk, both adapted to fix upon bodies by a sort of suction, and serving the leech as the principal organs of locomotion; for after extending itself, it fixes its anterior extremity, and approximates the other, which in its turn adheres, to allow the first to be carried forward. In several we observe, underneath the body, two series of pores, the orifices of as many little interior pouches, which some naturalists regard as organs of respiration, although they are usually filled with a mucous fluid. The intestinal canal is straight, inflated from space to space, as far as two-thirds of its length, where there are two cœca. The blood swallowed is preserved there, red and unchanged, for many weeks.

The ganglia of the nervous cord are much more separated than in the lumbrici.

The leeches are hermaphrodites: a large penis projects from under the anterior third of the body, and the vulva is a little farther back. Many of them collect their eggs into cocoons, enveloped in a fibrous excretion.

They have been subdivided according to characters chiefly derived from the parts of the mouth.

SANGUISUGA, *Sav.*, JATROBDELLA, *Blainv.* These are the leeches properly so called; the anterior sucker has its upper lip divided into several segments; its aperture is transverse, and it contains three jaws, armed, each on its edge, with two ranges of very fine teeth, which enables them to penetrate through the skin without making any dangerous wound there. Ten small points have been observed upon them, which are taken for eyes.

We are all acquainted with the common or medicinal leech, *Hirudo medicinalis*, L., so useful an instrument in local bleedings. It is usually blackish, striped with yellowish above; yellowish, spotted with black underneath. It is found in all dormant waters.

HÆMOPSIS, Sav., differ from the last only because their jaws have but a few and obtuse teeth. M. de Blainville calls them *Hypobdella, the Horse-leech; Hirudo sanguisuga*, L.; *Hæmop. sanguisorba*, Sav., Moq. Tand. pl. iv. f. 1.; Car. pl. xl. f. 7.; much larger, and altogether of a greenish black. It has been reported to be sometimes dangerous, from the wounds which it inflicts.

The difference of opinions as to the power of the horse-leech to draw blood is very singular. Linnæus says that nine of them can kill a horse. MM. Huzard and Pelletier, on the contrary, in a memoir lately presented to the Institute, and inserted in the Journal de Pharmacie, March, 1825, assure us that it never attacks a vertebrated animal. M. de

Blainville thinks that the difference arises from its having been confounded with a nearly-allied species, the *Black Leech*, which he makes the type of a genus that he calls PSEUDOBDELLA, and of which the jaws are merely folds of the skin without any teeth. I believe this fact would deserve a fresh investigation. Both species devour *Lumbrici* with avidity.

BDELLA, *Sav.*, have but eight eyes, and the mouth is completely destitute of teeth. M. Moquin-Tandon changes this name into LIMNATIS.

There is one in the Nile, *Bd. Nilotica*, Egg. Annel. pl. v. f. 4.

NEPHELIS, *Sav.*, have also but eight eyes, and their mouth internally has but three folds of the skin. M. de Blainville names them ERPOBDELLÆ, and M. Oken HELLUO. Such are *Hir. vulgaris*, L., or *H. octoculata*, Bergm., Mem. de Stockh. 1767, pl. vi. f. 5—8; *N. atomaria*, Caren. L. C. pl. xii. See also pl. vi. of M. Moquin-Tandon.

There are many small species in our waters, among which we distinguish TROCHETIA, *Dutrochet*, GEOBDELLA, *Blainv.*, which differ only by an enlargement at the seat of the genital organs.

We have one species of them, which frequently comes to land to pursue the lumbrici, *Geobdella trochetii*, Blainv. Dict. des Sc. Nat. Hirud. pl. iv. f. 6.

M. Moquin Tandon, under the name of AULASTOMA, describes a subgenus whose mouth is also furnished with nothing but longitudinal folds tolerably numerous, *Aulast. nigrescens*, Moq. Tand. pl. vi. f. 4.

After Nephelis are placed the BRANCHIOBDELLA, of M. Odier, remarkable for jaws, two in number, and the absence of eyes.

One species is known, which lives on the gills of the astaci. *Branchiobdella astaci*, Od. Mem. de la Soc. d'Hist. Nat. de Paris, tom. i. pl. iv.

All these sub-divisions have the anterior sucker but little separated from the body; in the following two it is clearly distinguished from it by a strangulation, is composed but of a single segment, and with a transverse aperture.

HÆMOCHARIS, *Sav.*, have, with this conformation, eight eyes, the body narrow, and the rings but little distinct; their jaws are projecting points, scarcely visible. They do not swim, but walk like those caterpillars called *geometrical*, and particularly attach themselves to fish. M. de Blainville, who had given them the name of PISCICOLÆ, adopted by M. de Lamarck, has again changed it into ICHTHYOBDELLA.

We have one species of them, pretty frequently found in the Cyprini.—*Hirudo piscium*, L. Rœsel, III. xxxii. Add.; *Piscicola cephalota*, Caren. pl. xii. f. 19.; and Moq. Tand. pl. vii. f. 2, *Piscic. tesselata*, Moq. f. 3.

ALBIONA, *Sav.*, differ from the preceding, because their body is bristled with tubercles, and their eyes are six in number. They live in the sea. They are the PONTOBDELLA, of *Leach* and *Blainville.*

Our seas nourish in abundance the *Albiona verrucosa, Hirudo muricata*, L., all bristling with small tubercles, Add. *Pontobd. areolata, P. verrucata, P. spinulosa*, Leach. Miscell. Zool. lxiii., lxiv., lxv.; *Hirudo vittata*, Chamisson, and Eisenhardt, Nov. Act. Nat. Car. t. x. pl. xxiv. f. 4.

BRANCHELLION.—This name has been given to a parasite of the torpedo, very similar to a leech in its two cups or suckers, its depressed body, and its transverse folds. Its anterior cup, which appears to have a very small mouth at its posterior edge, is supported on a part attenuated in the form of a neck, at the root of which is a small hole, for the organs of generation; there appears to be another behind. They are the POLYDORUS of Oken, the BRANCHIOBDELLION of Rudolphi, and the BRANCHIOBDELLA of De Blainville.

The lateral edges of its folds, compressed and projecting,

have been regarded as gills; but I can discover no vessels there. Its epidermis is ample, and envelopes it like a very loose sack. It is the *Branchellion torpedinis* of Sav.; but the species observed on the tortoise should not be associated with it. *Hir. branchiata*, Menzies, Linn. Trans. I. xviii. 3., which truly appears to have plumose gills, and which it would be necessary to examine again.

We also commonly range among the leeches,

CLEPSINA, *Sav.*, GLOSSOPERA, *Johns.*, which have a widened body, a posterior cup only, and the mouth in the form of a proboscis, and without sucker; but it may not be impossible that some of them rather belong to the family of the planaria.

M. de Blainville names them GLOSSOBDELLA; *Hir. complanata*, or *sexoculata*, Berg. Mem. de Stockh. 1757, pl. vi. f. 12—14; *H. trioculata*, ib. f. 9—11; *Hir. hyalina*, L., Gm., Trembley, Polyp. pl. vii. f. 7; *Clepsina paludosa*, Moq. Tand. pl. iv. f. 3, &c.

I believe them still more allied to the PHYLLINE, *Oken.*, and to the MALACOBDELLÆ, *Blainv.*, which have also broad bodies, and are destitute of proboscis and anterior sucker. They are parasite animals: the first are named EPIBDELLA by de Blainv.; *Hir. hippoglossi*, Mull. Zool. Dan. liv. 1—4. To the second belongs *Hir. grossa*, Mull. Zool. Dan. xxi.

GORDIUS, *L.*,

Have the body resembling a thread, slight transverse folds only marking its articulations, and neither feet, gills, nor tentacula are visible. Nevertheless, in the interior, a nervous system is still distinguishable in a knotted cord. Perhaps, however, it may be necessary to place them definitively with the cavitary intestinal worms, like the nemertes.

They inhabit fresh water, mud, and inundated grounds, which they pierce in all directions, &c.

The species are not yet very well distinguished: the most common, *Gor. aquaticus,* is several inches long, almost as fine as a hair, and brown, with blackish extremities.

SUPPLEMENT

ON THE

CLASS OF THE ANNELIDA.

It is of the utmost importance to the student of natural history, thoroughly to understand the application of the names, and the strict limits of its various divisions: to do this, becomes of still greater consequence, and we regret to say of infinitely more difficulty, as we descend in our researches to those beings which rank lower and lower in the great scale of animality. This arises from the multitude, the variety, the habits, and the localities of such beings, often presenting great impediments to observation and examination; but still more from the confusion and mutation of names and divisions, by careless investigation, or scientific vanity. This has been strikingly exemplified with respect to the animals to which our attention is now directed; and a few preliminary remarks in reference to their classification will not only be useful but indispensable.

In the infancy of the study of natural history, the name of Vermes, or Worms, was bestowed on all animals with long and soft bodies, from a comparison of them to the earth-worms, or *lumbrici*, to which this term had been specifically devoted. Consequently, the larvæ of insects were considered as worms, and are still considered as such, by the great majority of mankind.

The names of worms,—σκώληξ, εὐλαὶ, ἕλμινς, in Greek, and *vermes* in Latin,—were employed by the ancients to designate certain animals, which to a certain degree they suited; with much more reference, however, to the elongated form of their body than to the softness of their composition. But, as we have seen, the Greeks had three words for these beings, each of which had its peculiar signification. From what Aristotle tells us of his *scolex*,—a word, the root of which is indubitably *scolios*, which means *tortuous*,—it is evident that it applied to all the animals which exhibited the form of the common worm, or rather, perhaps, whose movements were tortuous, whatever might be the nature of the change which they were subsequently to undergo. It would seem, however, that it was more especially applied to the first degree of development in insects, to the state in which they appear on issuing from the egg of the parent. Aristotle certainly extends its application no further than to insects.

Such, however, is not the case with Ælian: in two places of his work on the nature of animals, where this expression occurs, he evidently intends the lumbrici; in a third, it is probable that he alludes to the caterpillar of the cabbage-butterfly; and in a fourth, he thus designates, after Ctesias, some fabulous animal, although he states it to belong to the genus of those which are nourished and engendered in wood.

Athenæus employs the word *Scolezia* to designate the small worms which live in the vulva of the she-mule.

The term *Eulai* appears to have been also employed to designate the form under which some insects exist, for a greater or less period of time, since we find it applied to animals which inhabit putrid flesh, and also wounds and ulcers. Its extension, therefore, was not very great. Ælian likewise employs it to designate what in all probability was a larva, when he tells us that in India the peasants remove the land tortoises from their shell with a mattock, in the same manner as they remove the worms from plants which are infested by them.

Finally, the word *Elmins*, which is frequently employed by Hippocrates in many of his works, and among others, in his General Treatise on Diseases, was applied by him to those animals which are at present known under the denomination of *intestinal worms*, of which he was acquainted with but a small number of species. Aristotle has employed it in the same manner, as well as Ælian, every time that he speaks of the substances which are used to rid dogs of the worms, to which they are subject.

The Latin authors, and Pliny among the rest, appear to have restricted the word *lumbricus* to the intestinal worms, and to have rendered the three Greek denominations by a single one, that of *vermes*, from which it has happened that the moderns have been led to the same confusion, by the word *worms*, which, as well as the French word *vers*, is evidently derived from the Latin. All the other animals, which they comprehended under the name of *exsanguia*, meaning by that term that they had not red blood, were divided under the three titles of *insecta*, *mollusca*, and *zoophyta*. The term *vermes* did not then possess that undue extension which it obtained among the naturalists of the last century, with whom it at last comprehended all animals, with the exception of the vertebrata, the insects and the crustacea. This extension was probably owing to the having united in the definition of the word the consideration of the softness of the body to that of its elongated form. To Linnæus unquestionably is attributable this unlucky innovation; for, previously to his time, we find that Isidore, of Seville, though he has employed the name of *vermes* for a sort of class, has only included under it such animals as might with some propriety be so called: in fact true worms, whether belonging to the Annelida of our author, or not, nevertheless, we find, in his definition of the slug, the word *vermes* employed as a principal name.

With respect to the *red-blooded worms*, with which we are most immediately concerned in this place, though some of our remarks may be found allusive to genera, thrown among the zoophytes by Cuvier, but comprehended by M. de Blainville in his classification of worms into *Chetopoda* and *Apoda*, (see text) the ancients seem to have known but very imperfectly the animals which constitute the class Annelida. Aristotle confines himself to stating, probably with regard to the nereides, and under the denomination of marine *Scolopendræ*, that they are similar to the land Scolopendræ, but a little smaller, of a redder colour, with a greater number of feet, and more weak; that they are born in the same manner as serpents, and in places full of rocks, not in the depths of the sea. Aristotle has also spoken of leeches, as well as of some intestinal worms. He adds, in another place, that they bite not with their mouth, but irritate by contact only, like the *acalaphæ*, or sea-nettles (a class of zoophytes); that they emit an unpleasant odour; and finally—which is somewhat less credible—that when they are caught with a hook, they reject, or vomit forth all their intestines, until they have expelled the hook, and then put them in again, and go about as well as ever.

Pliny has added nothing to the statements of Aristotle respecting this class of animals.

Ælian, Oppian, Dioscorides, and Galen, have likewise done nothing but copy and exaggerate what has been left us by Aristotle. The two latter have merely added the remedies which they conceived to be of suitable application in case of accidents from the contact of the marine scolopendræ, and those in the composition of which they employed the animals themselves.

Isidore, of Seville, is the first author in whose works we find a particular chapter, the fifth, under the title of *vermes;* but he confines his attention there to lumbrici, ascarides, leeches, and worms that inhabit flesh.

Albertus Magnus, in his book on the animals of the division *exsanguia*, speaks of the leech, and of the worm, in an alphabetical order.

Wotton has not extended the number of the animals of this class, and only speaks of the nereides under the name of *marine scolopendræ*, in his book upon insects; of leeches among the fish; and of earth-worms under the name of *intestini terræ;* as well as of intestinal worms under the generic denomination of *lumbrici; elmins* in Greek, among the insects.

Belon, in his history of aquatic animals, mentions, for the first time, under the name of *lumbricus marinus*, in opposition to the earth-worm, which he names *lumbricus terrestris*, the animal which we now call *arenicola.*

Rondelet went considerably farther: in fact, he not only described and gave figures of several nereides, still under the name of *sea-scolopendra*, but he remarked, for the first time, one of the *tubicolent* annelides, probably of the genus serpula. He also described and figured the common and the sea-leech, and also made known two species of *sipunculi* (zoophytes).

Gesner collected all that had been advanced previously to his time by the ancients and moderns respecting the *chetopoda**, and the worms in general. But he added nothing new, and speaks of them only in articles altogether detached

* We beg to use this term of M. de Blainville, as more conveniently expressing the *setigerous annelida*, or those which move by means of bundles of *setæ* or bristles. We may also occasionally employ his term *apoda* to express those which are destitute of such appendages, always, however, limiting the extent of those terms to the genera in the text.

and separate, and gives us less, perhaps, about them under the head *vermis*, than under that of *scolopendra*.

Aldrovandus, and his abridger, not having followed the alphabetical order, like Gesner, were necessarily forced to unite all these animals in a sort of class or group, under the common name of *vermes*. But it is remarkable enough that they did not comprehend among them any of the chetopoda, but merely the worms which live in the body of man, or that of other animals; those which live in plants, in the earth, as the earth-worm, which they call *lumbricus terrestris;* and finally, the slugs, probably after the definition of Isidore, who has characterized the slug as "*vermis limax dictus eo quod in limo nascitur, unde et sordidus semper et immundus habetur*," an etymology which had already been given by Varro.

The Chetopoda are, however, mentioned in the seventh book of those writers, when they speak of aquatic insects. In the sixth chapter, the nereides are comprised again under the name of *sea-scolopendræ*. In the seventh chapter, are placed the worms which live in cloths. In the tenth chapter is the gordius, which is named *seta, vel vitalis aquaticus*, and which has been called *gordius* from the habit of twisting itself up like the Gordian knot. The *Ololygon* of Theon appears to be the same animal. It is, he says, a palustral animal, simple, slender, oblong, indistinct, similar to a lumbricus, but thinner. In chapters xi. and xii. are the sea and land-leech; in xiii. the *lumbrici marini*, that is, the *sipunculi* of Rondelet, and the *arenicola* of Belon. In x. the *hippocampus*, a species of fish. In xviii. the *asteriæ*, which were then considered insects, according to the rigorous definition of the term *insecta*.

After this time, during almost the whole period, which preceded the regeneration of natural science, until Ray, and more particularly Linnæus, authors comprised under the term worms, the intestinæ, and the earth-worms, while the *chetopoda* were considered as insecta. Ray, whose method is very rigorous, divides his *insecta,* which comprehend all articulated animals, into insects which do not undergo metamorphosis, and those which do. The first section is divided into *apoda* and *phoropoda.* The division apoda comprehends the worms which live in the earth, as the lumbrici; those which inhabit the bodies of animals, as the intestinæ; and those which live in the water, as the fresh-water and sea-leech, &c. He observes, that among the terrestrial species, most authors range the limaces, whether naked or conchyliferous. The group of *phoropoda,* is then divided according to the number of feet, into *hexapods, octopods, tetradecapods,* and *polypods.* In this last section, under the name of *terrestria,* come the Juli, and the true Scolopendræ; and in the aquatic division the *sea-scolopendræ,* or nereides.

In the first edition of the *Systema Naturæ,* Linnæus extended the term *vermis* to all animals that were not mammalia, birds, reptiles, fishes, or insects, and consequently to mollusca, whether naked or shelled, and to zoophytes; but he excluded the *insect-worms* of Ray, from his class vermis. This class was then divided into four orders: 1. *Reptilia,* for the intestinal worms, comprehending, however, the leeches and lumbrici; 2. *Zoophyta* for the chetopoda, or setigerous annelida, the naked mollusca, the medusæ, and the echinodermata; 3. *Testacea,* for the conchyliferous mollusca, comprehending, however, the ascidiæ, under the name of microcosmus; 4. and last, the *Lithophyta,* for the madrepores and serpulaceæ.

After this first essay, appeared the genera *Amphitrite nereis,* and *aphrodita,* which belong to the chetopoda. In subsequent editions of Linnæus, the name *intestinæ,* was substituted for *reptilia,* for the first order. The denomination *zoophyta* was replaced by that of *mollusca,* for the second, while it replaced that of *lithophyta,* for the fourth, which was suppressed.

In the eleventh edition, the class of worms is divided into five orders: *Intestina, Mollusca, Testacea, Lithophyta,* and *Zoophyta,* and the genera which at present constitute the class of red-blooded worms, were parcelled out, some as *lumbricus* and *hirudo* in the first order; others as *terebella, aphrodita* and *nereis* in the second; and finally, some as *serpula* and *sabella* in the fourth, in consequence of the tube in which they live.

In the interval from this edition of the *Systema Naturæ,* to the last, which appeared in 1766, and which was very closely followed by Gmelin's, the thirteenth, some very important researches on the animals of the Linnæan class of worms, and especially on the chetopoda, or setigerous annelida, were given to the world. The labours of Pallas, in 1766, on the *aphroditæ,* the *nereides,* and the *serpulæ,* were the true origin of every thing judicious, which has been subsequently proposed concerning this class of animals. He made the very important observation, which he had already applied to the mollusca proper, that the presence or absence of a calcareous envelope did not constitute a sufficient ground for placing in two separate orders, animals which, in other respects, are similarly organized. Thus he approximated together the aphroditæ and the nereides of the order mollusca of Linnæus, and the serpulæ and amphitritæ of that of his testacea, saying, that they should form a distinct order, constituting the passage to the Zoophytes, and to which, he adds, may be joined the lumbrici, the hirudines, the ascarides, gordius, and the tæniæ or tape-worms; all which has been done since, has been founded on this observation of Pallas.

To two Danish naturalists, Otho Frederic Muller, and Otho

Fabricius, we are indebted for a considerable number of observations on the European species of nereis, as well as on many other animals of the Linnæan class of vermes. The first has proposed some changes in their distribution, introducing an entire order, and at the same time a tolerable number of genera.

By him, however, the class vermes is still divided into but five orders, as he has united the last two of Linnæus, 1. *Infusoria*, for a numerous group of animals which he supposed were produced in vegetable or animal infusions, and his labours are all that we yet possess on this subject; 2. *Helminthica*, or worms, in which he ranges, in two distinct divisions, the intestinal worms, and the hirudo in one, and all the chetopoda, comprising among them the lumbrici, in the other; 3. *Mollusca*, the same as Linnæus, abstracting the chetopoda, but leaving the planariæ, the fascioli and the medusæ, with the true naked mollusca; 4. *Testacea*; 5. under the name of *Cellularia*, the lithophytes and zoophytes of Linnæus; besides some very curious observations on the reproduction of the naides, and nereides, and the distinction of a great number of new species, science is indebted to this author for the establishment of the genera Nais and Amphitrite, and the more exact circumspection of those previously established. His researches were subsequently made available by Linnæus himself and several of his successors, in modifying the *Systema Naturæ*.

Though the celebrated naturalist, Blumenbach, has almost exclusively followed Linnæus, in his methodical distribution of worms, he has nevertheless introduced an observation, as characteristic of the class, namely, that it never possesses articulated organs of motion, in opposition to what he had said concerning the insects, in which those organs are articulated; "a character," he adds in a note, "which appears to me to be more precise, than that which has hitherto been employed to distinguish insects from worms."

Gmelin, in his edition of the *Systema Naturæ*, necessarily much augmented the number of genera, by collecting all that had been established since the last edition of Linnæus; but he made no great changes in the methodical distribution of worms, except by adding the last class proposed by Muller, under the name of *infusoria*. The chetopoda, and indeed all the red-blooded worms were still scattered through the three orders of intestina, mollusca, and testacea. He did not profit by the observations of Pallas, of Muller, or of Blumenbach.

In 1789, the part of the French Encyclopœdia, on worms, by Bruguières, made its appearance, with a table of the methodical distribution of those animals. But this writer, though he felt the necessity of establishing a new order, that of echinodermata (now in the zoophytes), effected no sensible amelioration. Nevertheless, in proportion as the study of the animals of the inferior class proceeded, the name of vermes, given to his last class by Linnæus, was reserved for the animals which the ancients thus distinguished, and ceased to be generally employed for the other orders, which were named mollusca, testacea, zoophytes, and infusoria. Thus, in the Synopsis of Animals, published by Baron Cuvier, in 1798, the seventh book treats of insects and worms. Under the name of *worms*, he then divided these animals into two sections, according as they were provided, or unprovided, with setæ or spines for locomotion; or, in other words, into chetopoda and apoda, as has been done by M. de Blainville. Among the first were grouped the species which live in tubes, and also the lumbrici; in the second were placed the leeches, and intestinal worms. Thus the observation of Pallas was appreciated, and put into execution, by Cuvier, respecting those animals, as it also was at the same time in regard to the mollusca.

Thus did M. Cuvier, by abandoning the views of Linnæus, return to those of the ancient naturalists, such as Aldrovandus, Mouffet, and Ray, by comprehending in one and the same division, the insects and worms; but he did more, by following, as he tells us himself, the ideas of Pallas, and uniting with his worms, the serpulæ, the sabellæ, and in general all the chetopodæ with tubes.

Two years after this, M. Cuvier, in the tables which form a sequel to the first volume of his Lessons on Comparative Anatomy, introduced a slight modification into the classification of the worms, which he subsequently carried farther. This consisted in comprehending under this name, in a definitive manner, only the chetopoda and apoda, which exist externally, the others, or the intestinal worms, being thrown into a sort of *incerta sedes*, with a view to a new order. As for the first, they are divided according as they have external organs of respiration or not; and the second, according as they are provided with lateral setæ, or destitute of them. In the first, are all the chetopoda, lumbrici excepted; in the second, those last animals, nais and thalassema, (now a zoophyte) and in the third, the hirudines, the planariæ, and even the fascioli, which are nevertheless intestinal animals.

In 1802, in a particular memoir read at the Institute, M. Cuvier, in giving his observations on the organization of the chetopoda, proposed to designate them as a class, by the name of red-blooded worms, comprehending in it the hirudines and lumbrici, without observing that the most common and thickest species of our seas has not the nutritive fluid of a red colour.

A short time after this, M. de Lamarck, determined most probably by the consideration put forth by Cuvier, also made a particular class of the worms, which he divided more easily, but perhaps less felicitously, according to their habitat being external, as is the case with all the chetopoda, and the hirudines, or in the interior of other animals, as the entozoa, or intestina. He admitted nearly the same genera as the Baron, or at least created but few new ones.

M. Dumeril nearly contents himself with giving new names to the divisions adopted by M. Cuvier. As to the intestinal worms, he cuts the difficulty short by making zoophytes of them, as Linnæus had done before, in the instance of the tænia.

In 1809, and afterwards in 1812, M. de Lamarck proposed for the class of Chetopoda, the new name of *Annelides.* He divided, as M. Dumeril had done before, this class into two orders, with reference to the situation of the gills, whether external or concealed (supposing the latter to have gills), under the names of cryptobranchia and gymnobranchia. He also then particularly established some new genera.

Notwithstanding these innovations of the French naturalists on the methodical distribution of Linnæus, the rest of Europe refused to follow the example, and adhered with obstinacy to the code of the Swedish Aristotle. In Germany, however, in 1815, M. Oken returned to the division of Aldrovandus, and though he did not in all respects follow the French naturalists, yet his arrangements were based upon the same principles. It is unnecessary to follow him through the details of his allocations, but though we cannot deny him the merit of extending just views, yet we must accuse him of those perpetual mutations of nomenclature which have proved, to such a signal degree, detrimental to the progress of natural science.

M. de Blainville, and especially M. de Savigny, have expended much and most meritorious labour upon this class of animals. But instead of pursuing this analysis of separate systematists any further, we shall sum up in a few words all that has been done for the arrangement of the *chetopoda* (leaving the others out for the present), from the time of Linnæus to our own day.

We have seen that these animals were scarcely known to the ancients, whose observations were confined to the nereides alone, under the name of *Sea-scolopendræ*. Linnæus, collecting the little which had been left by Belon and Rondelet, established them into four genera, *Serpula, Nereis, Sabella,* and *Amphitrite,* the relations of which he did not appreciate, placing them at a distance from each other, in four different orders of his class *vermes*. Pallas pointed out these relations, and did for them what he had done for the mollusca and testacea. Neither Gmelin nor Bruguières understood the value of his remarks, and profited very incompletely indeed by the detailed observations of Muller and Otho Fabricius. M. Cuvier was the first who executed what the genius of Pallas had devised, and who united all these animals under a single classic name. In this he was followed by M. de Lamarck, and by all zoologists who extended the application of the natural method to zoology. M. de Blainville, in applying his general notions of the methodical classification of animals to the class in question, which he restricted a little, introduced the consideration of the similitude of the rings of the body and of their appendages, for the establishment of orders and families, and even of a great number of genera, almost at the very moment when M. Savigny, after a long series of minute observations, was publishing a general system of these animals, adopted by MM. de Lamarck and Latreille, in which he made known a great number of new species of all seas, adopting always as the basis of his classification the jaws and gills, but still with a consideration of the nature of the setæ, and the division of the appendages into oars.

The small number of animals composing this division of the annelida (the chetopoda), has not permitted any very great differences among zoologists in their systematic distribution; for, after all, whether we give the name of family or even of order to the genera of Linnæus, pretty nearly the same distribution of species will always result. But such is not the case respecting the place which should be assigned to this class in the animal series.

Before the time of Linnæus, the proper situation of any group of animals was a question that but little disturbed the repose of naturalists. Linnæus himself, though his acute mind could have hardly failed to lead him to select the most suitable, yet in the present instance gave himself no trouble of the kind, since he has thrown the chetopoda into three or four different classes. But when natural methods were introduced into zoology, such anomalous distributions could no longer be admitted. M. Cuvier determined by the colour of the blood of these animals, which, without being of the same nature, has the relation with the blood of the vertebrata of being equally red, places them at the head of the articulated animals, and consequently before the crustacea, the arachnida, and the insects. Some objections have been made to this allocation, and with an appearance of reason. It has been said that animals in which the organs of sense are reduced to an obtuse touch, which move with difficulty, without complete limbs, which are hermaphrodite and cannot abandon the habitat of the waters, are placed before insects which enjoy all the organs of sense; which can execute all kinds of locomotion, even that of flight; whose nutriment is so various, or so select; who employ such a multitude of ingenious means to procure it, in which the sexes are constantly separated; and whose modes of assuring the development of their progeny excite our admiration the more highly in proportion as we become better acquainted with them: are the annelida to be raised to this rank because they possess a nutritive fluid of a red colour, or because they have what, after all, may be considered but as the semblance of a circulation?

In reply to all this we may briefly say that, anatomically and physiologically considered (and, after all, considerations of this kind form the proper basis for zoological classification), the annelida are more complicated in their organization than the subsequent classes; they form a more obvious link with the classes immediately preceding, and (be it remembered) that the objection against their being placed before the insects, if it prove any thing, proves too much, for the same objection is equally strong against the allocation of the mollusca in the animal kingdom. But the fact is, that this objection, like others of the same kind, is grounded upon the very untenable position, that the works of nature proceed in a linear series, a position which every true naturalist knows to be equally contradictory to fact and reason. As for the assertion relative to a circulating system, it will not bear examination. Analogy would lead us to conclude that in every organized being circulation must exist, and we are informed that its existence in insects has of late been demonstrated by an English naturalist *. If so, we cannot have any doubt concerning it in the annelida.

M. de Lamarck followed the example of Baron Cuvier, and placed this class between the cirropoda and the crustacea.

M. Dumeril, in his analytical Zoology, has not thought proper to follow the views of MM. Cuvier and Lamarck. He admits, however, with them, that the organization of the annelida is more complicated than that of the insects, and that, according to the natural scale of beings, they ought imme-

* J. Bowerbank, Esq.

diately to follow the crustacea, and conduct to the insects; but, in consequence of their form, of the trifling development of their organs of locomotion, and especially in consequence of the mode of respiration in insects, which appears to hold the place of the circulation of the blood, he thinks that the worms ought to be placed between the insects and the zoophytes; that is, between the class myriapoda, with which he finishes the former, and that of *Helminthii*, or intestinal worms, with which he commences the latter.

M. de Blainville intended to place this class between the myriapoda and the apoda, or intestinal worms. Thus we should pass, almost insensibly, to the sub-annelida, and by them to the holothuriæ, which would properly commence the zoophytes. But it was necessary to place between those two types the mollusca, which, after the vertebrated animals, also form a parallel line, proceeding to, and arriving at the zoophytes, through the ascidiæ.

Respecting the animals of which we are now writing, there is but little of any thing important in the study of them, except what may be derived from it by natural philosophy. Their utility to the human species is but slight indeed. The larger nereides and the arenicolæ are, however, much used as baits for catching fish, especially whiting and mackerel. They even constitute a small object of commerce with the inhabitants of the coasts of the Atlantic and the Mediterranean. The lumbrici, or earth-worms, are also employed to catch fresh-water fish, and among others, eels, and fish in general, that live in mud.

We must, however, now take a brief view of the organization of the chetopoda.

The organization of these animals has been studied in its external parts by Pallas, by Muller, and very particularly by Otho Fabricius, and M. Savigny. As to their internal organization, the little we do know is essentially owing to Pallas. Some genera have also been treated of by Baron Cuvier and Sir Everard Home, and among the rest the Arenicola, which is one of the largest species in our European seas.

The body of the chetopoda, in general elongated and vermiform, is sometimes, however, merely a lengthened oval, the longitudinal diameter not exceeding the transverse more than two or three times, as is the case, for example, in the genus aphrodite. It is also sometimes cylindrical, as in the lumbrici, and some other genera. But the permanent character of the body in this group of animals, is its being divided into a considerable number of rings, segments, or articulations, by transverse furrows, in which the skin, being softer, admits of the movements necessary for locomotion. The number of these segments varies considerably, and in some genera, as, for instance, the serpulæ, becomes an important character for the distinction of species. The rings too, at least in diameter, are never rigorously alike, but are gradually diminished towards the two extremities of the body. The head is constantly distinct, but it seldom happens that it is composed of a single ring. It is even difficult occasionally to decide where it begins, because it is never separated from the rest by a series of articulations, forming any thing like a neck. It is therefore necessary, in applying the denomination of *head*, to take its appendages into consideration.

In the majority of the species, such as the nereides, the amphinomæ, and still more in the lumbrici, not only is there no neck, but it is impossible to find means of separating the trunk into thorax, abdomen, and tail. But such is not altogether the case with the serpulæ and amphitritæ: in those groups a certain number of rings which follow the head are truly different from those which form the rest of the body, and a sort of thoracic region is perfectly distinguishable, and consequently an abdominal one. There is no caudal part, however, visible in any of the chetopoda, no more than in any other of the division articulata.

The rings or segments of the body of the chetopoda are constantly provided with a pair of appendages, either considerably complicated, or extremely simple. These appendages are never, even when most complex, composed of more than three parts: one proper for locomotion, another for sensation, and a third for respiration; and when most simple, have at all events the first. As to their position on the rings, the appendages generally occupy the extremities of the greatest transverse diameter; but it sometimes happens that they are situated lower, and more frequently higher, according to the peculiar uses for which they are intended. Generally speaking, they tend more to an upper, or dorsal position, as they appertain more to the anterior rings of the body; so that when placed upon the rings of the head, they may with propriety be termed *tentacula*. The exact reverse is the case when they are situated behind, their tendency being more and more to an under position, in proportion as the rings approach to that segment which contains the anus.

The extent of the lateral portions occupied by the appendage must, and in fact does vary, according to the complication of the latter. In the Naïdes we find it reduced to a mere point, while in the amphinomæ, and some of the nereides, it takes up more than a fourth of the circumference of the ring.

When the extent of the insertion of the appendage is considerable, it frequently happens that it is divided into two parts, one superior and the other inferior to the lateral line. To these M. Savigny gives the name of *oars*, a name deserving of adoption, as in fact they do serve the animal for the purposes of swimming. Each oar seems composed of the same parts, but arranged in an inverse way, the separation taking place in the fasciculi of setæ. But be this as it may, the appendage in the chetopoda may be composed of a gill, of cirri, of nipple, and of setæ, which we must proceed now to define successively.

The gill (for we have only occasion to consider a single side, as the animal is symmetrical), is always situated at the upper root of the appendage, whether simple or bipartite. This gill, which exhibits a character common to every organ of respiration, namely, that of being extremely vascular, with a very slender dermoid envelope, varies sufficiently in its form, as it may be either simply bifid or trifid, as in the nereides, or considerably ramified into arbuscula, as in the amphinoma; or, in fine, multifid, and longitudinally pinnate, as in the serpulaceæ.

Their position allows us to distinguish them into many kinds: in the regular state, they constantly occupy the upper root of a variable number of appendages, and they are dorsal, as in the amphinomæ, and the larger species of nereis. In that case, a number of them may be attached to all the rings without interruption. At other times they are anterior, and confined but to some rings.

The cirri, which we shall see might be termed *tentacula*, or *tentacular cirri*, on the *cephalic*, or *post-cephalic* rings, are species of filaments not vascular, in length and even in form extremely variable, which may be situated either at the upper part of the appendage, immediately under the gill, when that exists, and which sometimes seem even to take the place of it *(cirrus superior)*, or at the lower or ventral portion of the appendage *(cirrus inferior* or *ventralis.)* The form and proportional dimensions of the latter vary considerably, but in general it is smaller than the upper.

Sometimes we find at the root of the fasciculi of setæ, behind or before, dermoid prolongations of the nature of cirri, but which, shorter and broader, no longer deserve this name, yet are still worthy of notice. They may be designated by the name of *lobules*, *mammillary*, *cirrhous*, or *squamous*, according to their form.

The nipples, or *mammillæ*, are elongations, more or less considerable, of the sides of the segment, at the extremity of which the setæ are implanted. Sometimes they are almost nothing, and then the articulations of the body are very little sensible; at other times, on the contrary, they are excessively long, and then the body appears deeply incised throughout its entire length. It is those which support the lobules, of which we have been just speaking.

The *setæ* (whose structure we will give more in detail by and by), are stiff, hard, fragile parts, which are implanted more or less deeply into the skin of the chetopoda, and, in general, are considerably numerous. There is but a single one in some naïdes. They form simple or divided fasciculi, placed at the extremity of the nipple, and between the two cirri of the appendage.

Modern naturalists, such as M. de Blainville, Otho Fabricius, and Savigny, distinguish three kinds of setæ.

1. The simple setæ, which are slender, pointed, and straight at their extremity; these are the most common, and such as fasciculate the best.

2. Hooked setæ, which are still rather slender, and curved, and terminated by a hook at their extremity.

3. The needles, or spines *(aculei)*, which are straight, like the simple setæ, but which are always thicker, and much stiffer; such are to be found in some of the aphroditæ.

We have now stated all which can enter into the composition of the most complex appendage of a chetopoda.

When it is not divided into two parts, one superior and the other inferior to the lateral line, the appendage is composed of a single oar. In the other case, it is in the form of two oars.

We have already remarked, that even in the chetopoda, whose rings are most similar, there are, however, some differences; but in the genera, those differences are much greater. In general, on proceeding from the most complete segment, which is usually towards the anterior third of the body, and going towards the head, the gills and the setæ diminish gradually in length and strength, becoming, as we have said above, more and more dorsal, while, on the contrary, the cirri acquire a greater development. This may be observed in a very manifest manner in the cirri of the cephalic rings, and even in those which surround the anus in the nereides. These latter preserve the name of cirri; but such is not the case with those which accompany the rings which compose the head. Muller and Otho Fabricius have called them *tentacula.* M. Savigny designates them by the name of antennæ, a denomination which seems altogether improper, and calculated to produce confusion. The name of tentacula is preferable, though after all there is nothing in those organs which can cause them to be compared to the tentacula of the cephalous mollusca, nor to the antennæ of the hexapods, which rather appear to be organs of olfaction. Let this be, however, as it may, the tentacula of the head in the chetopoda are usually perfectly in even numbers; but it sometimes happens that there is one odd one, and that medial. It is, however, only in the aphroditæ that this singular disposition takes place. In our descriptions, we regard as cephalic, or belonging to the head, not only the cirri which are found upon the first ring, but also those which spring from some of the following, and which in general are very well distinguished from the cirri of the appendages by a greater degree of length.

It may be sometimes remarked, that those organs appear to be divided into segments by transverse folds, which has caused them to be termed *articulated* by some writers. But it rather appears that this effect, which is real, often takes place from the action of the liquor in which these animals are preserved, and is not remarked in the first state. Some species, however, constantly present this disposition.

We also find, in a tolerably great number of chetopoda, that the first two or three rings are provided with points, or rather with spots, very distinct, constantly arranged in one manner, and which have been honoured with the name of eyes, though we shall presently see that there is nothing of the sort in their structure, and that they do not in any wise serve the purposes of vision. It is proper, however, to remark, that these points are pretty nearly constant in their number, and in their disposition; so that very excellent zoological characters may be derived from them.

It only remains for us now to notice with respect to the exterior of the chetopoda, their colour, and the tubes or funnels which a great number of their species construct.

A character which appears peculiar to this class of animals, and which of itself alone might almost suffice to make them recognized, is, that besides their proper and fixed colour, the epidermis, or rather the skin, properly so called, appears tinted with colours, irradiated with magnificent reflections of gold or purple.

As to the external tube which the chetopoda often inhabit, although it is often sufficiently regular and solid, it cannot, however, in any manner be compared to the shell of the mollusca, not even when there is the greatest approximation, as in *dentalium* and *siliquaria*. These tubes of the chetopoda are always simple excretions from their body, which are by no means attached to it, and from which the animal may issue forth without dying immediately. We begin to observe something of this kind in the mucosity with which certain species line the hole, hollowed in the mud or sand which they inhabit, as in the arenicolæ and some lumbrici. This is analogous to the mucous pellicle of the tube of the amphitritæ and the sabellæ; but in the latter, surrounding this mucosity, is attached externally a stratum, more or less thick, composed merely of mud or very fine grains of sand, or, in fine, of debris, more or less thick, of shells and larger grains of sand. These tubes are constantly open at both extremities; there are also some of them more regular, which are completely calcareous. The double opening is a character whereby they are distinguished from tubular shells, the summit of which, on the contrary, is constantly imperforate. These last-mentioned tubes, however, appear constantly to grow after the manner of those shells, by laminæ or strata extremely thin, placed inside of and out-edging one another. From this result striæ marking the growth, more or less apparent outside, but we never remark longitudinal striæ on their surface, nor any thing indicating the delicate working of the edges of a mouth, as in the mollusca. This character alone might suffice to distinguish them from the true tubular *shells;* but to this we may add, that the constant perforation of the summit of the tube of the chetopoda never allows the animal, in growing and advancing in its tube, to form partitions there, whereas in the tubular shells the reverse is invariably the case. A final character which distinguishes the tubes of the chetopoda is, that they are adherent, and fixed flatly through a greater or less portion of their extent, on foreign bodies, which never takes place with the tubular shells.

The two last external parts which we have to examine in the Chetopoda, and which naturally conduct to the consideration of their internal organization, are the orifices of the intestinal canal, which constantly exist, and which are pretty nearly always terminal, but sometimes, however, a little oblique, under each extremity, especially the mouth.

This mouth is sometimes immediately at the origin of the body, or at the anterior extremity of the head, as in the lumbrici, and a great number of nereides; but also, in many cases, it is at the extremity of a long anterior elongation of the body, then without distinct articulations, without any other appendages than sorts of jaws, varying in number, and it is eminently retractile. This has been designated under the name of proboscis. We shall see that the mouth, when it is simple, is often accompanied with barbles, or cirruli very numerous, as in the serpulæ.

As for the anus, always very large, and transverse, it presents nothing very characteristic, except, perhaps, in the cirri, which may accompany it, but which belong to the appendages.

The internal organization of these animals has been much less studied than the form of the external parts. It is certain that it presented difficulties of a very different nature.

The envelope, which constitutes the external cylinder of which the body of the chetopoda is composed, is entirely soft, or is never supported either externally or even internally, by any solid part of a calcareous or corneous nature. We merely observe, that it is furrowed transversely, by striæ more or less deep, where the larger part of the body of the animal is situated; but it cannot be said that the envelope is there less hard, than on the rings themselves. This envelope is composed, as usual, of a skin, and a contractile, muscular stratum.

The skin does not, as in other animals, appear to be divisible into the usual parts. It seems to be formed only of a sort of mucous and transparent epidermis, doubtless decomposing the light, by a physical disposition of its parts, or perhaps even by its folds, so very fine that they form fissures. Underneath is sometimes found a true pigmentum, on which the natural coloration of the animal depends, which is very various. It has been found impossible to discover any nervous structure in this, though its sensibility, as we shall see by and by, is very great.

We are not acquainted with any special organ of sensation

in the chetopoda, unless we may consider as such, to a certain point, the cirri of the appendages, when any exist, and especially the tentacula, and the tentacular cirri. Their structure does not appear to differ from that of the general envelope, only they are filled with a substance, which by the action of alcohol, sometimes coagulates incompletely, and is divided into fragments, more or less regular, corresponding to the folds of the envelope, which sometimes gives them the appearance of being articulated. They are sometimes, however, really composed of globular articulations, regular, so as to be completely moniliform, as may be well observed in the species of nereides, which constitute the genus Syllis of M. Savigny.

As to the black points, or spots, which we have said exist at the upper part of the cephalic rings, and which are pretty generally regarded as eyes, they are evidently formed, each by a small flatted globule of a black colour, and lodged in a particular excavation of the dorsal muscular band, interposed between it and the skin, which appears more thin and transparent in this place than elsewhere.

The locomotive apparatus is essentially composed of the subcutaneous muscular stratum and the appendages, especially the setæ which enter into their composition.

The sub-cutaneous muscular stratum, only more thick underneath, and on the sides than above, exists through the whole extent of the body, and forms the greatest portion of its investing sheath. It is essentially composed of longitudinal fibres, divided into superior, lateral, and inferior fasciculi, each separated into two parts, by dorsal, ventral, and lateral lines. These fibres, however, are not extended without interruption, from one extremity of the animal to the other; but they terminate successively, at least in part, opposite to a variable number of rings anterior to that from which they have issued; but there is no more adherence to the skin in one place than in the other. Thus, in the common nereides, the two dorsal muscular bands, separated only by the dorsal vessel, are continued without interruption from one extremity of the body to the other, not, however, attaching themselves successively to the contraction of each ring, or to each transverse furrow. When arrived in front, they contract, and terminate at each tentaculum of the head.

The appendages, in their active, or contractile parts, are really composed like the rest of the skin, with the difference that the muscular stratum is there of necessity much less thick; but it has not appeared that those parts are provided with special muscles.

The passive parts of the appendage, or the setæ of what kind soever they may be, are always rigid and fragile. Their chemical composition would appear to be a mixture of calcareous and corneous matter. Each seta is hollow through its entire extent, at least if we are to judge of all, from the demonstration which has been made in some species. Ordinarily pointed, and harder on the summit, they are on the contrary truncated and soft at the base. We have already seen, that, according to their uses, they are straight, aciculated, or curved into hooks, at their extremity, in which last case, they are always much shorter. Sometimes they are even denticulated, as in the serpulæ.

These parts, usually retractile, and intractile, except in the final genera, are in fact capable of being withdrawn almost completely through proportional holes pierced in the skin. They do not appear, however, either particularly, or in fasciculi, to be provided with muscles which should produce movement. But their extremity, after having traversed the skin, pushes as it were within the longitudinal and lateral muscular fasciculus, which produces sorts of shrouds as to the masts of a vessel. By the contraction of the fibres, the seta is pushed outwards, more or less strongly, without which it re-enters into a state of repose. This is an arrangement which has not been remarked but in this class. There are, besides, some small basilary muscles, which are derived from the lateral contractile stratum, and which, according as they come before or behind the base of the fasciculus of setæ, must carry them forward, backwards, upwards, or downwards.

The other parts of the appendages of the chetopoda, mobile in all directions, extensible and retractile to an extremely remarkable degree, produce all their movements only by means of the subcutaneous, muscular stratum, which enters into their composition.

It is pretty nearly the same with the teeth or jaws. No peculiar muscles have been observed in them, and their movements are owing to those of that part of the subcutaneous muscular envelope in which they are implanted.

The setæ properly so called, of length and breadth so variable, that they are sometimes fine enough, and soft enough to be manufactured into felt, are often disposed in fasciculi. But they are also sometimes like a fan, and in a single rank.

The *aciculæ*, or pointed setæ, are sometimes divided in a tolerably fixed manner in the fasciculi, or bundles, most frequently but one or two in number; but sometimes also more numerous, as in the bristling aphrodite.

As to the hooks, they are always on a single rank, very much crowded one against the other, the hook being directed externally and forwards. The range which they thus form is supported on a linear nipple, not much projecting, and comprized between two lips of the skin, producing, when they are drawn in, a sort of stigma, analogous in appearance to those organs in insects, but in reality much more like the nipples or false feet of the caterpillars, as has been well remarked by M. Latreille.

The digestive apparatus of the chetopoda, is in general very simple, often composed only of a simple cylindrical canal, almost without enlargements, and extended from the mouth to the anus; but in some species a little more complicated.

The mouth is sometimes, as in the lumbrici, and neighbouring genera, a bivalve orifice or not, situated at the anterior extremity of the body; but in a great number of cases, it is no longer at the apparent extremity of the body, but rather at that of two or many of its rings a little modified, and which can re-enter or come forth from the part of the œsophagus which corresponds to a variable number of rings which follow. To this extensible part of the œsophagus in certain chetopoda, the name of proboscis has been given, perhaps erroneously; for there is no particular muscle which can serve to draw it in or put it forth. It is really formed like the double envelope which constitutes the rest of the body, with the difference that there never are any appendages properly so called, but only little accumulations of corneous tubercles, the use of which is not known. Sometimes we remark, besides, at the orifice of the first of these proboscidiform rings some papillary tubercles, or even short barbles, which no doubt assist the mouth in prehension. We also observe pretty often in the nereides a pair of teeth or corneous hooks, curved like a reaping hook, denticulated or not, on the concave edge. These hooks, which have sometimes been designated under the name of jaws, are in general two in number, forming a lateral pair. Four have been observed in a species of nereïs, each occupying an angle of the extremity of the proboscidal ring. Two are capable of being protruded to seize the food, which is then delivered to the other to be masticated.

These hooks are hollow in a great portion of their base, and this cavity gives insertion to some muscular fibres, which, without doubt, assist their movements.

In a tolerably great number of the chetopoda, the anterior part of the intestine is provided with a true buccal mass, much more complicated, and capable of being carried forward or backward, by fasciculi of muscles, which, belonging to the sub-cutaneous muscular stratum, are carried, either from the anterior margin of the buccal orifice, to the anterior half of its mass, or from the rings which immediately follow it behind, to its posterior half. We find, besides, that it is in a great measure composed by numerous transverse fibres, which must act strongly in mastication.

This mastication is performed by corneous or calcareous parts, which cover some longitudinal folds, corresponding to each other by pairs, and they often extend a considerable way into the entrance of the intestinal canal. The number and form of these various kinds of teeth varies sufficiently. No true odd one exists among them, that is to say, in the middle dorsal or ventral line; but almost always the two teeth which constitute the inferior pair, are contiguous, in the medial line, which forms a sort of under lip. The lateral pairs are variable in number, and most frequently composed of a sort of handle which is introduced into the muscular stratum, and of a free curved part, which may be either denticulated or not. It is this last part, which, being susceptible of being sometimes augmented at one of its angles by a corneous tubercle, has been the cause, that M. Savigny has defined, in a group or two of the nereides, some of the genera which he has established among the multidenticulated species, according to the odd or even number of these organs, which he has denominated jaws. M. de Blainville declares that he has found them always even, with the difference just noticed,—and which does not take place constantly on the same side, as M. Savigny supposed, but in an entire genus, that of aphrodite, the teeth which constitute the upper and lower pair, are contiguous, and touch each other in the middle line, from which it results that they act, like the jaws of bony animals, in a vertical direction, two against two.

At the end of this first part of the intestinal canal comes the œsophagus, which in the species where there is no buccal mass, is the direct continuation of the buccal cavity; but which, in the others, takes its origin at its upper part, so that the plane of the latter is much lower.

Accompanied in its traject by some salivary glands, often rather long, at least in the species where mastication takes place, or shorter in those in which this function does not exist, the œsophagus is more or less dilated, and constitutes the stomach.

This part of the intestine is sometimes enlarged, solidified by muscular fibres pretty thick, and constituting a sort of gizzard, as may be seen in the lumbricus terrestris. But most frequently it is membranaceous, and is continued, through the whole thoracic portion of the trunk, undergoing some dilatations, more or less marked, or being provided with sorts of cœca opposite to each interval of the articulations. These dilatations are often determined, in a great measure, by the plenitude of the stomach, and then its parietes are very thin. In the contrary case, they have a very great thickness, as much in consequence of that of the muscular stratum, as of that of the mucous membrane, which forms some considerable longitudinal folds. This disposition belongs essentially to the nereides.

In the amphitritæ, including the pectinariæ, the stomach is long, tolerably thick, but without any trace of cœcum. The same is the case with the terebellæ.

It does not appear that any distinct liver has yet been observed in the chetopoda, nor any organ which might perform its functions, otherwise than some granulations in the thickness of the intestine, as in the nereides, or which often terminate the cœca, and even form a projection under the skin; such may be considered the small denticulated crests which are seen behind the root of the appendages in the *aphrodite aculeata,* and which have been regarded as gills. In fact, in the family of the aphrodites, and especially in the principal species, A. Aculeata, the stomach, entirely membranaceous, is furnished throughout its whole length with long cœca, pedicled, or narrowed at the place of their communication with the intestinal cavity, and which proceed transversely, and dilating, even into the interval of the rings. Nothing similar seems to exist in any other genus.

The rest of the intestinal canal, most frequently without convolutions, and without any notable difference of diameter, proceeds directly to the anus, which is constantly terminal, and ordinarily very large and transverse; but in the pectinariæ it is not the same, the intestine making two convolutions of the length of the body before it terminates at the anus, and those convolutions being united by a sort of very slender mesentery.

The apparatus of respiration is not always specially marked in the chetopoda. In fact, in the final genera, it is impossible to find the modifications of the skin, proper to constitute lungs or gills.

When this apparatus is specially marked, it always forms true exterior gills, the form and position of which are sufficiently different.

In the serpulæ and amphitrites, those organs situated on the back of the labial ring, are formed by long cirri, furnished with two ranks of denticulæ, very short, and supported on a sort of pedicle, as it were lamellose. We find that they are almost entirely composed of a very thick vessel, which, seen with the microscope, appears a sort of trachea, analogous to what is observed in insects, its parietes being, in fact, supported by transverse fibres.

In the sabellariæ, the pectinariæ, and the terebellæ, the gills ramified like little shrubs, occupy the lateral portions of the cephalic rings, and it really appears that these are the vessels themselves, ramified, and covered by a skin extremely thin, so much so, that in the living state, they are of a blood-red extremely lively.

In the arenicolæ, they are again arbuscular, but they occupy the upper part of the root of the thoracic appendages.

In the amphinomæ, and the multidenticulated nereides, they are equally dorsal, and on a more or less considerable number of rings of the body, but they are merely pectinated and slender; those of the last rings of the body finish by even becoming unilobate or cirrhous.

Finally, none, unless indeed in a rudimentary state, are admitted to exist in the aphrodites, in many genera of nereides, in the lumbrici, naïs, &c.

The circulatory apparatus of the chetopoda, simply presents the vascular portion, without any heart, or true organ of impulsion. This vascular part is doubtless composed of two orders of vessels, but much less distinct, in any point of view, than in the higher animals.

We may nevertheless consider, as belonging to the venous system, a single thick vessel, somewhat flexuous, without swellings or dilatations, and which occupies the middle ventral line above the nervous system. It is the result, without doubt, of the branches which come to it transversely, on each side of each ring, in the whole length of the body, except in front, where it receives three thick branches, a medial, which is placed below the buccal mass, and which is really the continuation of the trunk, and two lateral, one on each side, much stronger, and which bring back the blood from the buccal mass itself by irregular branches, from its muscles, and from the skin of the first rings. All this process is very observable in the *nereis gigas*.

In the species which have the gills attached to the first rings of the body, as the serpulæ, the amphitritæ, and in general all those inhabiting sand, the vessels returning from the different ramifications which compose them, render complicated the anterior part of the venous trunk, or of its two branches, like a V. In the species in which the gills are dorsal, and on a great number of rings, the veins which return from them, belong to the corresponding transverse branches.

From the anterior extremity also of the bifurcation of the medio-ventral vein, spring the principal branches of communication with the arterial system. These branches, placed on the sides of the œsophagus, ascend towards the back, and end at the dorsal vessel.

The arterial system is formed by a thick medio-dorsal, or rather intestinal vessel, evidently swollen from ring to ring, at least in its action, by the blood which it contains, and furnishing on the right and left, some transverse vessels, which, when arrived at the root of each appendage, divide into two branches: one which proceeds forward, and the other behind. Each of these branches is itself divided into two branches; one which returns internally for the hepatic lobes, and the ovaries; the other which goes to the branchial part of the appendage. This, at least, is sufficiently observable on some nereides in the living state.

With respect to the species, whose gills are at the cephalic rings, and much more complex, it is easy to see that the lateral branches of the dorsal vessel, must be much thicker, and that they are continued in the branchial cirrus, and in its ramifications.

In the lumbrici, the thick dorsal vessel, with its pulsations and its lateral transverse trunks at each ring, are equally observable, but without the branchial ramifications. There have been merely remarked, in front, two vessels, occupying towards the end the medial line; one straight and smaller; the other flexuous and of a more considerable calibre, neither one nor the other having pulsation.

The apparatus of generation in the chetopoda has not yet

been studied in a sufficiently complete manner in all the principal genera, to render it possible to deduce any clear general conclusion concerning it.

It appears extremely probable that the two parts of the apparatus are distinct, but that they exist in the same individual, which produces what is termed an incomplete hermaphrodism. This seems at least to be absolutely true of the lumbrici.

In the terebellæ the ovary appears to constitute a white body, depressed, bifurcated behind, occupying the upper face of the abdominal muscular plane, from the head as far as the ninth articulation. Its communication with the exterior is made by a medial orifice, situated on the anterior part of the ventral disc.

In the *pectinariæ* (first division of the Amphitrite in the text, see note), the female organ is constituted by a pair of oval corpuscles, situated altogether at the anterior part of the body, on each side of the origin of the œsophagus, and which, according to the observations of Pallas, scarcely the size of a double seed of millet during the greatest part of the year, swell early in the spring, and form considerable masses, filling all the anterior part of the body, and composed of a great quantity of white grains.

The female part of the nereides consists in a series more or less considerable of globular masses, of a yellowish white, grained, placed between each gastric enlargement, without adherence or communication with it, and, on the contrary, terminating by an adherence to the skin, immediately at the upper root of the corresponding appendage.

In the terebellæ the spermatic vesicles are four pair in number, the anterior two being smaller. They are implanted in the interstices, which separate the peduncles of the appendages from the fourth as far as the sixth; and their external orifices, in the form of transverse chinks, are perceived, though with difficulty, in the furrows which separate these peduncles.

In the pectinariæ these vessels are placed in the same manner; but they are only two pairs in number, situated in front, and adhering to the peduncles of the second and third pair of feet, behind the gills. The liquor which they contain is, according to Pallas, of a yellowish colour, like bile.

The male part in the nereides is perhaps formed by a series of corpuscles, ranged by pairs on each side of the nervous cord, but only for the first three and twenty rings, diminishing in volume by little and little, in proportion as they approach from the middle part more and more to the extremities. It does not appear, however, that any canals arise from these organs, which should establish a communication between them, and still less any communication with the exterior.

The nervous system in all the chetopoda consists, as in all the articulata, of a series of ganglia, situated in the medio-ventral line, in as great a number as the body is composed of rings, often naked in the visceral cavity, but often also underneath a portion of the sub-cutaneous muscular stratum. Each of these ganglia is united to the following by a double cord, very distinct, which constitutes an uninterrupted thread from one extremity to the other of the animal, enlarged at intervals. It is from these enlargements that subsequently proceed in radii the threads which go to distribute themselves, especially to the fibres of the muscular stratum, whether it be sub-cutaneous or sub-mucous. The first, a little thicker than the others, does not appear to furnish more than two thick branches, which proceed from each side of the head, and which, arrived at the root of the tentacular cirri, are divided into threads for each of them. This is particularly observable in some nereides.

In the amphitritæ, we may very well trace along the branchial cirrus a nervous thread, which accompanies the vascular system.

One of the singularities presented by the chetopoda is, that their recrementitious fluid, or the blood, is almost constantly red, the aphrodites excepted. Of the cause of this we are entirely ignorant.

Their phosphoric property is altogether remarkable, at least in the smaller species. From this Linnæus long since denominated one of their species *nereïs noctiluca*. M. Viviani has likewise noticed one of them among the animals to which he attributed the phosphorescence of the waters of the sea of Genoa.

The chetopoda have in general been too little observed in the living state, and especially for a sufficient time, to enable us to know any thing with certainty respecting their physiology, or their natural history.

We only know from the experiments of Muller, that the nereides and the naïdes are capable of reproducing the parts of their body which have been amputated.

The chetopoda are almost all aquatic, the earth-worms excepted, and even those, to a certain point, may be regarded as such, so much need have they of humidity, and so much do they fear, and are injured by drought. A great part of these animals live in the waters of the sea. It may be even remarked, that there are few marine beings that perish so soon when they are put into fresh water. The majority of the naïdes live in fresh water. It would appear, too, that though the nereides generically belong to the sea, that some true species may be found in the lakes of North America.

Some of these animals are to be found in all quarters of the world, and, except the Amphinomæ, which have not yet been met with except in the seas of warm latitudes, and particularly in the Indian, all the other genera have some species in all seas. We may observe, however, that the largest species come to us from the Indian seas.

It is in general on the shores of the sea, in the midst of thalassiophytes, in the anfractuosities of the madrepores, of rocks, in the sand, and particularly in mud, that the chetopoda are to be found; and if some species are more commonly to be met with in the open sea, as, for instance, the amphinomæ, named by M. Savigny *Pleione vagans*, it appears that they may have been drawn along with marine plants by the currents, as is the case with many other animals.

A great many species are free and wandering, but others live in a tube without being fixed there.

The position of this tube is rarely horizontal, and when it is vertical, should any chance displace it, the animal torments itself until it has recovered its proper station, as Pallas has frequently proved by experiments on the pectinariæ.

The locomotion of animals of this class is in general rather slow, and may be compared to that of slugs, to a certain extent, though they have a great number of feet, or at least, as we have seen, of appendages that serve for locomotion. This is the case with the species which are most favoured in this respect, such as the nereides. In the aphroditus it is infinitely slower; and is reduced to a nullity in the serpulæ, the amphitrites, and the sabellariæ: in fact, they can only rise or sink in the tubes which they inhabit, by their fasciculi or hooked setæ.

The nereides not only creep in a serpentine manner on the surface of solid bodies at the edge of the water, but they often swim very well, either by successive undulations of their bodies, like eels and serpents, or by even agitating their appendages, which serve as oars.

The chetopoda appear to be for the most part carnassial, and feed upon smaller animals than themselves which come within reach of the species inhabiting tubes, or which they go in search of when they can move, as do the aphrodites, the amphinomæ, the nereides, &c. The species which have the mouth armed with corneous or calcareous teeth, whether trenchant or molar, must use as food living animals, and often such as are of a considerable size. Nereides have even been found in the holes of teredines, from which it has been concluded, that the former preyed upon them. The larger nereides with many teeth may doubtless attack pretty large animals, even fish perhaps not excepted.

There are chetopoda, on the other hand, which appear to feed only on organic molecules, or at least on the portions of organized bodies contained in the soil which they inhabit: such are the arenicolæ and lumbrici, whose intestinal canal is constantly found filled with earth and sand. It is true that their mouth is but a simple orifice, without any buccal apparatus.

The means which the animals of this class employ to procure their food cannot certainly be very inventive; for those which live in tubes, it is sufficient to agitate in all directions the barbles with which their heads are adorned, to draw towards their mouth a current of water, which must bring with it a certain number of little animals, or even to seek them and draw them towards the mouth by a sort of prehension, executed with the assistance of their branchial cirri, or of the barbles, when they are provided with them. Certain of those tubicolar species may thus remain in ambush at the entrance of their tube, and so much the more easily, as it is often composed of grains of sand, or of particles of shells, in all respects similar to those which constitute the surrounding soil.

The free and vagrant species may go in search of the objects which suit them, but in all probability they lay no ambush.

The lumbrici, thalassemæ, and the arenicolæ, have only to

excavate their habitations to find in the detritus of the materials which they swallow the substance which is to support them.

We know very little concerning the mode of reproduction in the majority of the chetopoda; it is in fact extremely difficult to make direct observations upon them in the bosom of the waters, and often in the soil itself which they inhabit; and it is not less difficult to institute experiments for the purpose of ascertaining any information on the subject.

We do not know whether all the species have need or not of the approximation of two individuals to propagate their species, as is certainly the case with the lumbrici.

It is only certain, that towards the commencement of spring, in our seas at least, we find the bodies of these animals filled with eggs, or with a milky, and as it were spermatic matter. In South America, we know from Darfel, quoted by Pallas, that the amphitrites at Curaçoa are in full vigour in the months of September and October; that at these times they deposit their eggs, and that the young ones come forth in the month of November.

The number of the young is immense; but are they deposited in the egg state or in that of the living animal, or in particular localities? These are facts of which we know nothing, and we are likely very long to remain in such ignorance.

The species which live in tubes are certainly not born with one, as is always the case, on the contrary, with the conchyliferous mollusca, because with the latter the shell in reality forms a part of their skin, which is not the case with the tubicolous chetopods. It is then probable that the very young chetopod has no tube, but that it forms one immediately after it has been expelled from the bosom of the mother.

We have still less information respecting the duration of life in these animals; we even know nothing in this respect about the earth-worms, which we find in such abundance in our gardens, and with which it might be useful for us to become better acquainted.

The chetopoda are of no great utility to the human species. It appears, however, that the larger species may be made available in the way of food. Pallas relates that some inhabitants of the coasts of Belgium eat the buccal portion or mouth of the *aphrodita aculeata;* but this must prove a very poor resource indeed.

The larger nereides, the arenicola, the clymenæ, the tipunculi, and even the lumbrici, are employed as baits with great advantage in hook or small net fishing. A tolerable number of fish are taken in this way alone; and it has been remarked, that the fishing is more successful when these worms can be employed in the living state.

These animals, notwithstanding the small number of cases in which they can be useful to us, are nevertheless more advantageous than hurtful. The earth-worms themselves, by dividing the earth, facilitate the development of the roots of the plants of our gardens.

As M. de Blainville, in his second class of worms *(apoda)* has thought proper to include along with the last division of our author's annelides (the *abranchia asetigera*), intestinal worms, we cannot in this general sketch avail ourselves of his observations, valuable as they otherwise may be. We must therefore reserve what we have to say concerning these animals, such as *Hirudo,* &c. until we come to observe upon them in their proper place. We shall now proceed to consider successively the orders and genera of the Annelides of the text.

We begin of course with the order

TUBICOLÆ.

Under the denomination of SERPULÆ (our first genus), coming doubtless from the word *Serpere,* to creep, Linnæus very early distinguished a pretty great number of marine tribes, which, adhering to immerged bodies, appear in fact to creep upon their surface, from the great number of inflexions which they form there.

Notwithstanding the labours of Linnæus and his editor Gmelin, it was only since the introduction of the consideration of the animal into the systematic disposition of shells and testæ in general, a consideration owing to Pallas, to Poli, and afterwards to MM. Cuvier and de Lamarck, that any kind of certain order could be established in the great genus Serpula of Gmelin. M. de Lamarck was the writer whom the nature of his labours most necessarily conducted to this task, and to him is owing the definitive establishment of the genera *Vermetus* and four others, which are now ranged under the type of the mollusca, and also of the genera *Spirorbis*, and some others, which have remained among the chetopoda or annelida, alongside of the true Serpula. Nevertheless, it would appear that science did not possess any just character whereby to distinguish the tribes which belonged to the mollusca from those of the chetopoda, until it was furnished by M. de Blainville. We shall briefly notice his distinction here. A tube or funnel of a molluscous animal, usually free through a great part of its extent, is never pierced at its extremity or summit, and its cavity often presents a series of partitions, which are formed in proportion as the animal in growing larger, has been obliged to abandon the most narrow part of it; sometimes this part of the shell is entirely filled and solid, as in mugil. The testa of a chetopod, or tubicola, however solid it may be, is, on the contrary, constantly, at least at its origin, fixed, applied by its ventral face, on a foreign body, and does not rise more or less, except towards its termination, which varies according to the form and the extent of the sub-posed body. It is always pierced obliquely towards its origin or its summit; and there are no partitions or divisions whatsoever in any part of its cavity. This is in strict relation with the organization of the animal, whose anus is always terminal and posterior, and which, besides, has no sort of adherence with its tube.

The genus Serpula may be thus characterized: animal moderately elongated, a little depressed, composed of a great number of articulations very much crowded, consisting of an abdomen and cephalothorax, tolerably distinct; the head, or first segment, is larger than the others, and has for appendages above a pair of elongated tentacula, dilated into an operculiform disk, radiated at the extremity, and one of which alone is completely developed. On each side is a tolerably large gill, in the form of an unilateral comb, composed of a variable number of long cirri, furnished with a double internal rank of mobile barbs; the thorax is short, with a sort of membranaceous sternal plate inferiorly; the appendages of the thoracic and abdominal rings are divided into two oars, the upper one provided with a fasciculus of subulated setæ, returning towards the back; the lower with a range of hooked setæ, for the thoracic rings, but the reverse for the second. The testa is in the form of a conical tube, solid, entirely calcareous, irregular convoluted, free in a portion of its termination, and fixed by its summit; the aperture is rounded.

The characters just assigned to the genus Serpulæ, after the most common European species, will suffice to make known the general form of the animal, and all that presents itself externally in relation to it, adding, however, that it is contained in a membranaceous envelope, as are all the tubicolæ—an envelope which, without doubt, produces the shell, always much larger than itself. As to the organization of these animals, it has been very little studied.

The intestinal canal commences by a buccal orifice, altogether anterior, provided with two lips, without any trace of teeth or proboscis. This orifice conducts into the intestinal canal, which proceeds directly to the anus, is always membranaceous, and even presents no very distinct gastric enlargement. The anus is altogether terminal, and very small.

The manners and habits of the serpulæ are extremely simple. Constantly fixed by their tube on bodies submerged in the sea, at tolerably great depths, all their movements are confined to advancing more or less out of their tube, so as to push forward the thorax, but rarely beyond that, and the gills more especially, which they develope like a fan, and agitate here and there. This partial sally from the tube, is doubtless performed by means of the hooked appendages, which are directed hindwards, taking their resting point on the parietes of the testa, somewhat in the manner that chimney-sweepers mount up our chimneys. On the least danger, which is indicated by the more rapid motion of the water, the animal sinks into its tube, deeply enough for its tentaculum, when dilated, completely to close the aperture, and thus serve as an operculum. We know but little concerning the sort of food used by the serpulæ, and still less concerning their mode of reproduction. It is said, however, that they are nourished by aquatic animalculæ, which they seize by the help of their branchial tentacula. The origin of the tube in a part which is open obliquely, leads us to suppose that the young animal is altogether naked, and that it forms no calcareous envelope, until some time after its birth.

It is known that species of this genus exist in all seas. Nevertheless it would be difficult enough positively to ascertain their geographical distribution, because it is possible that they might be confounded with the vermeti, or other animals with tubes. We find, however, from the researches of M. Savigny, on the species of this genus,—the only examination in which attention has been given both to the animal and the shell—that there are serpulæ in all our seas, in those of India, and in those of America, and that the largest come from the seas of warmer climates.

We have already said, that M. de Lamarck divides the serpulæ of Linnæus into four genera: these are serpula proper, spirorbis, vermilio, and Galeolaria. The spirorbes differ from serpula only because the entire testa is constantly applied to some body: vermilio, which belongs to the gasteropod mollusca of the "regne animal," and differs not from vermetus,—because the operculiform tentaculum of the animal is surmounted by a small testaceous piece; and galeolaria, the second division of serpula, in the text, because this testaceous piece, operculiform, is not simple, but very complex. M. Savigny has not adopted this arrangement, nor indeed mentions any species with calcareous operculum: his disposition of the species is deserving of all attention in consequence of the above-mentioned principle which he has followed, of taking the animal into consideration, as well as the envelope.

The genus SPIRORBIS was established by Daudin, for the species of Serpulæ by Linnæus and Gmelin, whose testa, adherent in its entire extent, is rolled flatly, in a manner almost regular, and thus forms a sort of planorbic shell. As for the rest, the Spirorbes have all the characters of the true Serpulæ, the animal perhaps differing still less than the shell. Accordingly, M. Savigny has rejected this genus from his distribution of Annelida.

The manners and habits of the Spirorbes do not differ from those of the other Serpulæ: they are always very small; some of them are to be found in all seas, fixed upon every species of marine body, dead or living. M. de Lamarck characterizes five living species, but it is not improbable that there is a greater number in existence.

The name SABELLA has been employed by Linnæus in the tenth edition of the Systema Naturæ, and afterwards by Gmelin, in the thirteenth edition of the same work, to designate a genus of his order of testaceous worms which he defined: animal similar to the nereides, open mouth, with two thick tentacula behind the head, contained in a tubular shell, composed of grains of sand retained in a vaginal membrane. But as at this period, the animals which form the arenaceous tubes, were but very ill known, it is not surprising that attention was given rather to the tube than to the animal, for the purpose of increasing the species of this genus. To this was doubtless owing the addition made by Gmelin to the species of Linnæus, of a great number of the cases or sheaths of the *phryganeæ*, or of neighbouring genera of insects, the larvæ of which are found in fresh water, and of which, in imitation of Schröter, whom he has unluckily copied, he makes as many species, as there are different bodies which enter into the composition of those tubes. Some modern zoologists, and particularly the French authors, have adopted this genus with some modification, and others have taken no notice of it.

Thus, M. de Lamarck, in the first edition of his Invertebrated Animals, makes no mention of the Sabellæ. It appears, according to his definition, that he arranged the known species under Amphitrite; but in his last work, he established under the name of *Sabellariæ*, a new genus for one of the species of the Sabellæ of Linnæus.

M. Cuvier, in his "Tableau Elementaire," appears to have intended to place the true Sabellæ of Linnæus in the genus Amphitrite; but in the Regne Animal, he has named Sabellæ, those chetopods, whose tube is usually clayey, whose gills are fan-like, and without appendages in the form of a comb. Of the habits of these animals there is nothing to remark.

To pursue this genus through the subdivisions of the text would be useless. We should find nothing to entertain our readers, but the controversies and blunders of naturalists, and the everlasting revolutions of nomenclature.

Terebella is a genus of Chetopoda, with artificial tubes,

established by Linnæus for a tolerably great number of vermiform animals, belonging to our author's red-blooded worms, or annelida, which live upon our coasts, sometimes imbedded in the sand. This genus, adopted by all zoologists, but rectified, and purged of the species not naturally belonging to it, may be thus characterized: body elongated, subcylindrical, inflated in its anterior third part, attenuated behind; head not very distinct, but yet formed of three segments; thorax of twelve, and provided underneath with a sort of sternal scutel, lengthened as far as the twenty-first ring; abdomen cylindrical, and composed of a great number of articulations; mouth subterminal, bilabiate; upper lip advanced, and furnished above with a great number of unequal and filiform barbles, cleft underneath, and prehensile; no tentacula; gills in the form of arbuscula, two in number, or four, or six, disposed in pairs on the first, second, or third thoracic ring; feet dissimilar; the thoracic, with two compound oars, the dorsal with subulate setæ, the ventral, with a double rank of hooked setæ, and the abdominal consisting of hooked setæ simply; the tube cylindrical, open at the two extremities, membranaceous, and invested with large grains of sand, and fragments of shells.

The organization of the *T. conchilega*, has been examined by Pallas under the name of *nereis conchilega*. The exactness of his description has been verified by M. de Blainville. The body of this animal, of a whitish colour, with a rose-tint arising from the irradiation of the red colour of the sanguine vessels, is rather elongated, lumbriciform, a little depressed, nevertheless rather more convex above than below, and gradually attenuating behind. Under the belly is a narrow, flat, small band, which commences immediately after the head, and which terminates by being narrowed and attenuated a little beyond the middle of the body. It is divided into many parallelograms, by the scissures of the segments.

The anterior third of the body, a little broader than the rest, forms a sort of thoracic region, composed of seventeen segments: they are provided on each side with a pencil of setæ, above and underneath, with a sort of vertical cleft, with two lips, pretty similar in appearance to a long stigma, but which is really produced by the re-entrance of two very crowded ranks of setæ, very short, and curved hook-wise. In the rest of the body, composed of more than one hundred and fifty segments, there are no other appendages but those ranks of hooked setæ.

The head of the terebella is but little distinct, and appears as it were truncated. We may nevertheless recognize there, by means of the appendages, several segments. The first, or labial, which advances more or less above the mouth, in the form of an epiglottis, and which supports at its upper face, a variable, but often a considerable number of cirrous barbles, of different lengths, and which are formed by a sort of shred, bending longitudinally underneath. The second ring constitutes the posterior lip; it is much more narrow above than underneath. The third supports, on each side, a sort of membranaceous lobe, and forms a kind of neck.

After the head, come the first three thoracic gills; the gills are in the form of arbusculæ, more or less ramified, and rather dorsal. The third pair is accompanied, underneath, with a pencil of simple setæ, but without any range of hooks.

The mouth very large, is anterior and inferior.

The anus is terminal, folded, and circular.

The organs of generation (or, to speak more properly the ovary), are terminated, according to Pallas, by a medial orifice, situated at the edge of the first segment of the abdominal band.

In the organization of the terebellæ, we may remark, as in all the chetopods, that the sub-gelatinous epidermis, is recovered with the greatest facility, from the rest of the cutaneous envelope. The dermis is confounded with the muscular stratum, which is divided into five longitudinal bands; a single dorsal, and two inferior pairs. It is between the inferior edge of the dorsal, and the superior edge of the external abdominal band, that the setæ are implanted.

The intestinal canal is retained to the parietes of the visceral cavity, only by a small number of fibrillæ. It is extremely narrow at its commencement at the mouth, and without any traces of teeth in the buccal cavity, but it increases by degrees, in bulk, and after a sensible and sudden contraction, it forms the stomach. The latter, corresponding pretty nearly to the eighth articulation, is membranaceous, a little swelled in the middle, and at the twelfth, or thirteenth articulation. It becomes a little more narrow, but more fleshy in the place where the intestine, properly called membranaceous, exists, which is continued directly to the anus, surrounded immediately by the external envelope, to which it adheres in all parts.

The respiratory apparatus, altogether exterior, is formed by gills above described, which are evidently only ramified vessels enveloped with a skin considerably attenuated.

The latter, proceeding no doubt from the ramifications which return from the gills, follow, as usual, the inferior and medial part of the abdomen.

The apparatus of generation consists of a female, and of a male part.

The female part is composed of a single medial ovary, occupying the whole inferior face of the visceral cavity, as far as the ninth ring. Pallas tells us that it terminates behind in a bifurcation. This mass, which presents irregular, transverse strata, is white, and is composed of a great number of oviform grains. It is terminated by a single orifice, of which we have just spoken.

The male part consists of four pairs of small pyriform vesicles, the base being behind, and the neck forward, placed on each side of the anterior moiety of the ovary. The anterior pair is much smaller than the others, and especially than the two middle ones. Their termination takes place externally, by very small orifices situated between the fifth and sixth pair of feet, for the fourth pair; between the fourth and fifth for the third; between the fourth and second for the second; and between the third and last, for the first.

These animals live often assembled in great numbers, in those parts of our coasts where plenty of sand is to be found; and especially argillaceous sand, for, although their tube often contains sand, properly so called, fragments of shells enter still more into its composition. These tubes often exceed, by two or three inches, the surface of the ground in which they are implanted. When the sea covers them, we see the animal put forth its barbles and gills, and agitate them in all directions. The last are of a fine blood-red, and have so much sensibility, that if they are touched, they contract, the blood is expelled from them, and they are no longer red. The barbles of the mouth are not coloured by the blood. They are in a perpetual movement of contortion, of extension, and through a property dependent, no doubt, on the mucous matter secreted by their surface, as well as on the particular form of their inferior face, they adhere with a great facility to all the bodies which they touch. It is, in fact, with the assistance of these organs, that the terebella constructs its tube; its base is a glutinous membrane, to the surface of which the foreign bodies are cemented. They are pretty nearly cylindrical, a little narrowed, however, behind, but broadly opened upon that side, as well as the other. They are uncovered for the space of an inch from their origin. Some appear to have been found totally so, as Pallas notices in the case of individuals, which he has observed on the masses of eggs of the buccinum undulatum.

When the terebella is withdrawn by force from its tube, it

seems probable that it can proceed to form another. Pallas, in fact, informs us, that when it is very much annoyed there, it will issue forth spontaneously, and make its escape by a vermicular kind of movement.

The AMPHITRITE is a genus of marine worms, whose character consists in having the head furnished with two pieces of a metallic lustre, and similar in form to combs. These animals, for the most part, inhabit tubes, which they form by agglutinating small grains of sand or fragments of shells, and which are sometimes fixed, and sometimes free, according to the species. The body of the amphitrite is in the form of an elongated cone, and is ordinarily terminated by a long and tubular tail. The rings which compose the body support all, or in part, on each side, a fasciculus of some stiff setæ, which the animal employs for motion. There are also at each of these rings some fleshy filaments. The gills of the amphitrite are attached to the anterior part of the body only, as happens with all the worms that inhabit tubes, because gills attached to the other parts of the body, which are sunk in the tube, would be useless for respiration; these gills are in the form of a plume of feathers, and their greatest number is four pair. The head of the amphitrite is as it were truncated: around its mouth are filaments, more or less numerous, which serve as tentacula; the two brilliant combs are placed underneath. It is probable that the animal makes use of them to attract towards it the bodies which are to become its prey, or the grains of sand which it desires to attach to its tube.

The amphitrites have, for the most part, a skin so fine, and so transparent, that all their interior is perceptible through it; this interior generally consists only in an intestinal canal, as thin as the skin itself, and almost always filled with sand. There are, however, some species which have a muscular gizzard. We also observe in them a vessel filled with red blood, longitudinally directed, and which in the living animal exhibits very marked contractions.

M. Cuvier originally attached both his Sabella and Terebella to this genus.

The SIPHOSTOMÆ is a very remarkable genus of these animals, established by Dr. Otto, in a dissertation printed at Breslau in 1820, on an animal discovered and observed first on the coast of Naples in 1818, and which he has thus characterized: body cylindrical, articulated, elongated, attenuated at the two extremities, enveloped in a skin extremely thin, diaphanous, provided on each side with a double series of setæ, directed forwards, and the anterior of which approximated, forms two sorts of advanced combs; the mouth inferior, subterminal, with a mass of cirri extremely numerous in front, and a pair of tentacular cirri behind, composed of two orifices, placed one before the other, the first smaller, canaliculated at the base, with an advancement in the form of a proboscis, and the second much broader and more rounded behind. However extraordinary may be this character of a double mouth, an arrangement not known in any other animal, M. Otto has described and figured it with so many details, that it would be difficult to deny its existence, however one might be at first disposed to do so. Some doubts, however, may attach to the use of these two orifices, as one of them might very well be supposed to belong to the apparatus of generation. Be that as it may, we shall present our readers with M. Otto's description of this curious animal, which he names *S. diplochaites*.

Its body cylindrical, elongated, flexuous, and about three inches long, is attenuated at the two extremities, but especially behind; at the distance of about half an inch from the anterior, it presents an enlargement, an indication of the place occupied by the viscera; the number of segments of the body is about forty, but they are not very distinct, except on the side of the belly which is flatted; the sides of the body are bristled with a great number of stiff setæ, long, thick, especially in the middle, of no great lustre, and whitish, forming two longitudinal distant ranges, each ring supporting two of these setæ at each side. A singularity which they present is, that they are all turned forwards, contrary to their arrangement in all the other annelida; the setæ of the rings which compose the anterior extremity are very large, crowded against each other horizontally, so as to imitate on each side a sort of comb, directed forwards, as in M. Lamarck's division of amphitrite *(pectinaria)*, and furnished at its root with a considerable quantity of tentacular cirri, extremely short and labial; between these two fasciculi, and at the inferior face, is the head, properly so called, of a conical form, adherent to the body by the summit of the cone, and elongated anteriorly into a small proboscis; it is at the base of this elongation that the first buccal orifice is placed, which is continued in a sort of gutter for its whole length, and which M. Otto considers to serve as a sucker; the second mouth is farther back; it is much larger, and surrounded by a labial pad, in the form of a horse-shoe, at the posterior part of which is a pair of tentacula, sub-compressed, mobile, sub-articulate, and with a deep furrow on the edge; the anus is rounded, large, and altogether terminal. No other orifice has been observed on the exterior of this animal.

The cutaneous envelope, rather thin, and transparent enough to let the nervous and vascular system be seen through it, is formed of two laminæ, one of which is the skin, properly so called, and the other, which M. Otto names the peritoneum, still thinner, is very little adherent; the latter, at the anterior third of the body, separates the anterior cavity into two very unequal parts, by a sort of diaphragm, pierced only by the intestine; it is in the anterior part that lie the principal viscera. The two mouths have each an œsophagus, about an inch long, which communicate by a lateral orifice partly with a large bladder, which M. Otto thinks to be the stomach, and partly with a single intestine, which continues it; the first œsophagus, more narrow, has been most frequently found empty, but sometimes filled with a whitish juice, while the second was always filled with the same brown matter as the rest of the intestine; the latter, placed under the other, has also more ample and more solid parietes. At the place where they approximate to form a common intestine, both adhere with the gastric bladder and with the intestine, but the first more with the former, and the second with the latter; so that it appears that the bladder should be the continuation of the superior œsophagus, and the intestine of the inferior. This bladder is large, spheroïdal, very thin, diaphanous, most generally empty, but sometimes filled with a yellowish juice. M. Otto does not think that this is a true stomach, but a sort of bladder proper for suction. On each side of the anterior mouth is an organ, in the form of a cylindrical cœcum, an inch in length, flexuous, and full of a viscous juice. M. Otto makes this to be a large salivary gland. The continued intestine is very narrow and cylindrical; it makes many convolutions round the bladder, traverses the membranaceous partition, assumes the dimensions and cellular form of a bulky intestine, and proceeds directly to the anus, surrounded in its course, as well as the narrow intestine, by a hepatic mass, thick, and of a yellow colour.

The apparatus of generation consists of many small sacs of eggs, situated in the anterior part of the visceral cavity, and attached together, and with the peritoneum, by filaments extremely fine. In the month of January, some individuals had neither these little sacs nor ovules, while others had them very much developed. M. Otto was not able to find the orifice through which they could pass.

He was easily able to convince himself of the existence of

a vascular system, by many vessels which proceeded from the intestine to the cutaneous envelope, and from the latter to the former, and especially towards the bladder, and the hepatic mass of the upper œsophagus, and under the large intestine; but he was unable to follow the distribution of them, in consequence of their fineness, their great number, and their colour being yellow throughout.

The nervous system consists in a filament extended from one extremity to the other of the middle ventral line, and swelling into a ganglion, giving out filaments for each articulation. The first appeared thicker than the others; accordingly, the nerves which it furnishes for the sides of the head are thicker.

The *S. diplochaites* thus named by Dr. Otto, on account of the double series of setæ with which its body is furnished, was found by him in tolerable abundance on the coast of Naples, cast ashore by a tempest, but dead. At other times the fishermen brought him living specimens in their nets. We have followed his mode of considering this animal; but might it not be possible, says M. de Blainville, that he took one extremity for the other, and that the two orifices, which he considered as mouths, were one of them, the inferior, the anus, and the other, the superior, the orifice of the generative apparatus? The direction of the setæ would seem to favour this opinion, and then the extraordinary anomaly of an animal with a double mouth would vanish.

The DENTALIUM is a genus of animals probably articulated, and belonging to this class, but rather imperfectly known, and the classification, as our author observes, is doubtful. The body is a little conical, terminated behind by a sort of basement, which issues from the tube, and superiorly by a cephalic enlargement, in the middle of which is found the mouth, at the extremity of a sort of button, and having at its base a frill, the nature of which is unknown, but which very probably is branchial or vascular. It is contained in a calcareous tube, tolerably thick, solid, slightly arched, open at the two extremities, smooth, striated, or even polygonous at its superficies.

We have very few details, indeed, concerning the manners and habits of these animals. It is nevertheless very probable, that they live sunk perpendicularly in the sand or ooze, into which they can no doubt penetrate more or less. But it seems, at least, very doubtful that they can change place, transporting their tube along with them. The species of this genus are however very common on the sandy coasts of the seas of warm climates, and even on those of the Mediterranean. It appears that they were equally abundant in the ancient sea, for many of them are to be found in the fossil state.

We are acquainted with a great number of tubes belonging to species of this genus, which are named *Dentalia,* in consequence of their resemblance to the tusks of the elephant, and which are divided into smooth, or striate, angular, or polygonous.

We come now to the second order of annelida, the DORSIBRANCHIA.

ARENICOLA is a genus of marine worms established by Lamarck, and which appears even yet to comprehend but a single species, which was designated by Linnæus under the name of *lumbricus marinus,* and by Pallas under that of *nereis lumbricoides,* while the animal is, in fact, neither a lumbricus nor a nereis. The generic characters are the want of tentacula and jaws, and the possession of gills only in the middle of the body, the two extremities being destitute of them. This worm is about eight or ten inches long, and a little thicker in the middle, than at the two ends. Its skin is marked with a multitude of annular wrinkles. For every five there is one bulkier, and more projecting than the others, and on this are attached the feet and gills. This distinction of large and small wrinkles does not exist in the posterior third of the body, where no gill nor foot is discernible. In front the feet advance, close up to the head, which itself is distinguishable only by the presence of the mouth, but the gills do not proceed so far. There are six of these large wrinkles in front, which have only feet without gills. There are fourteen which support gills, very small at first in front, enlarging by degrees as far as the tenth or twelfth, and afterwards diminishing. The feet, like those of the other genera of this order, are only fasciculi of stiff and brilliant setæ, which the arenicola causes to enter, or issue forth at will, by means of certain muscles, of which we have already given a sufficient notion in our general article.

The gills are those parts which most quickly attract the eyes of the observer, in the living animal, by the beauty of their structure and the changes of colour which are remarked in them. These gills are sorts of aigrettes, composed of eight or ten principal slips or blades, which proceed from a common base, and separate like the radii of a circle, at the same time curving gently. Each of these blades or stalks supports a dozen of small branches, which are subdivided two or three times into smaller branches. All this apparatus cannot well be seen but for a very short instant, during which it is extended in every direction, and of a fine red colour. In a second it sinks in upon itself—the branches are folded, it grows pale, and becomes altogether grey. These two states thus alternate with each other as long as the animal is in good health, and are caused by the blood, which is carried into the gills to be respired, that is, to undergo the action of the circumambient element, and which afterwards returns into the interior of the body. It is thus that respiration takes place in all the marine worms with red blood, such as amphinoma, &c. &c.

The œsophagus of the arenicola can, at the will of the animal, be unrolled externally, and form a sort of proboscis, all bristling with small tubercles. There are neither jaws nor teeth. The stomach extends towards the tenth gill. Its membrane is of a fine yellow colour, on which the vascular net-work of which we have spoken, is very agreeably drawn. At the anterior part of the body are on each side five blackish pouches, which probably serve as testicles, and on the junction of the œsophagus and stomach are two other conical and muscular pouches, of the use of which we are ignorant. The eggs are small grains of a yellowish colour, which swim in the interior of the body.

This worm is very common on most sandy coasts. Fishermen employ it as the best bait for catching sea-fish. It even forms an object of commerce, and is sold dear enough in those places which do not produce it. It is found in the sand, at about a foot and a half or two feet in depth. Its retreat is discovered by small cordons of sand which it has voided, and which close the orifice of its hole. Its external colour is reddish, and it changes into a dark green. When it is touched it emits a liquid, of the colour of bile, which causes spots upon the fingers difficult to remove; but in the month of August it only sends forth a milky fluid. On drawing it rather slightly by the tail, the latter separates into several articulations, without any appearance of tearing.

The Amphinome is a genus of marine worms, originally established by Bruguières. The characters consist in an elongated body, more or less flatted, each articulation of which supports a pair of gills in the form of tufts, or little plumes, and in a mouth without jaws. Each articulation supports besides two tufts of hairs, or stiff setæ, most frequently accompanied, each of them, with a fleshy filament more or less long. The head supports a certain number of similar filaments, and sometimes an ornament in the form of a crest.

The mouth is a longitudinal cleft, which can re-enter or issue forth from a rounded cavity. The anus is at the posterior extremity of the body. The intestinal canal is usually straight and without any great convolution. Nothing is found there but sand or mud. The stomach is a sort of gizzard, fleshy and robust. The interior of the body exhibits numerous vessels, which doubtless, in the living animal, were filled with red blood, like those of the other articulated worms. These animals belong either to the East Indies or South America.

We shall include what we have to say respecting EUNICE, &c. under the head of

NEREIS. Linnæus was the first naturalist who applied this name, derived from that of a family of nymphs of the court of Neptune, to a tolerably numerous group of elongated and depressed worms, usually composed of a great number of rings or segments, provided with tentacular appendages, and which are commonly found on all the sea-coasts, concealing themselves in the holes or anfractuosities of all the bodies which they can find, and penetrating even into the sand and mud. This name we find employed even in the first editions of the *Systema Naturæ*. At a later period Linnæus introduced under it some species which do not belong to this genus, as Pallas demonstrated in his memoir, on the aphroditæ; but the latter also, in his turn, united animals to it, which really do not appertain to it. Muller was more happy in his work upon worms, and in the Prodromus of his Fauna of Denmark; although he greatly augmented the species of this genus, he separated however the fluviatile species, of which he made the genus Naïs. Otho Fabricius described a greater number of them, and in a manner frequently still more complete, so that Gmelin, in collecting all that had been done upon the subject, has carried the number of the Nereïdes to more than twenty-nine species. From that time until late years the systematic writers adopted this genus, such as it is in Gmelin, almost without making any other change than adding a very small number of species. This may be easily seen in Bruguières' "Tableau des Vers," in the Encyclopédie Methodique, Blumenbach's Manual of Natural History, that of the Baron Cuvier, the first edition of Invertebrated Animals, by M. de Lamarck, and even in M. Bosc's Treatise of Worms, although he indicates some new species. In the first mentioned work, however, we find a new established genus upon a species equally new, and the observation that the three sections of Linnæus and Gmelin should form so many distinct genera, when the mouth of the animals composing them should be better known. M. Oken also established a small section in this genus, but without any new observations. The naturalists of our country, in the mean time, and among others MM. Donovan and Leach, made known some species of Nereïdes not before observed. The nature of the labours of M. de Blainville also led him to endeavour to throw some light on this great genus, by carefully studying the small number of species which he could procure. It was thus in considering the disposition or composition of the appendages of these animals, that he was led to this result, of some importance to the philosophy of the science, that in the animals articulated externally, to which he has given the typical name of *entomozoaria*, or *entomozoa*, each ring of the body in its complete state is provided with appendages, formed of three parts, one respiratory, or branchial, the other locomotive, and the third sensitive; that all the rings are not necessarily provided with these parts of the appendage; that it may have three, two only, or one, or there may be none whatever in some of the lowest species, or in some parts of the body of an animal modified by some other cause. From this point the transition was easy to show how in the Nereides the general respiratory and locomotive parts of the appendages, diminish in proportion, as from the middle of the body, they approach the two extremities; and how, in front, the attachment of these two modified appendages, rises by little and little, so as to open into the first rings of the lateral tentacula, and then of the superior tentacula, more or less elongated, according to some local cause. He also succeeded in showing that the teeth themselves, in such as are provided with them, are nothing but fasciculi of hard setæ, approximated together, and analogous to those of the other rings of the body.

This mode of considering the external parts of the nereides, naturally conducted M. de Blainville to the attempt of rendering their systematic arrangement more perfect. He proposed to divide the nereides into eight subgenera, according to characters derived from the presence or absence of teeth, from those of the tentacula, and from their number; from the form of the mouth, furnished with proboscis or not; from the existence, or absence of gills, from the form of the parts of the appendages, and even from the absence or presence of the black spots which have been regarded as eyes. To all these genera he gave different names, but at the period to which we allude, the defect of materials did not permit him to perfect or finish his labours.

While M. de Blainville was thus occupied in revising the genus nereis, M. Savigny was employed in the same manner, and as materials were furnished him in great abundance, he was enabled much more completely to perform his task, as may be seen by consulting his system of annelida, forming part of the great work on Egypt, by the French Savans. He considered these animals in a much more detailed manner than had been done previously to his time, at least as far as all the external parts are concerned, and established a great number of genera, which have since been adopted by M. de Lamarck, in his New System of Invertebrated Animals.

Leaving systematic distribution, however, for the present, we must now give a succinct view of the structure and history of the nereïdes.

The body of the nereïdes is in general extremely elongated, slender, attenuated, cylindrical, sub-cylindrical, or even sometimes depressed, especially underneath. Finally, a little broader in the middle, it becomes gradually attenuated towards the extremities, but much less so towards the anterior extremity, which is always more or less truncated, than towards the posterior. It is formed of a great number of rings, or segments, broader than long, very distinct, very mobile, one upon the other, in their whole circumference, and the broadest and longest of them are in general towards the anterior third of the body. The posterior ones decrease insensibly as far as the last. On the first or second anterior ring we may pretty frequently distinguish one or two pairs of black orbicular spots, which have been regarded as eyes. These first rings, in certain species, do not appear to be complete, but rather cleft, or open in their inferior part, so that the aperture of the œsophagus is preceded by an oblique in ferior cleft, to which the name of mouth has been given. In other species these rings are complete, but they are wide enough to permit the re-entrance of a considerable cephaloïd proboscis; which itself is sometimes composed of one or two rings.

To suit to this disposition of the cephalous rings, it is evident that the mouth must present different forms. In the first case, it is a cleft more or less long; in the second, it is a round hole, sometimes accompanied in its circumference with papillæ, or papillary tubercles disposed in radii.

The posterior extremity of the body of the nereïdes terminates by being very much flatted. Sometimes the last ring is in the form of a dagger point, which alone extends beyond the anus. This last is always very broad and transverse.

The appendages which furnish the sides of the rings of the nereides are always much more complicated than in the true lumbrici, and even than in the naïdes, but less so than in the amphinomæ, and neighbouring genera. They constitute, in general, a small lamina, compressed from front to back, and placed vertically on each side of the ring, of which it occupies the entire upper part. But this lamina, or this species of fold, is sometimes almost nothing, while at other times it is longer than even the diameter of the ring, and forms a true pedicle. This lamina, in its greatest state of complication, is divided by an emargination, or a bifurcation, into two portions, more or less distinct, placed one above the other. M. Savigny designates them by the name of *oars*. The upper one is composed of a soft, flexible, tentacular part, more or less elongated, sometimes giving out at its base a bifurcation, by way of branchial appendage; and of a fasciculus of hard, rigid, corneo-calcareous setæ, situated at the superior base of another tentaculary nipple, which, after that, is constantly inferior. In the two parts of the fasciculus of setæ, there are almost always two or three harder and stiffer setæ, which M. Savigny names *aciculi*. Sometimes, however, the setæ are not divided.

The simplification of this appendage may be considered to commence by its non-bifurcation, but afterwards the denticulations of the branchial lobe disappear; then the branchial lobe itself; the setæ afterwards become attenuated by a diminution of their number, and are reduced in length. There remain then only the tentacular lobes. They subsequently diminish, either both or one of them, and the appendage is sometimes represented only by one or two small tubercles. We may very well conceive too, that what are named eyes in the nereïdes, are perhaps but the extremities of these rudiments of tentacula.

But this *rudimentation* (if we may use such a term) of certain parts of the appendage in the nereïdes, is sometimes accompanied with an augmentation of another part, which produces what have been named *antennular cirri*, *tentacula* and *antennæ*, by the great development of the sensitive part of the appendage, on the cephalic rings, and by their superior position, in the same wáy as the agglutination of the setæ, or of the aciculi on the lateral parts of the anterior rings, has constituted what has been named jaws, or teeth, at least in the species which have them corneous and simple. Finally, to this same augmentation of the tentacular cirri of the appendages, is owing the long pair of setaceous filaments which seem to terminate the body behind, forming a sort of tail. They are always double, and constantly belong to the ring before the last.

The attachment of the appendages in the nereides still presents something remarkable enough, in this, that being nearly lateral in the medial or normal rings, it descends a little in the more posterior ones, and ascends again, much more evidently in the anterior, so that what remains on the cephalous rings, namely, the tentaculary filaments, are almost altogether superior, and become frontal. In certain species, even the two tentacula which are most approximated, form but one, which is then odd and medial.

From all this, it is evident that we may distinguish, and with just reason, to a certain point in the body of a nereïd, 1. a proboscis, which is but one or two entire anterior rings, in the interior of which, there may be teeth, which, without any doubt, become exterior, or at least marginal in their action. 2. A cephalous enlargement, formed of two broader rings than the proboscis, and disposed obliquely, so that the anterior produces a sort of advance, or front, under which the proboscis may withdraw, and the posterior, a complete ring, receives it when it is more deeply retracted. This it is which gives to the incomplete appendages of these cephalous rings, a disposition more or less tentacular, according as, from being lateral, they become more and more dorsal, or frontal. We may then very well apply a common denomination to those tentacula, since their origin is the same, and distinguish them only, as they are altogether superior, or lateral. 3. A species of neck, as some authors have made of it, formed of the narrowest and shortest rings, which immediately follow the cephalous rings, and the entire appendage of which is almost rudimentary. But these rings are something less distinct than the others, the trachelian rings passing by little and little into the thoracic, and those again being less distinct from the abdominal. We may, however, consider as such those whose appendages are the most complete, and especially in the branchial part. As for the coccygeal, or post-anal rings, there are never more than one, terminated by a short point, a little dagger-wise.

The appendages may be equally divided into cephalous, cervical, thoracic, abdominal, and præanal.

Considering the cirriform, or tentacular appendages of the cephalous rings, as of the same nature and of the same origin, we may, for the purpose of rendering ourselves better understood, divide them into superior or frontal, and into lateral.

The teeth, more or less deeply placed, will be distinguished from the masses of corneous tubercles, equally unciform sometimes, which arm the rings of the proboscis, by their size and their development, and even sometimes by their calcareous nature.

There is no objection to preserving the name of eyes, (although it is very doubtful whether they be really organs of vision,) for the black orbicular spots which are remarked upon the principal cephalous ring of a tolerably great number of nereides, and their consideration should not be neglected, as their presence or absence, their number and disposition, appear to furnish tolerably good zoological characters.

We may term cervical appendages, the first, in greater or less number, which do not present the complete composition of those of the rest of the body, and especially are destitute of gills.

We may reserve, on the contrary, the denomination of thoracic appendages, for those which are perfectly complete, at least in relation to the particular species which may happen to be under consideration; for it may occur, that in a group of species, no appendage is absolutely complete, that is to say, formed of its two parts, divided by a lateral line, and both composed of a fasciculus of setæ and of aciculi, of tentacular filaments, and particularly of a primate or branchial tentaculum.

Finally, the name of abdominal appendages suits those of the rings, which have lost something of their complication, from the most perfect to the pair which precedes the anus, and which may be termed præanal.

In the different parts of the appendages, some differences may be found, on which it is proper to remark. Those of the absolute, or proportional length of the tentacular filaments, or of the nipples which support them, are of but little importance. Such, however, is not the case respecting the simplicity, or the complication of form of the filaments, or superior tentacular cirrus. In the first case, there is no gill, properly so called, but in the second there is, and then the number of digitations and their forms become characteristic of the species, and often of the rings.

The fasciculus of stiff, corneo-calcareous setæ, which completes the appendages, is sometimes like the latter, divided into two fasciculi, more or less distinct, by a lateral line. But, moreover, they are themselves formed of two sorts of setæ; some finer and more flexible, to which this name may be preserved, and the others stiffer, more resistent, and constantly of a black colour, which M. Savigny has named *aciculi*. The setæ also exhibit differences in their mode of termination; but it requires magnifiers of considerable power to enable us to perceive this character.

The skin, or general envelope of these animals, in most instances extremely thin, presents a character common to the entire class, of exhibiting the different colours of the prism, according to the inclination of the luminous rays; in other respects it offers no great differences of thickness and of structure, according to the parts of the body to which it belongs. It is, however, always thinner on the tentacular cirri, whether branchial or not.

These cirri, which constitute the only supposable organ of touch in these animals, are of different sizes, often very long, and placed throughout the whole extent of the body; the longest are in general the upper ones of each appendix, and especially those which terminate it in front and behind, when we have considered them as tentacula. They seem at times to be as it were articulated, although they are never in reality so, the skin being of the same thickness throughout. This should seem to be owing to the manner in which the substance which fills them is divided, and it may be that this sub-articulate disposition exists only in the nereides which have been preserved in alcohol.

The black points which the authors who have observed the living nereides regard as eyes, are tolerably large, comparatively to the thickness of the cephalic ring which supports them. Of number and position constant in each little group, they are sessile, and rather independent of the skin. The eye itself forms an elongated spheroid, altogether black on one side, and shining on the other, leaving in the muscular stratum on which it is supported a slight excavation; this side is white, and evidently in communication with the rest of the system.

The locomotive apparatus of these animals is extremely simple, although considerable: it is composed by two strata of muscular fibres, situated under the skin; the external stratum, much thinner than the other, is formed by transverse fibres, while the internal one, very thick, is entirely formed by longitudinal fibres; this last stratum is much thicker underneath, where it. constitutes two or four narrow bands, extended from one extremity to the other, and sometimes perceptible externally; on the upper part, or in the dorsal line, the longitudinal band is perhaps not thicker, but it is not much divided except into two, by a medial furrow; on each side the muscular bands, superior and inferior, are thicker, so as to fill almost all the interval comprised between the skin and the intestinal canal. It is in fact from these bands that proceed the small oblique muscles which go to the root of each appendage; and it is in their thickness that are deeply sunk the fasciculi of setæ, and of aciculi, of those appendages, which are provided with them so as to be drawn forward or backward by the fibres which are attached to their base, or which penetrate even into the nipples which support them; these are true specific muscles. In the species in which the appendages are deeply divided into two branches, there is besides a longitudinal muscular fasciculus which follows the whole lateral line, and to which the transverse fibres adhere.

It is not necessary to say that all the longitudinal muscles are divided into as many parts as there are rings to the body of the animal: in the species which have a pair of teeth, each of the latter has peculiar muscles to approximate and remove it; in those whose masticatory apparatus is more complicated, the muscles are more complicated: at first the upper lip, or the cleft ring, which immediately precedes the mouth, has a strong lateral muscle, which is carried from the lateral base to the side of the general muscular sheath; another, more considerable, belongs to the second cephalic ring: it attaches itself to the lateral face of this ring, and is elongated pretty far backwards into the visceral cavity, between the intestinal canal and the longitudinal cutaneous stratum; one or two small oblique fasciculi, and much shorter, proceed from the first cervical to the posterior cephalic ring. The buccal mass, or mouth, with its appendages, is drawn back by a very powerful muscle, which is inserted at its dorsal and lateral face. What is more remarkable is, that the enlargements of the intestinal canal, at least the anterior ones, have also retractor muscles; those which carry the buccal mass forward, are altogether dorsal, and much less strong: the teeth are as it were implanted by a sort of pedicle, in the very parietes of the œsophagus, or rather of the buccal cavity, the fibres of which are longitudinal; and they are surrounded from their base to a little distance from their far extremity, by a thick stratum of transverse fibres, which must act in the manner of a powerful sphincter.

The mouth of the nereides, in the species provided with a true proboscis, and even in those which have one formed by proboscidian rings, is always terminal, and consisting of a rounded orifice, and sometimes of a transverse cleft; but in the multidentated species, it is a sort of oblique cleft underneath and behind the first ring.

The proboscis, properly so called, may be considered as a very long ring, sometimes filiform, and at other times more or less claviform, and enlarged at its extremity. The skin which invests it is always much thinner than on the rest of the body; in fact, this organ is often retracted into the interior part of the intestinal canal, pretty nearly like the tentacula of the limacinæ. It is rarely armed at its extremity; sometimes, however, there are two pairs of little hooks, one at each corner of the enlarged extremity of the proboscis; more frequently it is furnished with folds, with tubercles, or even with tentacular papillæ at its orifice.

Finally, in the multidentated species, the mouth, as has been already mentioned, presents the disposition of a cleft, at the orifice of which the whole of the jaws may be supported by the muscles which move the whole buccal mass; these jaws appear to be nothing but the corneous or calcareous encrustment of longitudinal folds to the number of three or four on each side, wider, and more open in front, narrower, and convergent one towards another behind. Accordingly, they always seem to be of the same number on each side; but the first two folds, the most anterior and inferior, much shorter than the others, are sometimes composed of two corneous or calcareous particles, placed one at the end of the other. This anomalous disposition has occasioned MM. Savigny and Lamarck to attribute to these animals jaws of uneven numbers, and always more numerous on the left than the right. In an individual of the *nereis pinnata*, examined by M. de Blainville, there were four teeth, the first anterior and inferior one was semi-lunar, very small, simple on the right hand, and with a small tubercle at the posterior angle on the left; the second, a little larger, and of the same form, but denticulated on its free edge, was in like manner undivided on the right, and divided on the left, the accessory part being in front of the principal. Finally, the other two were perfectly similar on both sides: one broader, stronger, and denticulated on its free or internal edge, was joined to the fourth, which was longer, more slender, and more posterior; for its posterior third part, and beyond that this double tooth, united on a common pedicle, is continued with its corresponding one of the opposite side, something like a pair of sheers, and it was this part which was principally seized by the thick annular muscle. In an individual of the *nereis gigas*, whose jaws had pretty nearly the same form, with the difference of being calcareous, the same gentleman found that at the right the second tooth was equally ditomous, the accessory piece being also in front. But besides these sorts of lateral jaws in this section of nereides, there is a sort of lower jaw, composed of two symmetrical pieces, a little widened into a palate at the extremity, and approximated in the medial line; their posterior extremity is implanted in the transverse fasciculus of muscular fibres of the buccal mass.

The communication of the mouth with the rest of the intestinal canal is direct in the proboscidian nereids, but in the unidenticulated species, and particularly in the multidenticulated, this is not the case. The buccal cavity forms a sort of cul-de-sac behind, and the continuation is made with the œsophagus, by an aperture in the form of a cleft, situated at the anterior part of the superior paries. This orifice conducts into a sort of pharynx, or enlargement, from which issues the intestinal canal. The latter is always extended in a right line, from one extremity to the other of the body of the animal. We often distinguish there a somewhat short œsophagus, but much more narrow than the rest, as in the unidentated species, and moreover some salivary glands, pretty long, and a little twisted, and they open considerably behind the buccal mass. The stomach, which is of a pretty considerable diameter, and almost equal to that of the abdominal cavity, is sometimes without very distinct enlargements, and corresponds to the rings of the body. It is sometimes pretty markedly strangulated towards its grooves, and more or less dilated into cœcal appendages in the enlargements of the rings. Its parietes, often very slender, are sometimes, as in the *Nereis gigas*, almost as thick as the skin, so that we may easily distinguish in them, as in the latter, two strata of fibres, one longitudinal and the other transverse. Often also these two parts of the general envelope are strongly connected together by means of numerous cellular frœna, which are carried from one to the other, forming numerous transverse cells.

The termination of the intestine is made, as has been already mentioned, altogether at the posterior extremity of the body. It is usually very wide and transverse.

No liver, properly so called, has been distinguished round any part of the intestinal canal of the nereides, but the salivary glands have been found to be particularly distinct, at least in the unidentated species.

The portion of the exterior envelope, modified so as to be converted into an organ of respiration, is constantly external in the nereides, and consequently constitutes gills, but those gills are not always perfectly distinct from the tentacular cirri of the appendages of the rings, in which case it is probable that the latter take the place of them. In the multidentated nereides there are some very distinct, and which are often even more or less pinnated. In the unidentated species the lateral cephalic tentacula may doubtless be regarded as gills, as well as the little cirrous tongues, which are found underneath the superior and inferior cirri. Finally, in the species with proboscis, it is very rare that there is any thing else but simple tentacular cirri, sometimes, it is true, augmented at the superior part with sorts of cirri compressed into little leaves, which may also be regarded as respiratory organs, but from no proof except analogy of position.

The circulation of the nereides appears to be extremely simple; and the blood is constantly of a fine red colour, comparable to that of the arterial blood of vertebrated animals, with warm blood, and that in the whole vascular system.

The apparatus of decomposition in these animals is still less known than that of composition. The cutaneous exhalation is sometimes sufficiently abundant, since certain species can form from it a sheath, or at least line their dwelling-place, such as holes or anfractuosities, which they inhabit.

The organs of generation are still more obscure. It appears, however, that we may consider as ovaries some small whitish utricles, granular, which are found on each side of each ring, and between the cœca of the stomach, the orifice of which appears to exist at the base of its appendage, so that these animals should have a large number of ovaries. They do not, however, possess their full development but in the extent of about three-fifths of their total length. In front they are very small, and still more so behind.

The nervous system consists in a long abdominal and medial thread, extended from one extremity to the other, and often concealed by fasciculi of the inferior longitudinal muscle. In certain species it does not merely form a thread, but it swells a little in the middle of each ring, and it is from this enlargement that the threads come forth which go to the muscles and the skin. In the *nereis pelagica* we find even that these ganglions, about the second anterior third of the body, are very thick and very distinct, while they are so little so elsewhere, as to lead us to doubt of their nature.

No naturalist has yet studied the different functions of the organs which constitute the nereides. The faculty, however, of continuing to live, has been recognized in them, after a considerable portion of the posterior extremity of their bodies has been cut off, the amputated part being reproduced in a tolerably short time.

Their eyes appear to be but of very small utility to them; but it is not so with the tentacular cirri of their body, and especially with those of the head, which they direct in all ways, as it were to examine and scrutinize those obstacles which may occur to them.

Their locomotion on a resisting soil is very lively, and is

made in a serpentine manner from right to left. Otho Fabricius tells us that he has even seen the two extremities moving at once in the same direction. It appears that certain species can swim equally well in this serpentine way, without doubt, by means of the oars, formed by the pencils of setæ of their appendages, and especially by the aid of their foliaceous cirri.

Those which bury themselves in the sand or mud, which doubtless are the unidentated species, appear to do this by the assistance of the tuberculous points with which their proboscidian rings are armed.

We know but little concerning the natural history of the nereides, except what we have received from Otho Fabricius and M. Bosc. It is generally admitted that the species of this genus are found only in the waters of the sea.

It is equally evident that there are species of nereides in all parts of the world, although they are very far from having been sufficiently studied. From the little that we do know concerning their division, it would appear that a species of each group may exist in the different zones of the globe, but the largest belong to the torrid zones of the two sides of the equator.

The nereides most usually live in the excavations of littoral rocks, in the hollows of sponges, in certain alcyones, in univalve or bivalve shells, in madrepores, in the interstices of the radicles of thalassiophytes, under stones, and in general in all bodies which present fissures more or less profound. There are some which bury themselves in mud or sand, where they excavate a lodge proportional to the dimensions of their body, and sometimes they line this dwelling with a mucous matter issuing from their body in sufficient abundance to construct a tube or sheath. From this they put forth a greater or less portion of their body, but rarely the posterior extremity, so that they may be able to re-enter on the slightest indication of danger.

They all appear to feed upon animal substances, whether in the living state, or in a state of putrefaction more or less advanced. M. Bosc, who has observed the manners of some species on the coasts of the United States, tells us positively that these animals feed upon polypi and small worms, on which they throw themselves, by darting the anterior part of their body, which they have first contracted. Otho Fabricius tells us of some species of spio, or nereides with tubes, that they seize the planariæ, on which they feed, by means of their long tentacula.

No observations appear to have been made on the mode of reproduction in animals of this genus, and it is easy to conceive that such observations must be very difficult to be made. It appears merely probable that the eggs must be very numerous, and that they issue from each ring, for the ovaries have been remarked to have filled the emarginations made by the gastric enlargements.

It has not, we believe, been at all observed that the nereides were hurtful animals; the larger species are, on the contrary, regarded by fishermen as forming an excellent bait, which causes them to be sought after pretty carefully on the coasts of the Channel. It is generally women and children who at low tide gather them in the muddy or sandy places, and in the intervals of shingles, with a sort of trident.

We have already had occasion to mention that the distinction of the species of the nereides was but little advanced before the labours of M. Savigny, notwithstanding that Pallas, Muller, and especially Otho Fabricius, had given us some excellent descriptions of those which live on the coasts of the northern seas. It must be allowed that it is difficult enough when a careful study is not given to each part, composing the appendages, which is not always so easy to be done, as certain of these parts may be more or less retracted, when the animal appears in our collections.

The arrangement of the species of nereides, according to M. de Blainville, should indicate the passage from the aphrodites, which he places at the commencement of the class of Chetopods, to the lumbrici, which are at the end. It should be established on the development of the appendages in general, and particularly on those of the head. More regard, however, has been paid by naturalists to the mode in which the mouth is armed.

The genus Eunice of the text is put at the head of the Nereides by M. de Blainville; the anterior extremity of the digestive canal is provided with a buccal mass, armed at the interior with corneous folds or teeth, denticulated, and lateral, the two inferior of which are very much approximating the medial line, and constitute a sort of lower jaw.

This group encloses the largest species of nereides; many appear to exist in the European seas, but they are never of so large a size as those of the seas of warm climates.

The corneous or calcareous armour which invests the lateral folds of the buccal cavity in these nereides, would lead us to suppose that their food was more solid than that of the others, and that they might even attack the smaller fish.

We shall conclude our remarks on the present order of annelida with a brief notice of the aphrodites.

This is a genus of marine worms, the character of which is to have membranaceous plates in the form of a scale, which form two ranges on the back, where they are attached by their middle, one pair fastened to two of the rings. The gills are placed on those rings which have no scales, and so small that they can hardly be perceived; each ring, moreover, supports feet composed of stiff setæ, which vary in number according to the species.

The body of the aphrodites is generally broader and flatter in proportion to its length, than that of other worms of the same family. It has been supposed that they have articulated tentacula round the mouth, but this is erroneous; they are nothing but simple and fleshy filaments, similar to those which are attached to the rings, and only a little larger; their œsophagus can re-enter and come out so as to represent a species of proboscis. When it is altogether elongated externally, the aperture of the gizzard presents itself, and with it four small teeth that are there attached, two above and two below. This gizzard is fleshy, and very strong: the intestinal canal is straight, and gives out on each side a multitude of cœca which terminate sometimes by a simple dilatation, sometimes by some ramifications.

The blood vessels of the aphrodites are a little smaller in proportion than those of the nereides and the arenicolæ; nevertheless, it is easy to perceive them, and also to demonstrate that they are filled with a red fluid. Their nervous system is very apparent, and consists in a medullary cordon, which predominates through the whole length of the belly, and is swelled into as many ganglia as there are segments in the body.

It is said that the sexes are separated in the aphrodites, and that they are oviparous. In certain seasons the body of the female is found full of eggs, which swim in a liquid, and that of the males full of milt; but no internal organs have been found destined to produce them, nor external apertures to evacuate them.

We shall now proceed to the third and last order—ABRANCHIA.

The genus LUMBRICUS was indicated originally by the authors of antiquity, and admitted successively, under the same denomination, by all the modern zoologists, except by M. Savigny. Linnæus, Gmelin, and his followers, who are very numerous, place this genus in the division of external worms. M. Cuvier at first imitated Linnæus; but he gave to the division of the worms in which he placed the lumbrici, the name of red-blooded, which was changed by M. Lamarck into that of *Annelida*. The generic characters of the lumbrici are: body elongated, very extensible, attenuated at the two extremities, but especially at the anterior, composed of a great number of articulations, having for appendages only spines or setæ, forming longitudinal striæ; mouth, terminal, simple; anus likewise terminal, and longitudinal; the organs of generation terminating towards the anterior third of the body, near a sort of pad or swelling, more or less considerable, which may be remarked there.

The organization of the lumbrici has been studied by several persons, and among others by Willis, Redi, Montegre, and Sir Everard Home. Their body, perfectly round, is terminated in a manner more obtuse behind than in front, where it is considerably attenuated, and becomes very pointed; the furrows which divide it into articulations are so much the deeper and more crowded, as they approach more to the posterior extremity; also the articulations are much more marked in front than behind; they are especially so in a place situated towards the anterior third of the body, where we remark an enlargement of a redder colour, formed by six rings not so distinct as the rest. At the sixteenth ring, at its inferior and lateral part, is a sort of ovaliform tubercle, transverse, whiter than the rest of the body, which is pierced by a cleft equally transverse. This is particularly evident when the animal is elongated. At the thirty-sixth ring we see equally on each side a part more flesh-coloured than the rest, and which represents an elongated tubercle, occupying the space of three rings: no trace of aperture is visible there. O. Fabricius, in his description of the common earth-worm, puts this pad at the twenty-seventh and twenty-eighth rings, and says that in front, that is to say, at the twenty-fourth, he has seen a pendant, soft appendage, whose envelope being very thin, suffered to escape a limpid fluid, through an orifice by which it was pierced; at the upper part of the back on each side is a series of pores very symmetrically placed : it is from those orifices that the humour which invests the body of the lumbrici comes forth. Some authors think that at the same time they are sorts of stigmata for respiration.

The general envelope of the lumbrici is eminently contractile, in consequence of the thickness of the muscular stratum, which doubles it. As for the skin itself, it presents this character of irritation which is found in all the animals of the class. The skin is more slender, and softer in the intervals of the rings, but these latter are more inflated and more resisting. Each of them is provided, to the right and left, with a number, variable, as it would appear, according to the species, of small calcareo-corneous setæ, of a golden yellow, disposed in pairs; one latero-superior, and the other latero-inferior, and the succession of which, on each ring, forms four longitudinal series on each side of the animal, or eight in all. These setæ, rough and resistent, are more or less short, and strongly directed backwards. This is what the appendages become reduced to in this genus of animals. There is, in fact, no trace of tentacular parts, not even around the mouth. The intestinal canal is simple, extending from the mouth to the anus. The former is very small, for it is pierced in the first ring, which is remarkably pointed; but as it opens a little obliquely at its inferior part, there result two kinds of lips, the upper of which is oval and much longer than the other, which really possesses but little sensibility. There is not, at the anterior part of the canal, any buccal dilatation, or teeth, or lingual enlargement. The œsophagus, when arrived to about the sixteenth ring, terminates in a true gizzard, almost as thick as a pea, of a fleshy and tendinous tissue, with fibres a little oblique. All the rest of the intestine goes directly without enlargement to the anus, which is pierced in the form of a longitudinal cleft in the last ring.

In the passage of the intestinal canal, the muscular fibres which pass from one ring of the body to another, attaching themselves to the interval, form sorts of diaphragms, which proceed to terminate at the parietes of the intestine. No author speaks of the liver, properly so called. Some, however, have considered as holding its place, a thick flexuous vessel, which predominates all along the inferior face of the intestinal canal. But this is probably an erroneous opinion, this vessel being more likely to be a mesenteric vein. The apparatus of the circulation of the lumbrici appears to be very simple. From all the parts of the exterior envelope, and from the intestinal canal, springs a very close net-work of small veins, which unite in a single thick trunk, situated in the medial line of the ventral face. This trunk, when arrived near the head, ascends by five pair of lateral canals, to the dorsal surface of the body. These canals soon unite in a very long heart, occupying all the middle line of the back, broader in front, and growing more narrow in proportion as it goes backwards. The heart may then very well be considered as an aortal artery, from which, subsequently issue the divisions which repair into the different parts of the body. Its motions of systole and diastole are very perceptible. From this distribution of the circulatory apparatus, it is extremely probable that there is no special organ of respiration, and that the entire skin is modified for this purpose. There are, however, many authors who regard as sorts of lungs, the little follicles, to which the dorsal pores conduct, as the same has been supposed with regard to the leeches. The organs of generation appear to have sufficient relations with those of the same animals. Like them, the two sexes exist in the same individual, and the apparatus are situated towards the anterior third of the body. They are composed, behind, of a double series of yellowish pores, situated above the stomach, into which repair a great number of blood-vessels, and in front;

of those other pairs of white vesiculæ, the posterior of which is thicker and more oblong. It would seem that these last communicate with the exterior by the vertical clefts which we have observed on each side of the sixteenth ring. It is a query whether the posterior bodies are the ovaries, the eggs from which, previous to issuing forth externally, should pass through the anterior vesicles, that would thus perform on them the office of spermatic organs. Of this we must be the less certain, as Montegre assures us that the young ones, which come forth in the living state, do so by the anus, and that the eggs from which they proceed have descended between the external envelope, and the intestinal canal, as far as the circumference of the rectum, a disposition which is, to say the least of it, very singular. It seems, however, quite certain that the lumbrici are ovoviviparous.

The nervous system of the lumbrici is composed of a brain extremely small, situated above the mouth, and of a gastric, or abdominal cord, which is formed by a series of a great number of small ganglia, much crowded, one against the other.

The lumbrici cannot be supposed to taste, smell, see, or hear, in any wise, since they have no specific organ of sense. But, in compensation, their touch appears to be extremely delicate. Accordingly, it is sufficient to strike, or even slightly to stir the earth which they inhabit, to make them come out quickly. The mucous nature of their skin causes them to be partial to humidity, whether in the earth or atmosphere. They are, therefore, much afraid of the drying action of light, sun, and air. If, by any cause, they find themselves exposed to it they quickly endeavour to withdraw from its influence by sinking into the earth, or getting under some shelter, and if they cannot do so they soon dry up, and are deprived of life. They move with tolerable swiftness on the surface of the earth, by the alternate extension and approximation of the rings of their body, a part of which is more or less fastened to the soil by means of little hooks, and that in all directions. They doubtless walk more generally forward, but they can likewise move a little in a contrary direction. For entering the earth they always use the upper lip, which they contract, so as to give it solidity, and a terebrant form, but they never can perform this operation, except in a very loose and humid soil. The canals which they form in the earth, have always at least a double issue, one by which they enter and another by which they come forth. It is by the first that they eject in a vermicular form, the earth which they have swallowed in excavating their galleries, and they issue forth through the other. To ascend thus in their hole it appears that they make a little use of their spines. It is generally supposed that these animals feed only upon the animal and vegetable substances found in the earth which they traverse. But it appears that they join with those evident portions of organized bodies. It is quite certain that the lumbrici seek, in preference, unctuous soils, such as the holes where dung is deposited, the strata of our gardens, &c.

Although these animals are really endowed with hermaphrodism, that is to say, they have both sexes at once, it does not appear to be complete, but the above approximation of two individuals is necessary for reproduction, without, however, any reciprocal penetration of an exciting organ. It is at the commencement of the spring that coupling takes place, and this always occurs during the night, and half out of the ground. The two individuals adhere so closely together, by a sort of agglutination of the enlarged ring, that they sooner suffer themselves to be crushed than separated. Montegre tells us, however, that this adherence is not so great as to prevent the animals from sinking into their hole, on the least sensation of approaching danger. At the end of a space of time of greater or less duration, but the exact extent of which we do not know, they deposit their young in the earth. We are no better acquainted with the time which they take to acquire the necessary development for reproduction, and for attaining their largest size. Neither do we know any thing concerning the duration of their life.

The lumbrici are in full possession of their faculties only during the seasons of spring, summer, and a part of the autumn. In proportion as the cold approaches, they sink deeper and deeper into the earth, where, according to M. Latreille, they form a sort of lodgment, or case, probably with the mucous matter which issues from their bodies. Under certain circumstances, the nature of which does not appear to be sufficiently understood, the lumbrici become phosphorescent. Experiments have been tried upon them, respecting their capacity of reproducing parts removed. Some authors tell us that they have even seen the parts of a lumbricus cut in two, become each of them a complete animal.

This is conceivable respecting the anterior half, because it contains almost all the essential parts of the organization, and that, so to speak, there is nothing to form but an anus; but it is not probable that the posterior moiety can repair the loss of the stomach, the organs of generation, &c. The lumbrici are scarcely of any other use to the human species than as bait for fish. They are procured by searching for them, with spade or fork, in the unctuous or loose soils of our kitchen-gardens, &c. or even by stamping on and disturbing the soil in which, from the multitude of holes by which it is pierced, we may recognize that there is an abundance of these animals. What comes to the same thing, is by sinking the spade or a stake in the earth, to cause all around a considerable commotion and pressure. If this operation be continued for a while, especially in warm and humid weather, there will soon appear a great quantity of lumbrici, which may be preserved until it is necessary to employ them.

However abundant these animals may be in our gardens and fields, it seems certain that they do no harm to our gardening or agriculture; and even, as they divide and turn the earth, some persons are of opinion that they are more useful than injurious.

We have but little knowledge concerning the distribution of the species of this genus on the surface of the earth. They have as yet been studied in a satisfactory manner only in Europe. It is however extremely probable that they exist in North America, and the north of Africa and Asia. But on this subject we have no positive certainty, and still less so respecting their existence in the southern portions of those mighty continents, or in Australasia.

The NAIDES have evidently many relations with certain species of Nereides, and strongly so with the lumbrici. Their intestinal canal is simple, extended from one extremity to another of the body, and adherent to the external envelope by cellular bridles. The mouth is round and terminal, without any trace of tentacular, or masticatory apparatus. The anus is likewise terminal and rounded. We see all along the back of the animal a flexuous vessel, filled with a red fluid, as in the nereides. There is no trace of gills on any of the rings; but all of them, or almost all, are provided on the right and left with spines, or calcareo-corneous aciculi, simple, and sometimes fasciculate, but always few in number, and pretty nearly as in the lumbrici. The nervous system is almost unknown. These animals live almost constantly in fresh waters, running or stagnant, in the mud or soft clay which border them, and are seldom visible. It appears that they feed on small animals, whether infusory or not, which they probably swallow entire. M. Bosc has seen a naïs, whose intestine was filled with Daphniæ still living, which it had taken very dexterously when swimming. Their mode of generation is pretty nearly unknown. It is said, however, that they are oviparous, and that towards spring, is perceptible, towards the two thirds of the length of their body, and underneath, an elongated mass of a different colour from the intestine, and which, seen through the microscope, appeared to contain an innumerable quantity of eggs. This is to be seen for a longer or shorter time, which depends upon the heat of the season; but in general it has disappeared at the commencement of summer. The naïdes can be multiplied artificially, by cutting their body transversely into many pieces. Such, at least, was the result of the experiments of Trembley and of Rœsel; M. Bosc, however, tells us that he has repeated them without success.

The ancients appear to have been acquainted only with the most common species of HIRUDO, and these animals were not yet employed in medicine, at the time of Hippocrates. Pliny designates them very well under the name of *Hirudines*, and of *Sanguisugæ*, distinguishing two species. Many new species have been added in modern times, and they have been divided into generic groups. Many scientific men have studied their organization, the knowledge of which yet appears hardly complete, if we may judge from the very great and singular differences of opinion existing on many points of consequence respecting it.

A part of their history, which had been much neglected until later times, but the study of which has been necessitated by their dearness, and the difficulty of procuring them for medicinal purposes, is their mode of reproduction and preservation. M. Lenoble, a physician of Versailles, appears to be the first who noticed that the medicinal leeches formed a sort of cocoon, which Bergman had before observed in relation to another species. In our observations respecting the hirudines in general, we shall take as a type, the medicinal leech, which is by far the best known.

The body of a leech, in a state of moderate extension, is elongated, a little depressed, more convex above than underneath, attenuating insensibly in front, and much less so behind, where it is rounded. Therefore its greatest diameter is towards its posterior third or fourth. It is formed of a variable number of rings or articulations, very regular, very equal, separated by interstices a little more narrowed, and sub-linear. The anterior extremity is obtuse, although sub-angular. In a state of inaction it presents a large orifice, oval, depressed, and oblique, because the upper lip, composed of incomplete rings, advances sensibly more than the lower, formed by the edge of the first complete ring. Thus there is no distinct sucker, or cupper, although these lips perform the functions of one. On the first rings are observable some black points which have been called eyes, but which, at most, are but very imperfect rudiments of such. They are to the number of five pairs, very regularly disposed like a horse-shoe. In all the rest of the back, nothing is perceptible but some irregular mucous pores. Finally, altogether backwards, is an aperture much more evident, and perfectly medial, for the anus. The ventral face of the body presents, towards the first fourth of its length, two large medial orifices, at some distance from each other, the anterior of which serves as we shall see, for an issue to the male organ, and the posterior is the female organ of generation. In its whole length, we see on this inferior face of the body, some lateral pores pretty much swelled or tuberculous, ranged by pairs, one at every fifth ring. Finally, at the posterior extremity is a muscular disk, perfectly circular, a little concave, forming a sort of cupper. Sometimes we observe that each ring is provided on each side with a small, not projecting, and retractile tubercle; doubtless the rudiment of an appendage, but in which no setæ are perceptible. M. de Blainville appears to be the only author who has noticed this.

The envelope of the leeches is soft in all its parts, and in all directions, so that the animal can easily pass from a semi-globular to a sub-linear form. The skin, properly so called, is adherent in all its points, and even almost confounded with the subjacent contractile tissue. An epidermis may be there distinguished, or rather a sort of varnish extremely thin, applied on a pigmentum tolerably thick, granular, and coloured in various ways. The dermis itself is not very thick, it is adherent, and subtuberculous, in consequence of the great number of crypta with which it is sown, which give it a porous aspect. Each ring separated from the others by a tolerably deep furrow, is itself divided into two by a transverse fold, on which there are numerous longitudinal fissures.

The crypta of the skin are larger, and more developed on each side of the belly, at every fifth ring, and form a pretty considerable projection or tubercle, pierced with a large pore. These organs, M. Thomas considered as sorts of lungs, but with small reason, as they never contain any air, and are filled with the same sort of mucosity as the other crypta of the skin, nor is there any thing in their position analogous to that of respiratory organs. Neither have they any real relation to the male organ of generation, as was supposed by M. Spix.

This very soft and contractile skin of the leeches, is, in all probability, the only organ of sense which can be recognized in them. There seems to be no appendage, no cavity, which can be regarded as the seat of smell. There certainly may exist an organ of taste, probably in the sort of lips which precede the dentiferous tubercles; but of this we can be by no means certain.

We remark, as we have already mentioned, at the upper

face of the first rings of the body, certain black points, which have some apparent resemblance to the simple eyes of hexapod insects, and still more so to the similar organs which exist in the nereides; but whether they are really eyes is more than doubtful, for on examination they do not appear to possess either the structure or the uses of such organs.

As for an organ of hearing, the place of the leeches in the animal series will not permit us to believe that they possess any.

The apparatus of locomotion consists in its active part alone, and it is entirely confounded with the skin which covers it; thus it is erroneously that some authors have admitted that each ring of the body is supported by a cartilaginous band. The contractile fibre has a shining satiny aspect.

The distinct part of the muscular system forms a stratum moderately thick at the internal face of the skin, on its whole extent, but a little thicker underneath than above. This stratum is composed of two planes—the external, formed of circular or transverse fibres, is much more slender than the internal. The latter is tolerably thick, especially below; it is entirely composed of longitudinal and fasciculated fibres, the most external of which terminate from one ring to another, while the internal have an extent much more considerable. We remark, moreover, certain fasciculi of transverse fibres within the plane formed by the longitudinal fibres.

At the anterior extremity of the body these two planes of fibres appear, as it were, to be confounded, from which results a contractile tissue, not distinct from that of the dermis, and which constitutes the two lips or edges of the anterior aperture, there susceptible of assuming any form.

At the posterior extremity there is also a sort of confusion of the two planes of muscular fibres, but they assume a new and peculiar disposition; in fact, the longitudinal fibres, approximated in consequence of the absence of the viscera, proceed from a central point, and form radii at the circumference of the discus, which we have said terminates the body of the leech, while the circular fibres preserve their usual disposition.

The apparatus of digestion is pretty similar to that of most of the class: there is no evident serous lacuna between the exterior and that which constitutes the intestinal canal; on the contrary, these two parts are united by numerous cellular and vascular bridles, which pass from one to the other, and thus produce a sort of strangulations, and then a series of pouches or dilatations of the intestine.

We have already remarked, that the anterior extremity of the leeches presents an aperture more or less considerable, conducting into a cavity sometimes conformed like a cupper; at the bottom of this cavity exists a labial fold, composed of three lobes, not very distinct, rather narrow, two lateral and one ventral, leaving between them a triangular space, the summit of which is upwards. In this space appear projecting more or less the dentiferous tubercles; these tubercles are the same in number as the labial folds, but in an inverse order. Their form is sublenticular. They are placed on a level, so that the free part of their edge, very obtuse, converges towards the longitudinal axis of the body, and the adherent part is confounded with the contractile stratum of the external envelope. They are of a yellowish and shining white, and appear to be of a denser tissue and more compact than the rest of the contractile stratum of which they are a dependence. They seem entirely contractile, though there is at their dorsal base a muscular fasciculus, more distinct, forming a part of the longitudinal stratum, extending under the œsophagus, and, besides, another transverse bridle, very visible, particularly between the two inferior tubercles. It is on the very blunt edge, however, that one can distinguish, by the assistance of a glass of very short focus, a double series of corneous teeth, extremely fine; sometimes they are so much so that it is impossible to perceive them. In the middle of a space comprised between the internal roots of the dentiferous tubercles, is a round orifice, extremely small, and conducting into the intestinal canal.

We have already said, that this last is not free and floating in the visceral cavity; in fact, its parietes are adherent by their external surface, almost immediately to the muscular sheath, by means of a stratum of cellular tissue, of a spongy aspect, of a deep brown colour, and which may be very well considered as hepatic. Be this as it may, the intestinal canal has its parietes extremely slender; the mucous membrane is scarcely less so: it forms longitudinal folds, very little marked; they are much more so in the whole length of the œsophagus, which is very short, and all its parietes are distinct. Beyond this is the stomach, which extends almost to the posterior sixth of the total length of the body; its most singular peculiarity is, that in the leeches which are gorged with blood, we find it divided by a considerable number of strangulations, and of lateral pouches, most usually of a sigmoïd form; these pouches, which some authors have regarded as stomachs, are variable in number, according to the report of different authors. The stomach of the leech, when completely empty, which is the proper state in which it should be studied, and not in a state of enormous distension, presents a great number of longitudinal folds, which converge or approximate at the entrance of the sinuses. At the place where this stomach terminates, it is again continued in a vast pouch on the right and left, which extends as far as the extremity of the body, occupying its entire breadth, and without its parietes being more separated from the skin than the rest of the stomach. There are also perceived strangulations formed by the transverse muscular fibres, which have been named cœcums. The intestine, properly so called, is very short; its communication with the stomach is made by an extremely narrow orifice; it has a similar opening into the rectum, which latter proceeds directly to the anus, which is very small, and pierced in the last ring of the body.

No author has spoken of a liver in the leech; yet M. de Blainville regards as such the cellular system, already mentioned, of a deep brown, which clothes externally, in the form of a membrane, the greatest part of the intestine and stomach.

The respiratory apparatus, according to the generality of observers, does not exist in a specific manner in the leeches.

The circulatory system is considerable and very complicated in these animals: it is composed as usual of a venous and arterial system, but there is no heart, properly so called.

The venous system is formed of two very long vessels, with their distinct parietes, situated underneath each side of the body, between the intestinal canal and the longitudinal stratum of muscles of the external envelope. These vessels, which are evidently larger in the middle than at the extremities, receive in their passage a great number of transverse branches, some of which return from the tissue itself of the animal, and others come from the vessel on the opposite side, from which it follows that these two large veins and their ramifications form a net-work with broad meshes on the back of the leech.

Towards the anterior extremity these two veins are continued in branches which curve upwards, and unite in the middle and dorsal line to a smaller vessel, but with parietes a little thicker, placed in a longitudinal furrow, hollowed in all the length of the intestine. This is the aorta, from which spring, at a right angle from the two sides, the vessels destined by their ramifications to carry the blood into all the parts of the body of the animal, but especially to the parietes of the intestinal canal.

The reproductive apparatus, or system of generation, is

also very complicated in this genus of animals, first, because the two sexes exist in each individual, and secondly, because each of the sexual organs is very much developed.

The female sex is composed of several parts, all accumulated at a little distance from the posterior genital orifice, which belongs to it. There are at first two ovaries, ovoid, or sub-globular, placed in front of the orifice; from each of these springs a very short oviduct, but by its union to its congener is formed a single canal, directed from behind forwards, and projecting almost entire into a pouch or sort of matrix with distinct parietes, mucous and contractile, and the neck of which is extended as far as the external orifice.

The male sex is still more complicated, and much more extended. It is formed of a complex secretory organ, of an excretory canal with epididymis, and finally of an excitatory organ with its sheath.

The secretory organ is composed of a series of small white globular masses on each side of the intestinal canal, between the sinuses of the stomach, contained in the cellular tissue, but certainly without any adherence to the skin. Each little mass is nothing but a white vesicle with very slender parietes, and contains a whitish fluid. Each of these furnishes a small white canal, which is soon united to a common canal, situated at the external side of the series, and which advances directly, making, however, many sinuosities from rear to front. When this canal has arrived towards the genital region, its diameter diminishes in a sensible manner, and it comes in connexion with a white ovaliform mass, which seems formed by the crowded convolutions of this canal, so as to represent the disposition of the cerebral convolutions in the mammifera. From this mass issues a distinct canal, cemented against it, and which terminates at the root of the sheath of the excitatory organ. This sheath, considerable and very long, is directed backwards for its first half, then forward for the second. These two parts are cemented one against the other, and towards the place where the ovaries are, the extremity gives issue to the excitatory organ. The latter very long, slender, and cylindrical in a part of its extent, claviform at its extremity, issues forth by the anterior genital orifice, probably from the contraction of the sheath, which appears to be of a muscular tissue.

We have to state that this description, differing considerably from what we find in the works of many authors, is taken from M. de Blainville, who had the fortunate opportunity of dissecting a leech taken in the act of coupling, and in which consequently all the parts were in the greatest state of development, so that at all events we may rely upon its general truth. This part, however, of the anatomy of this animal involves much difficulty. The nervous system of the leech is pretty nearly similar to that of the lumbrici, and the rest of the class. Placed on the middle abdominal line in the cellular tissue, which separates the intestine from the sub-cutaneous muscular stratum, it is composed of a certain number of ganglia placed in file, and furnishing, beside the cords of communication in front and behind with each other, some transverse threads for the external envelope. The number of these ganglia appears to be very variable. The first, much larger than, and differently composed from, the others, is immediately on the under lip. Besides the threads which it furnishes to the surrounding parts, there issues forth on each side a thick cord, which is continued with a cephalic or epilabial ganglia scarcely thicker than itself, and which gives out the nerves of the upper lip.

Each of the following ganglia is of a lozenge form ; the anterior and posterior angles furnish the double cords which continue the nervous system along the belly and from the exterior angles proceed the extremely fine threads, to distribute themselves to the different parts; the last ganglion supplies those which proceed to the posterior disk.

The leeches feed on animal or vegetable matter, and on the juices or substances of the former only, according to the species. The medicinal leech, when feeding by suction, fastens itself by the posterior disk, and having selected the spot to bite, applies the anterior disk to it, which acts like a cupping-glass; the mouth is then advanced, its lips widened, and the three dentiferous tubercles which carry the hooks are erected and stiffened by a strong contraction of their muscular tissue. By the alternate contraction and slight expansion of these tubercles, a combined action of pressure and rubbing of the hooks is produced, and the smaller blood vessels thus become ruptured, and the wound, slight as it is, sometimes produces a degree of inflammation, which would not be likely to result from a simple cut produced by a smooth-edged instrument.

On opening the leech shortly after it has gorged itself with the blood of its prey, it will be found that none of this blood has passed into the intestines. The operation of digestion is extremely slow, notwithstanding the rapid and excessive manner in which the leech fills its stomach: a single meal of blood will suffice for many months, nay, more than a year will sometimes elapse before the blood has passed through the intestines in the ordinary manner, during all which period so much of the blood as remains undigested in the stomach continues in a fluid state, and as if just taken in, notwithstanding the vast difference in the heat of the body of a mammiferous animal and that of a leech.

The species which swallow animal matter in an entire state take it only when alive or very recently dead. These do not present the singularities in the structure of the œsophagus and stomach found in the blood-sucking leeches, but digestion is extremely slow in all.

The leeches are essentially aquatic animals, most of them living in fresh water, but some in the sea; a few occasionally quit the water, and as nature seldom acts on a single rule, one species is said never to go into it—humidity, however, is as necessary even to these as to the earth-worms.

We have nothing to add on the genus Gordius, and proceed therefore to the translation of Cuvier's description of such of the articulated animals as have articulated feet.

ARTICULATED ANIMALS,

AND

PROVIDED WITH ARTICULATED FEET;

OR,

CRUSTACEA, ARACHNIDA, AND INSECTS.

THESE three last classes of articulated animals, which Linnæus united under the name of *Insects*, are distinguished by articulated feet, at the least six in number. Each articulation is tubular, and contains in its interior the muscles of the following articulation, which always moves by ginglymus, that is to say, in one direction only.

The first articulation, which attaches the foot to the body, and which is most frequently composed of two pieces, is named the *haunch*; the following, which is usually in a situation nearly horizontal, is the *thigh*; the third, most usually vertical, is named the *leg*; finally, there remains a series of little ones, which rest upon the ground, and properly form the *tarsus*.

The hardness of the calcareous, or corneous envelope of the greater number of these animals, is referrible to that of the excretion which interposes between the dermis and the epidermis, which in man is termed the *rete mucosum*. It is also in this excretion that the colours, frequently so brilliant and so various which adorn them, are deposited.

These animals always have eyes, which may be of two

sorts: the simple eyes, which present themselves under the form of a very small lens, commonly three in number, and disposed triangularly on the summit of the head; and the composite, or eyes with facets, whose surface is divided into an infinity of different lenses, called *facets*, and to each of which a thread of the optic nerve corresponds. These two sorts may be united, or separated, according to the genera. We do not yet know whether, when they exist simultaneously, their functions are essentially different or not; but in both vision is performed by means very different from those by which it is caused in the eye of the *vertebrata.*

Other organs which appear in this division for the first time, and which are found in two of these classes, the crustacea and the insects, namely, the *antennæ,* are articulated filaments, infinitely diversified in form, often so even according to the sexes, attached to the head, appearing eminently adapted for the purposes of a delicate tact, and perhaps of some other species of sensation of which we have no idea, but which possibly may refer to the state of the atmosphere.

These animals enjoy the sense of smell, and that of hearing. Some place the seat of the former in the antennæ. Others, like M. Dumeril, at the orifices of the tracheæ; others again, like M. Marcel de Serres, in the palpi. But these opinions are by no means founded on positive and conclusive facts. With regard to hearing, the decapod crustacea, and some orthoptera, are the only animals of this division which have a visible ear.

In the mouth of these animals there is a great analogy, which, according to the observations of M. Savigny, even extends, relatively at least to the hexapod insects, to those which can only suck fluid aliments.

Those which are called *grinders,* because they have jaws proper for the trituration of food, always have them in lateral pairs, placed one before the other. The anterior pair are especially named *mandibles;* the pieces which cover them in front and rear, are named *lips;* and that of the front in particular receives the name of *labrum.* This remark belongs more especially to the hexapod insects. Certain articulated filaments attached to the jaws, or the under lip, and which appear to serve the animal for the purpose of recognizing its food, are termed *palpi,* or *antennulæ.* The forms of these various organs determine the nature of the aliment as precisely as do the teeth of quadrupeds. To the lower lip the tongue, or *ligula,* commonly adheres. Sometimes (the *bees* and many other hymenopterous insects,) it is considerably prolonged, as well as the jaws, and forms a sort of false proboscis *(promuscis)* having the pharynx at its base, often covered by a species of sub-labrum, called by M. Savigny *epipharynx.* Sometimes *(hemiptera* and *diptera,)* the mandibles and jaws are replaced by scaly pieces, in the form of bristles or lancets, received in a tubular elongated sheath, either cylindrical and articulated, or more or less elbowed, and terminated by sorts of lips. These parts then compose a true proboscis. In other sucking insects *(lepidoptera)* the jaws alone are considerably prolonged, and unite to form a tubular body, in the form of a thread, having the appearance of a long tongue, very fine, and rolled up spirally, *(spiritrompe,* Lat.); the other parts of the mouth are very much diminished. Sometimes, as in many crustacea, the anterior feet approach the jaws, assuming their form, exercising a part of their functions, and one might then say that the jaws are multiplied. It may even happen that the jaws shall be so much reduced, that the maxillary feet, otherwise *jaw-feet,* may replace them altogether. But whatever may be the modifications of these parts, there is always a means of recognizing them, and of reducing these changes to a general type.

FIRST CLASS OF ARTICULATED ANIMALS, AND PROVIDED WITH ARTICULATED FEET.

THE CRUSTACEA (CRUSTACEA,)

ARE articulated animals with articulated feet, respiring through gills, covered in some by the edges of a testa, or carapace, external in others, but which are not inclosed in special cavities of the body, receiving the air through apertures placed at the surface of the skin. Their circulation is double, and analogous to that of the mollusca. The blood repairs from the heart situated on the back, to the different parts of the body, whence it comes back to the gills, and from thence it returns to the heart. These gills, situated sometimes at the base of the feet, or on the feet themselves, sometimes on the lower appendages of the abdomen, form either pyramids composed of plates piled together, or bristling with barbs, or tufts, with simple laminæ, and even appear in some to be solely constituted by hairs.

Some zootomists, and especially M. le Baron Cuvier, have made us acquainted with the nervous system of many crustacea of divers orders. The same subject has lately been treated profoundly by MM. Victor Audouin and Milne Edwards, in their third memoir on the anatomy and physiology of animals of this class (*Ann. des Scienc. Nat.* xiv. 77,) and we want nothing to complete these researches but the publication of those made by M. Straus, on the branchiopoda, and particularly on the limulæ, of which these two naturalists have not spoken.

" The nervous system of the crustacea, submitted to their

observation, presents itself, as they inform us, under two very different aspects, which constitute the two extremes of the modifications which it exhibits in the crustacea. Sometimes, as takes place in the *talitrus,* this apparatus is formed by a great number of nervous enlargements, similar among themselves, disposed in pairs, and united by cords of communication, so as to form two ganglionic chains, distant one from the other, and occupying the whole length of the animal. Sometimes, on the contrary, it is merely composed of two ganglia, or knotty swellings, dissimilar in form, volume, and disposition, but always simple and uneven, and situated one at the head, and the other at the thorax. This is what we meet with in the maja.

"Certainly, at the first glance, these two modes of organization appear to be essentially different, and were we to limit the study of the nervous system of the crustacea to those two animals, it would be very difficult to recognize in the central nervous mass of the thorax of the maja the analogue of the two ganglionic chains, which occupy the same part of the body in the talitrus. But if we recall to mind the divers facts which we have mentioned in this memoir, we must necessarily arrive at this remarkable result."

They were conducted to this by the exact study of the nervous system of various intermediate crustacea, forming so many links of this series, such as *cymothöe, phyllosoma, palemon,* and *palinurus;* they also rest on the observations of M. le Baron Cuvier and M. Treviranus: they deduced from them this consequence, that in spite of these differences of disposition, the nervous system of the crustacea is nevertheless formed of the same elements, which, isolated in some, and uniformly distributed through the whole length of the body in others, present various degrees of centralization, at first from without inwards, afterwards in a longitudinal direction; and that, finally, this approximation is carried to an extreme, when there exists only a single nucleus at the thorax (the crabs proper, or *brachyuri.*) Of all the macrourous decapods observed by MM. Victor Audouin and Milne Edwards, the palinurus would be that whose nervous system is the most centralized, and in fact, in our method, this crustaceum is but little removed from the brachyuri; but it would not be the same with Palemon and the lobster, for according to them the first would approach more in this respect to Palinurus than the lobster does, while in our distribution this last crustaceum precedes palemon, an arrangement which appears to us to be founded on many very natural characters.

The crustacea are apterous, or without wings, provided with two eyes, with facets, but rarely with simple eyes, and commonly with four antennæ; they have, for the most part (the pœcilopoda excepted) three pair of jaws, (the upper two which are designated under the name of *mandibles* comprehended); as many of jaw-feet, but the last four of which become, in a great number, real feet; and ten feet, properly so called, all terminated by a single claw; when the last two pair of jaw-feet fulfil the same functions, the number of feet is then fourteen. The mouth also presents, as in the insects, a labrum, a ligula, but no lower lip, properly so called, or comparable to that of these last. The third pair of jaw-feet, or the first, closes the mouth externally, and replaces this part.

The sexual organs, or those at least of the males, are always double, and situated under the breast, or at the lower origin of that posterior and abdominal part of the body which is commonly called tail; but they are never placed behind. Their teguments are usually solid, and more or less calcareous; they change skin many times, and generally preserve their primitive form and their natural activity; for the most part, they are carnivorous, aquatic, and live many years. They do not become adult, or fit for generation, until after a certain number of moultings. With the exception of a small number, in which the changes of the skin exercise a trifling influence on their primitive form, modify or augment their locomotive organs, these animals are, when born, size excepted, such as they will remain for the whole of their existence.

DIVISION OF THE CRUSTACEA INTO ORDERS.

The situation and form of the gills, the manner in which the head is articulated with the trunk, the mobility or fixedness of the eyes, the masticatory organs, and the teguments, will form the basis of our divisions, and give rise to the following orders:—

We shall divide this class into two sections, the MALACOSTRACA and the ENTOMOSTRACA. The first have generally very solid teguments, of a calcareous nature, and ten or fourteen feet, usually unguiculated; the mouth, situated in the usual way, is composed of a tongue, a labrum, two mandibles (often bearing a palpus), two pairs of jaws covered by the jaw-feet; in a great number, the eyes are carried on an articulate and mobile pedicle, and the gills are concealed under the lateral edges of the testa or carapace; in the others, they are usually placed under the post-abdomen. This section is composed of five orders, the DECAPODS, the STOMAPODS, the LÆMODIPODS, the AMPHIPODS, and the ISAPODS; the first four embrace the genus *Cancer* of Linnæus, and the last that which he names Oniscus.

The Entomostraca, or insects with shells, of Muller, compose the genus *Monoculus* of Linnæus. Here the teguments are corneous and very thin, and a testa in the form of a buckler, of one or two pieces, or in the form of a bivalve shell, covers or encases the body of the great majority; the eyes are almost always sessile, that is fixed, and often there is but one; the feet, the number of which varies, are in the majority exclusively adapted for swimming, and without any claw at the end. Some, having an anterior mouth, composed of a labrum, of two mandibles (rarely furnished with palpi), of a tongue, of from one to two pair of jaws at most, the external ones, naked, or not covered with jaw-feet, approach the preceding crustacea. In the other entomostraca, and which appear in many respects to border on the Arachnida, sometimes the masticating organs are simply formed by the haunches of the feet, advanced and disposed in the manner of lobes, bristling with small spines, around a large central pharynx; sometimes they compose a little siphon or bill, serving as a sucker, as in many Arachnides and many insects, or do not show themselves at all, or scarcely show themselves externally, whether the siphon be internal, or that suction is performed after the manner of a cupping-glass. Thus the Entomostraca are either denticulated or edenticulated; the first form our order of BRANCHIOPODS, and the second that of PŒCILOPODS, which in the first edition of this work were only a section of the preceding order.

The singular fossils called TRILOBITES, and of which M. Brogniart, our fellow-member of the Royal Academy of Sciences, has given an excellent monograph, being considered by him, as well as by other naturalists, as crustacea which border on the entomostraca, we shall treat of them succinctly at the sequel of the latter.

FIRST GENERAL DIVISION.

THE MALACOSTRACA.

The Malacostraca are divided naturally into those whose eyes are on a moveable pedicle, and those in which these organs are sessile and immoveable.

MALACOSTRACA WITH EYES SUPPORTED ON A MOVEABLE AND ARTICULATED PEDICLE, OR DECAPODA AND STOMAPODA IN GENERAL.

Eyes supported on a mobile pedicle, of two articulations, lodged in fossets, distinguish these crustacea from all others. Considered anatomically, they appear to be still more remote from them, inasmuch as they are the only ones which present sinuses where the venous blood collects, before repairing to the gills to return to the heart.

The decapoda and stomapoda resemble each other in many common characters: a large shell, sometimes divided in two, called testa or carapace, covers in front a more or less extended portion of the body. They all have four antennæ, of which the middle are terminated by two or three threads; two mandibles, with a palpus near the base of each, divided into three articulations, and usually inclined upon the mandibles; a bilobate tongue; two pair of jaws; six jaw-feet, but the four posterior of which are in some transformed into claws; ten or fourteen feet, in those in which the four jaw-feet have this form.

In the greater majority, the gills, seven pair in number, are concealed under the lateral edges of the testa; the two anterior pair are situated at the origin of the last four jaw-feet, and the others at that of the feet properly so called. In the other crustacea, they are annexed, under the form of tufts, to five pairs of feet like fins, situated under the post-abdomen; the under part of this posterior portion of the body is equally furnished in the others with four or five pairs of bifid appendages.

FIRST ORDER OF CRUSTACEA.

DECAPODA,

HAVE the head intimately united to the thorax, and covered with it by a test or carapace, entirely continuous, but most frequently presenting some deep lines dividing it into divers regions, which indicate the places occupied by the principal interior organs. Their mode of circulation offers some characters which distinguish them from the other crustacea. The heart, very much circumscribed, of a figure approaching oval, and with muscular parietes, gives birth to six vascular trunks, three of which are anterior, two inferior, and the sixth posterior. Of the three anterior arteries, the median (*the ophthalmic*) is distributed almost exclusively to the eyes; the two others (the *antennary*) spread themselves over the carapace, the muscles of the stomach, a portion of the viscera, and over the antennæ; the two lower, (*the hepatic*) carry the blood to the liver; the last, (or the *sternal* artery) the most voluminous of all, and which sometimes originates on the left, sometimes on the right of the posterior part of the body, is principally destined to carry the nutritious fluid to the abdomen, and to the organs of locomotion. It furnishes a great number of vessels of considerable volume, among which we must particularly remark that which MM. Audouin, and Milne-Edwards, name the upper abdominal artery, because it issues from the posterior part of this artery, (a little before the articulation of the thorax and abdomen, vulgarly called the tail) and because it penetrates into the abdomen, (the tail) where it divides itself into two thick branches, continuing its passage backward, becoming more and more slender in its progress, and terminating at the anus. The blood which has served for the nutrition of these divers organs, and which is thus become venous, flows from all parts into two vast sinuses, one on each side, above the feet, and formed of venous gulfs, united in a longitudinal series, in the manner of a chain. It discharges itself into an external vessel of the gills, is renewed there, becomes arterial again, passes into an internal vessel, and afterwards directs itself towards the heart, traversing some canals (*branchio-cardine*) lodged under the vault of the flanks. All the canals of the same side are united in a wide trunk, have a common opening into the lateral and corresponding part of the heart, through one aperture, whose folds form a double valve, and open to allow the blood to proceed from the gills to this viscus, but close to shut it out from an opposite direction, or to hinder it from passing from the heart to the respiratory organs. Examined internally, the heart exhibits a great number of bundles, and of muscular fibres intercrossed in various directions, and composing many small lodges in front of the orifices of the arteries. These lodges are so many small auricles, which communicate together with facility, when it is dilated, but which appear to form, for each vessel, when the heart is contracted, an equal number of little cells, whose capacity is in relation with the quantity of the blood of the vessels which are proper to them. These vessels open into the interior of the heart, by eight apertures, the two lateral ones of which we have spoken, comprized. Such is, some few modifications excepted, the general system of the circulation of the decapods.

The upper surface of the brain is divided into four lobes, the middle ones of which, furnish each, from their anterior edge, the optic nerve, which is carried directly into the pedicle of the eye, and is divided there into a multitude of threads, proceeding each to a corresponding facette of the cornea of these organs. The inferior surface of the brain produces four other nerves which go to the antennæ, and give out some threads to the neighbouring parts. From its posterior edge spring two nervous cords, very much elongated, embracing the œsophagus laterally, and uniting underneath. In some, as in the brachyuri, this union takes place only in the middle of the thorax, and the medullary substance afterwards assumes the form of a ring, and in proportions eight times larger than the brain. This ring gives rise, on each side, to six nerves, the anterior of which repairs to the parts of the mouth, and the five others to the five feet on the same side. From the posterior edge proceeds another nerve, repairing to the tail, without producing sensible ganglia, and appearing to represent the ordinary nervous cord.' In others, as in the macrouri, the two nervous cords, before they unite under the œsophagus, give birth each at the middle of their length, to a thick nerve, repairing to the mandibles, and their muscles. United, they form a first median ganglion, (the *sub-cervical*) furnishing nerves to the jaws and the jaw-feet. Approximated afterwards, in their entire length, they present, successively, eleven other ganglia, the five first of which give, each of them, nerves to as many pairs of feet, and the six others furnish those of the tail. That of the paguri has some ganglia less, and these crustacea thus appear to form the passage, from the brachyuri to the macrouri. We shall add, that M. Serres believes that he has recognized in these decapod crustacea, some vestiges of the great sympathetic nerve.

The lateral edges of the carapace or test, refold underneath, to cover and protect the gills, but leave a vacancy anteriorly, for the passage of the water. Sometimes even (see *Dorippus*) the posterior and inferior extremity of the thorax presents for this purpose, two peculiar apertures. These gills are situated at the origin of the last four jaw-feet and feet. The anterior four are less extended. The six jaw-feet are all of a different form, applied on the mouth, and divided into two branches, the exterior of which has the form of a little antenna, composed of a peduncle, and a setaceous, and many-articulated stem. It has been compared to a whip, *(palpus flagelliformis).* The two anterior feet, sometimes even the two or four following ones, are in the form of talons. The last articulation but one is dilated, compressed, and in the form of a hand. Its inferior extremity is prolonged into a conical point, representing a sort of finger, opposed to another, formed by the last articulation, or the tarsus proper. This last is mobile, and has received the name of thumb *(pollex)* the other, or the one which is fixed, is thought to be the *index*. These two toes are also called *mordentes*. The last is sometimes very short, in the form of a simple tooth, and the other folds underneath. The hand, as well as the fingers, will form for us the forceps, properly so called. The preceding, or antepenultimate articulation is denominated *carpus*.

The respective proportions, and the direction of the locomotive organs are such as to enable these animals to walk sideways or backwards.

Except the rectum, which opens at the end of the tail, all the viscera are enclosed in the thorax, so that this portion of the body represents the thorax and abdomen of the insects. The stomach, supported by a cartilaginous skeleton, is armed internally with five osseous, and denticulated pieces, which complete the mastication of the aliments. We may see there, at the time of moulting, which occurs towards the end of spring, two calcareous round bodies, convex on one side, and plane on the other, which are vulgarly called *crab's eyes*, which, as they disappear after the moulting, give grounds to presume, that they furnish the matter for the renovation of the testa. The liver consists of two large clusters of vessels, closed at one end, and filled with a bilious humour, which they pour into the intestine near the pylorus. The alimentary canal is short and straight. The sides present a range of holes, placed immediately at the insertion of the gills, but which are not discovered until these organs are removed. The breast-piece, seen internally, presents, at least in many large species, some transverse lodges formed by crustaceous laminæ, and separated in their middle by a longitudinal crest of the same consistence.

The sexual organs of the males are situated near the origin of the two posterior feet. Two articulated pieces, of solid consistence, in the form of horns, of stylets, or of setaceous antennæ, placed at the junction of the tail and thorax, and replacing the first pair of sub-caudal appendages, are regarded as the generative organs of the males, or at least their sheaths. But according to our observations on various decapods, they should consist each of a small membranaceous body, sometimes in the form of a bristle, sometimes filiform, or cylindrical, issuing from a hole situated at the articulation of the haunch of the two posterior feet, with the breast-piece. The two vulvæ are placed on this piece, between those of the third pair, or at their first articulation, dispositions which depend on the enlargement or diminution of the breast-piece. Copulation takes place belly to belly. The growth of these animals is slow, and they live a long time. It is among them that the largest species are found, and those that are most useful to us as an article of food; but their flesh is difficult of digestion. The body of some decapods attains a length exceeding three English feet. Their forceps, as is well known, is very formidable, and so powerful in some large individuals as to raise and trip up a goat. They remain habitually in the water, but do not perish directly on being exposed to the air. Some species even pass a portion of their lives in this latter element, and visit the water only during the season of their amours, and for the purpose of depositing their eggs. They are nevertheless obliged to make their sojourn in burrows, or in fresh and humid places. The nature of the decapod crustacea, is voracious and carnivorous. Certain species even proceed to cemeteries to devour carcases, and to feed on them. Their limbs are regenerated with great rapidity, but it is necessary that the fractures should take place at the junction of the articulations, and they know how to cause them there, if the breach should accidentally occur elsewhere. When they are desirous of changing skin, they seek a retired place, so as to be in shelter from the pursuits of their enemies, and to remain there in tranquillity. When the moulting is done, their body is soft, and, according to some persons, is then of a more delicate flavour. From a chemical analysis of the old testa, we learn that it is formed of carbonate and phosphate of lime, united in divers proportions to gelatine. On those proportions depends the solidity of the testa. It is much less thick and flexible in the last genera of this order, and farther on it becomes almost membranaceous. M. de Blainville has observed, that that of the palinuri is composed of four superposed strata, of which the two lower ones, and the upper are membranaceous; the calcareous matter is interposed between them, and forms the other stratum. From the action of heat, the epidermis assumes a tint of red, more or less lively, and the colouring principle is decomposed in boiling water. But other combinations of this principle produce in some species, a very agreeable mixture of colours, often bordering on blue or green.

The greater number of fossil crustacea, which have been discovered up to the present day, belong to the order of decapods. Among those of Europe, some, and the most ancient, approach the species now actually existing in the zones which neighbour the tropics. The others, or the more modern, have a great affinity with the living species, which are proper to our climates. But the fossil crustacea of the tropical regions appear to me to have the greatest relations with many of those found there at the present day, in the living state; a fact which would be interesting to geology, if the study of the fossil shells of those countries, and collected from the deepest strata, should present us with a similar result.

The first family, or that of

BRACHYUROUS DECAPODS, (KLEISTAGNATHA, *Fab.*,)

Has the tail shorter than the trunk, without appendages or fins at its extremity, and folding underneath, in a state of repose, to lodge itself in a fosset of the chest. Triangular in the males, and furnished only at its base with four or two appendages, of which the upper are larger, in the form of horns, it becomes round, broader, and gibbous in the females. Its under part presents four pair of double filaments edged with hair, destined to carry the eggs, and analogous to the natatory subcaudal feet of the macrourous crustacea and others.

The vulvæ have two holes, placed under the breast, between the feet of the third pair. Their antennæ are small. The intermediate usually lodged in a fosset under the anterior edge of the testa, terminate, each, by two very short threads. The ocular pedicles are generally longer than those of the macrourous decapods. The auricular tube is almost always petrous. The first pair of feet is terminated by a claw. The gills are disposed on a single range, in the form of pyramidal tonguelets, composed of a multitude of little leaves, piled one upon the other, in a direction parallel to the axis. The jaw-feet are generally shorter and broader than in the other decapods. The two external ones form a sort of lip. Their nervous system again, differs from that of the macroura. (*See* the generalities of the decapods.)

This family might, as in many methods anterior to the distribution of these animals by Daldorf, form but a single genus, that of

CRAB, (CANCER.)

The greatest number have the feet all attached to the sides of the breast, and always exposed. The first five sections are in this predicament. The first, or the SWIMMERS (PINNIPEDES) unites to this character, that of having the final feet at least terminated by a very flatted articulation like a fin, (oval or orbicular, and broader than the same articulation of the preceding feet, even when they also are fin-like.) They remove more frequently from the shore, and proceed into the high sea. If we except the orithyiæ, the tail of the males presents but five distinctly marked segments, while that of the females has seven. We shall commence with those, all of whose feet, the claws excepted, are natatory.

MATUTA, *Fab.*,

Have the testa almost orbicular, and armed on each side with a very strong tooth in the form of a spine. The hands are denticulated above, in the manner of a ridge, and bristling at their external face, with pointed tubercles; and the third articulation of the external jaw-feet is without any apparent notch, and terminates in a point, so that it forms, with the preceding articulation, an elongated and almost rectangular triangle. The external antennæ are very small. The ocular pedicles are a little arched. *Cancer latipes*, Degeer.

POLYBIUS, *Leach*.

Approximate to the *Portuni;* but their testa is proportionably less broad, and more rounded. Its sides only present the ordinary teeth. The third articulation of the external jaw-feet is obtuse and emarginated. The eyes are much thicker than their pedicles, and globular. We know as yet but a single species which has been found on the coast of Devonshire, and which M. Dorbigny, the correspondent of the Museum of Natural History, has also observed on our maritime coasts of the western departments. (*Polybius Henslowii*, Leach.)

In all the following swimmers, the two posterior feet alone are fin-like. We may at first detach from them those whose testa is almost ovoid, narrowed and truncated transversely in front, whose tail presents distinctly in the males (the only individuals known) seven segments. Such are,

ORITHYIA, *Fab.*,

Only species known, *O. Mamillaris*, Fab. *Cancer bimaculatus*, Herbst.

The testa of the last swimmers is notably wider in front than behind, in the form of a segment of a circle, narrowed towards the end, and truncated, so as either to represent a trapezium, or approach very nearly the shape of a heart. Its greatest transverse diameter generally exceeds its opposite diameter. The tail of the males presents but five segments instead of seven, the number of those in the female, and which is generally proper to the tail of the decapods. The third and the two following are confounded together, or form but one, nevertheless we discover the traces of them, at least on the sides.

We shall separate at first those whose eyes are supported on slender and very long pedicles, proceeding from the middle of the anterior edge of the testa, prolonged as far as its lateral angles, and lodging in a groove formed under the edge.

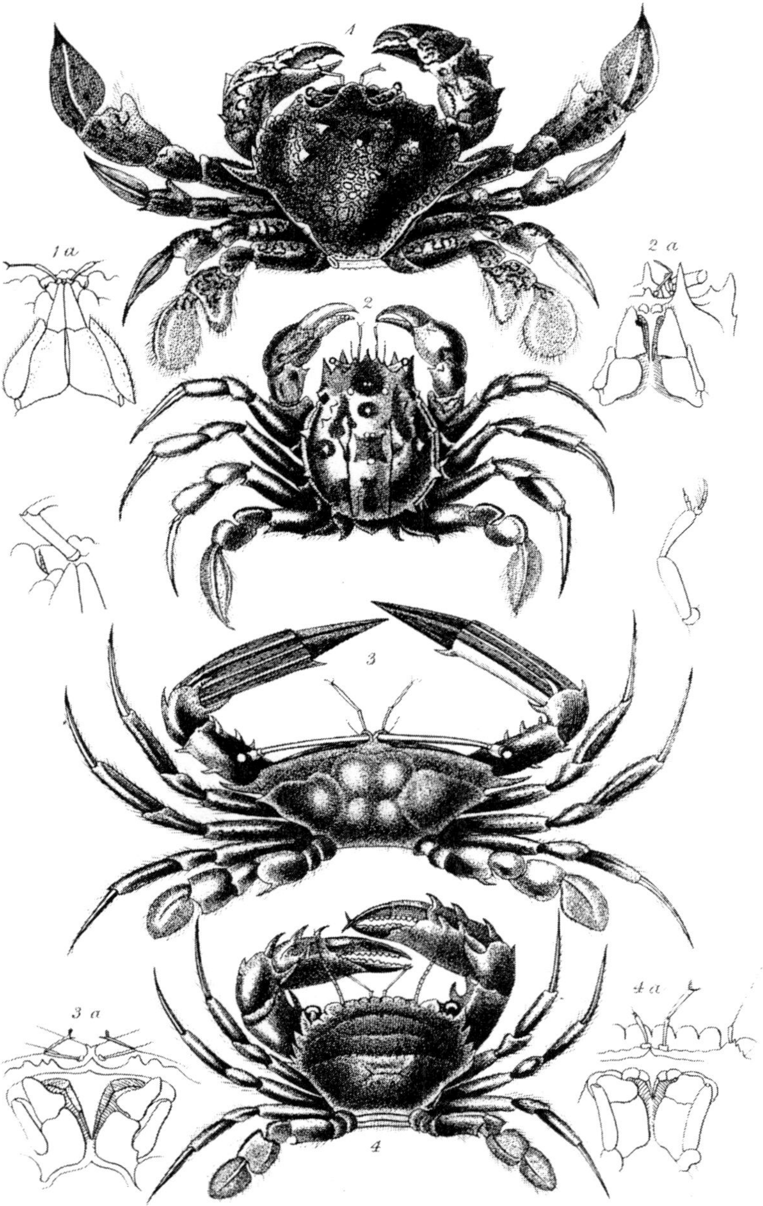

1 *Matuta Peronii.* 3 *Podophtalmus vigil.*

2 *Orythia mamillaris.* 4 *Thalamites admetes.*

London, Published by Whittaker & C.° Ave Maria Lane, 1833.

Such are,

PODOPHTHALMUS, *Lam.*

The test is in the form of a transverse trapezium, more broad and straight in front, with a long tooth in the form of a spine, behind the ocular cavities. The claws are elongated, spiny, and similar to those of the majority of the species of the genus *Lupa* of Dr. Leach. The only living species known inhabits the coasts of the Isle of France, and those of the neighbouring seas.

The ocular pedicles of the other crustacea of this section are short, occupy but a very small portion of the transverse diameter of the testa, lodge in oval cavities, and resemble in general those of the ordinary crabs, with which those swimming crustacea are almost insensibly connected.

These crustacea may be united into a single subgenus, that of,

PORTUNUS, *Fab.*

Some species proper to the seas of the East Indies, such as the *Admetus* of Herbst, are distinguished from all the following by their testa in the form of a transverse quadrilateral figure, narrowed posteriorly, and whose ocular cavities occupy the anterior lateral angles. The eyes are thus distant one from the other, by an interval almost equalling the greatest breadth of the testa. The insertion of the lateral antennæ is very remote from these cavities.

Other species, whose testa is in the form of the segment of a circle, truncated posteriorly, and wider in the middle, are remarkable for the length of their claws, which is at least double that of the testa. Its sides present each nine teeth, of which the posterior one is much larger, in the form of a spine. The tail of the males is often very different from that of their females. These portuni compose the genus *Lupa* of Dr. Leach, and are, for the most part, tolerably large, and exotic. The Mediterranean presents us with one species—*Portunus Dufourii*, Lat.

A third division shall be composed of species analogous to the last in the form of the testa, but whose lateral teeth, commonly five in number, are almost equal, or the posterior one of which differs but little from the preceding. The length of the claws but little exceeds that of the testa.

Those which have from six to nine teeth on each side are all exotic. (*P. Tranquebarius*, Fab. Herbst.) is the only known species, having nine teeth, and all equal at each lateral edge.

The following species, all of the European seas, have five teeth at each lateral edge of the carapace. *Cancer puber*, L.—*C. corrugatus*, Penn. Zool. Brit.—*C. mænas*, Lin. Fab. (*Common crab of our coasts.*) This appears to us to belong to Portunus, rather than to crab properly so called.

We shall form a fourth division with the subgenus

Platyonichus,

Whose denomination has replaced that of *portumnus* of Dr. Leach, too much approximating to the word *portunus*, already employed. Here the testa is at least as long as broad, and almost in the form of a heart. All the tarsi of the feet, the claws excepted, are terminated by a small semi-elliptical lamina, elongated and pointed; the index is very much compressed. This division again comprehends but a single species, which is the *cancer latipes* of Plancus. The front presents three teeth, and each lateral edge of the testa five.

From the swimming crabs we pass to those, all of whose feet terminate in a point, or by a conical tarsus, sometimes compressed, but not forming a fin, properly so called. Those among them whose testa is broader, cut in front like the arch of a circle, narrowed and truncated behind, whose claws are identical in the two sexes, in which the tail presents the same number of segments as in portunus, and which, with the exception of the tarsi, almost entirely resemble the last-mentioned genus, shall compose our second section, that of ARCUATA.

THE CRABS, properly so called, (CANCER, *Fab.*,)

Have the third articulation of the external jaw-feet notched or marked with a sinus, near the internal extremity, and almost squared. The antennæ, but little exceeding the front, and with but few articulations, are folded, smooth, or but little furnished with hairs. The hands are rounded, and do not present the appearance of a ridge above.

Some have the radical articulation of the external antennæ much larger than the following, in the form of a lamina, terminated by a projecting and advanced tooth, closing inferiorly the internal corner of the ocular cavities. The fossets of the middle or internal antennæ are almost longitudinal. Such is *C. pagurus*, Lin. which Dr. Leach separates generically from other crabs.

In the others, the inferior articulations of the antennæ are cylindrical. The first, although a little larger, does not differ from the following as to form and proportions, and does not pass the internal canthus of the ocular fossets. Those of the intermediate antennæ extend rather in the direction of the breadth of the testa, than in that of its length.

There are some among them whose toes have their extremity hollowed like a spoon, (*C. dentatus*, Fab.) These are the *clorodius* of Dr. Leach. Many of these species in which they terminate in a point are remarkable, inasmuch as the arching of the edges of the testa terminates posteriorly by a fold, and a projection jutting out in the manner of an angle. Those whose front is tridenticulated, and whose testa presents on each side only this projection or posterior tooth, compose Dr. Leach's genus *Carpilius*. The species of this subdivision (*C. corallinus*, Fab.; *C. macalatus* ejusd.) have marblings, or round spots, of the colour of blood, (*Xantho* of the same.) Other considerations might induce us to augment the number of these sections. But we have thought proper to confine ourselves to an indication of the principal ones.

PIRIMELA, *Leach*,

Altogether resemble the crabs, but their external antennæ are remarkably extended beyond the forehead, and their stem, longer than the pedicle, is composed of a great number of articulations. The fossets of the intermediate are, as well as in *C. pagurus*, rather longitudinal than transverse. But one species known, (*P. denticulata*, Leach.)

ATELECYCLUS, *Leach*,

Have, as well as the pirimelæ, the fossets of the intermediate antennæ longitudinal. The lateral antennæ are elongated, projecting, and composed of a great number of articulations. But they are very setaceous, as well as the claws; these claws are strong, with the hands compressed. The third articulation of the jaw-feet is sensibly narrowed above, in the manner of an obtuse, or rounded tooth. The tarsi are conical, and the ocular pedicles are of the ordinary size. The tail is more elongated than in the preceding crustacea.

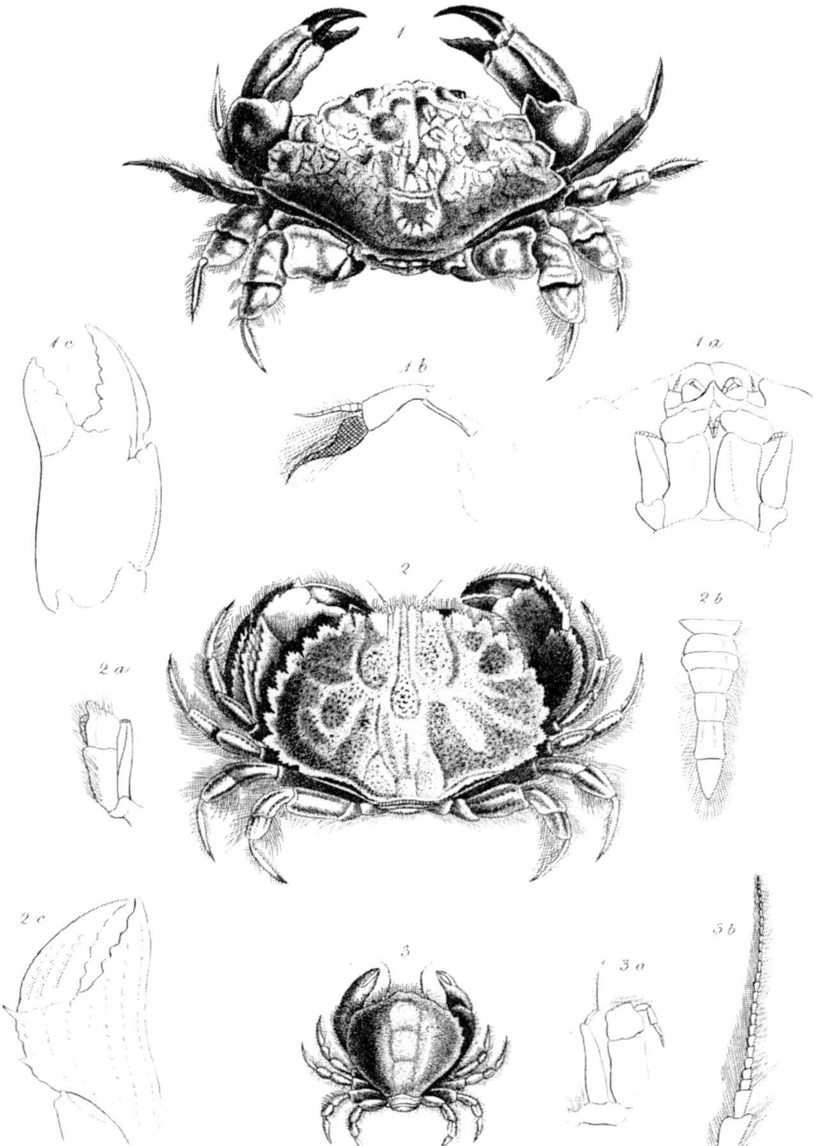

1 *Cancer Rhumphii.*

2 *Atelecyclus cruentatus.*

3 *Thia polita.*

London, Published by Whittaker & C.o Ave Maria Lane, 1832.

THIA, *Leach*,

Approach the last by reason of their lateral antennæ, of the direction of the fossets lodging the intermediate ones, of the form of the third articulation of the external jaw-feet, and of their sub-orbicular testa; but their eyes, as well as their pedicles, are very small and scarcely project; their tarsi are very much compressed and sub-elliptical; the front is arched, rounded, and without marked denticulations. The pectoral space comprised between the feet is very narrow, and of the same width throughout; the claws are proportionally much less strong; the testa is smooth, and in some other respects these crustacea approach *leucosia* and *Corystes*. The prototype species is *Thia polita*, Leach.

MURSIA, *Leach*,

Of which but a single species is yet known, proper to that part of the ocean which encompasses the southern extremity of Africa; it approaches Matuta and many portuni, by reason of the long spine with which each side of the testa is armed posteriorly. It also approaches the crabs, properly so called, in the form of the testa, and the external jaw-feet, with this difference, that their third articulation is in the form of an elongated square, narrowed and truncated obliquely at its superior extremity; but as in Calappa and Hepatus, the hands are very much compressed superiorly, with a short and denticulated edge.

HEPATUS, *Latr.*,

Have, as to the widened form of their testa, the shortness of their lateral antennæ, a great affinity with the crabs, properly so called, and approach Mursia and Calappa, by reason of their hands being compressed, and terminated above in the manner of a ridge: but the third articulation of their external jaw-feet is in the form of an elongated triangle, narrow, and pointed, without apparent emargination, a character also observed in Matuta and Leucosia. The type on which this section is founded is *Hepatus fasciatus* (Latr.), which Fabricius has confounded with Calappa.

A third section, QUADRILATERA, has the testa almost square, or formed like a heart, with the front generally prolonged, inflected, or very much inclined, and forming a sort of hood. The tail of the two sexes is of seven segments distinct in their whole breadth; the antennæ are generally very short; the eyes of the majority are supported on long or thick pedicles. Many live habitually on land, in holes which they excavate; others frequent fresh water. Their course is very rapid.

A first division will comprehend those in which the fourth articulation of the external jaw-feet is inserted at the internal upper extremity of the preceding articulation, either on a short and truncated projection, or in a sinus of the internal edge. These are they which approach the nearest to the crabs proper.

Some have the test sometimes square or trapezoïd, but not transverse, sometimes in the form of a truncated heart; the ocular pedicles are short, and inserted either near the lateral and anterior angles of the test, or more interior, but always at a sufficiently great distance from the middle of the front. Here come

ERIPHIA, *Latr.*,

Which have the lateral antennæ inserted between the ocular cavities and the medial antennæ; the test is almost always heart-formed, truncated posteriorly, and the eyes are remote from its anterior angles. (*Cancer Spinifrons*, Fab.)

1. *Eriphia lævimana.*

2. *Pilumnus aculeatus.*

3. *Thelphusa indica.*

4. *Front of Thelphusa fluviatilis.*

London, Published by Whittaker & C°. Ave Maria Lane, 1832.

TRAPEZIA, *Latr.*,

Similar to Eriphia, in the insertion of the lateral antennæ, but whose testa is almost square, depressed, smooth, with the eyes situated at its anterior angles, and the claws very large, in comparison with the other feet. *Cancer cymodoce*, &c. all exotic.

PILUMNUS, *Leach*,

Different from the two preceding subgenera, by reason of their lateral antennæ being inserted at the internal extremity of the ocular cavities, below the origin of the pedicles of the eyes; they are more approximate, as to the form of the testa, to the crustacea of the preceding section, than the other quadrilatera, and ambiguous, in this respect, between the two sections. As in the majority of the arcuata, the third articulation of their jaw-feet is almost square or pentagonal; the lateral antennæ are longer than the ocular pedicles, with a setaceous stem, longer than the peduncle, and composed of a great number of small articulations; the tarsi are simply furnished with hairs.

THELPHUSA, *Latr.*,

With lateral antennæ, situated as in the pilumni, but shorter than the ocular pedicles, of but few articulations, with the stem scarcely longer than the peduncle, and cylindrico-conical; the testa is almost in the form of a truncated heart, and the tarsi are furnished with spiny or denticulated ridges. *Cancer fluviatilis* of Belon-Rondelet, and Gesner.

Other quadrilatera, having, like the preceding, the fourth articulation of the exterior jaw-feet inserted at the internal extremity of the preceding articulation, are removed from them by the trapezoïdal form of the testa, transverse and widened in front, as well as by their ocular pedicles, which, like those of podophthalmus, are inserted near the middle of the front, long, slender, and reaching the anterior angles; the claws of the males are long and cylindrical. Such are

GONOPLAX, *Leach*,

Our seas furnish two species, *Cancer angulatus*, Linn., and *Cancer rhomboïdes*, Ejusd.; the principal difference between which is, that the former has spines at its anterior and posterior angles, the letter at the anterior only.

In the second division of quadrilatera, the fourth articulation of the external jaw-feet, or of those which cover inferiorly the other parts of the mouth, is inserted at the middle of the end of the preceding articulation, or more externally.

Sometimes the testa is trapezoïd or ovoid, or in the form of a heart, truncated posteriorly; the ocular pedicles, inserted at a little distance from the middle of its anterior edge, extend as far as the anterior angles, or even go beyond them.

In commencing with those whose testa has the form of a transverse four-sided figure, widened in front and narrowed behind, or has the form of an egg, we first find

MACROPHTHALMUS, *Leach*.

As well as in gonoplax, the testa is trapezoïd, the claws are long and narrow; the ocular pedicles are slender, elongated, and lodged in a groove, under the anterior edge of the testa; the first articulation of the intermediate antennæ is rather transverse than longitudinal, and the two divisions which terminate it are very distinct and of middle size; the external jaw-feet are approximate inferiorly at the internal edge, without vacancy between them; and their third articulation is transverse.

The following, forming the three next subgenera, have the fourth pair of feet, and then the third, longer than the others; the intermediate antennæ are excessively small, and scarcely bifid at the end; their radical articulation is almost longitudinal. These animals are proper to hot countries.

In the following the testa is solid, of a quadrilateral or trapezoïd figure, wider in front.

GELASIMUS, Lat., *Uca*, Leach.

The eyes terminate their pedicles in the manner of a little head; the third articulation of their external jaw-feet is in a transverse square; the last segment of the tail of the males is almost semi-circular, that of the females is almost orbicular.

The lateral antennæ are proportionally longer and more narrow than the same in the *ocypodes*. One of the claws, sometimes the right, sometimes the left, (for this varies in individuals of the same species) is usually much larger than the other; the toes or fingers of the small one are often in the form of a spatula or spoon.

OCYPODE, *Fab.*

The eyes extend into the major part of the length of their pedicles, and form a sort of knob; the third articulation of the external jaw-feet is in a long square. The tail of the males is very narrow, with the last articulation in the form of an elongated triangle; that of the females is oval.

The claws are almost similar, strong, but short, with the forceps almost in the form of an inverted heart. Thus, as the etymology of the generic name announces, these crustacea run with very great velocity; it is so great, that a man on horseback would find some difficulty in keeping up with them. From this faculty they received the name of *eques* among the ancient naturalists. By the moderns they are sometimes named *land-crabs*, (*Cancer cursor*, Linn.) In some others of them the eyes terminate the pedicles, and form a sort of knob. Some of the old continent (*O. Rhombea*, Fab.), and all of the new, are in this case; but the latter have a peculiar character, announcing that they visit the water more frequently, or swim with greater facility; their feet are more even, more flatted, and furnished with a fringe of hairs.

In the next, the testa, at least in the females, is very slender, membranaceous, and flexible; the body is almost round or sub-ovoïd. The ocular pedicles are sensibly shorter than in the preceding subgenera.

At first come,

MICTYRIS, *Latr.*

Their body is sub-ovoïd, very much inflated, more narrow and obtuse in front, truncated posteriorly, with the hood much lessened, and narrowed into a point at its extremity. The claws are elbowed at the junction of the third and fourth articulation; this latter is almost as large as the hand. The other feet are long, with the tarsi angular. Add to these essential characters, that the ocular pedicles are curved, and crowned with globular eyes; that the external jaw-feet are very ample, very hairy at the internal edge, with the second articulation very large, and the following almost semi-circular.

Immediately after Mictyris we shall place

PINNOTHERES, *Latr.*

Very small crustacea, living for a part of the year in divers bivalve shells. The testa of the females is suborbicular, very slender, and very soft, while that of the males is solid, almost globular, and a little narrowed into a point in front. The feet are of moderate length, and the claws are straight, and conformed in the usual way. The external jaw-feet present distinctly but three articulations, the first of which is large, transverse and arched, and the last provided at its internal base with a small appendage. The tail of the female is very ample, and covers all the under part of the body.

We now arrive at crustacea, which, analogous to the last by reason of the insertion of their ocular pedicles, are nevertheless remote from them as respects the testa: it has the form of a heart, truncated posteriorly; it is raised, dilated, and rounded on the sides, near the anterior angles. The ocular pedicles are shorter than those of the preceding subgenera, and do not altogether reach the lateral extremities of the testa. The intermediate antennæ are always terminated by two very distinct divisions.

In some, such as

UCA, *Latr.*,

The size of the feet, beginning inclusively at those of the second pair, diminishes progressively; they are very hairy, with the tarsi simply furrowed, without denticulations, or remarkable spines. (*Cancer uca*, Lin.)

In the others, the third and fourth pair of feet are longer than the second and fifth; the tarsi have denticulated or very spiny ridges. These crustacea form two subgenera.

CARDISOMA, *Latr.*,

Having the four antennæ, and all the articulations of the external jaw-feet exposed; the first three articulations of the same jaw-feet are straight, the third shorter than the preceding, emarginated superiorly, almost in the form of a heart. Finally, the first of the lateral antennæ is almost similar to the rest, and broad.

Gecarcinus, *Latr.*,

Whose four antennæ are covered by the hood; the second and third articulation of the exterior jaw-feet are large, flatted, as it were foliaceous, arched, and leaving a vacancy between them on the internal side; or the last of these articulations is in the form of a curvilinear triangle, obtuse at the summit. It reaches the hood, and covers the three following articulations (4, 5, and 6.) *Cancer ruricola*, Lin.

Sometimes the testa is almost square, subisometrical, or but little broader than long, flatted, with the front diminished in almost its entire breadth; the ocular pedicles are short, and inserted at the lateral anterior angles; the two ordinary divisions of the intermediate antennæ are very distinct; the external jaw-feet are separated internally, and form by this separation an angular vacancy; their third articulation is almost as long as broad; the claws are short and thick, and the other feet are very much flatted; the fourth pair, and then the third, are longer than the others; the tarsi are spinous.

Plagusia, *Latr.*,

Have their middle antennæ lodged in two longitudinal and oblique fissures, traversing the entire thickness of the middle of the hood. *P. depressa*, Latr.

They are inferior, or covered by this part in

Grapsus, *Lam.*

Their testa is a little wider in front than behind, or at least not more narrow, while in plagusia it widens a little from front to rear. *Grapsus varius*, Latr.

Our fourth section, Orbiculata, has the testa either sub-

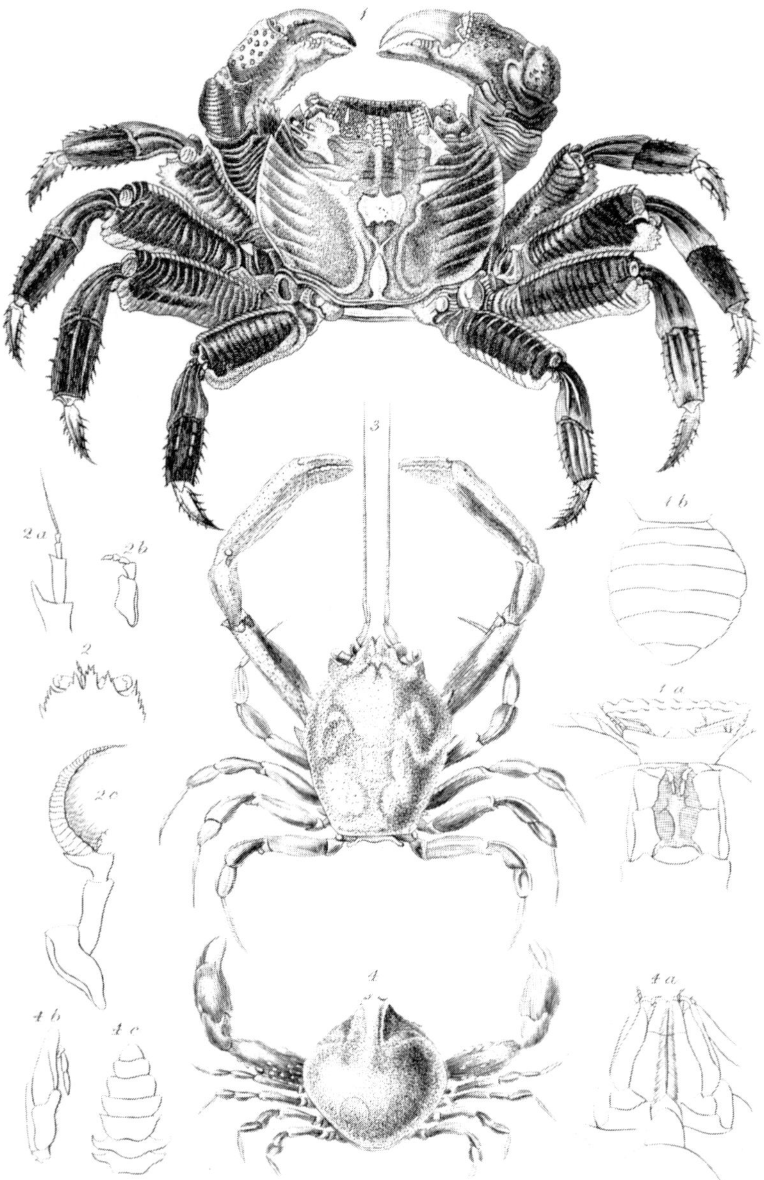

1. *Grapsus variegatus.* 3 *Corystes personatus.*

2 *Details of Plagusia depressa.* 4. *Leucosia urania.*

London: Published by Whittaker & C°. Ave Maria Lane. 1833.

globular or rhomboïdal, or ovoïd, and always very solid; the ocular pedicles are always short, or but little elongated; the claws are of unequal size, according to the sexes (larger in the males); the tail never presents seven complete segments; the buccal cavity proceeds narrowing towards its superior extremity, and the third articulation of the external jaw-feet is always in the form of an elongated triangle. The posterior feet resemble the preceding, and none of the latter is ever very long.

CORYSTES, *Latr.*,

Have the testa in an oblong-ovoid figure, crustaceous, with the lateral antennæ long, advanced, and ciliate. The ocular pedicles are of middle size, and apart; and the third articulation of the external jaw-feet is longer than the preceding, with an apparent emargination for the insertion of the following articulation. The tail consists of seven segments, but two in the middle are obliterated in the males, *Cancer personatus*, Herbst., Leach.

LEUCOSIA, *Fab.*,

Have a testa, whose form varies, but is most generally almost globular or ovoïd, and always of a very hard and stony consistence. The lateral antennæ and the eyes are very small; the eyes are approximate; the third articulation of the external jaw-feet is smaller than the preceding, and without apparent internal sinus. These parts are contiguous inferiorly, along the internal edge, and form an elongated triangle, whose extremity is received in two upper lodges of the buccal cavity. The tail very ample, and suborbicular in the females, usually presents but four or five segments, but never seven.

Dr. Leach has divided this genus into several others, but which we shall present as simple divisions.

The species whose testa is transverse, with the middle of the sides greatly prolonged, or dilated in the manner of a cylinder, or cone, form his genus *Ixa*, (*Leucosia cylindrus*, Fab.)

Those whose testa is rhomboïdal, with seven conical points, in the form of spines, on each side, compose that of *Iphis*.

If the testa, having always the same rhomboidal form, presents only angles or sinuses on the sides, we shall have his genus *Nursia*; and that of *Ebalia*, if these lateral edges are smooth.

The leucosiæ, with ovoïd, or almost globular testa, and distinguished besides from many of the preceding, by having the claws always longer than the body, and thicker than the other feet, and the tarsi sensibly striated, may be thus divided.

Some have the point advanced beyond, or at least not outedged by the superior extremity of the buccal cavity. The external branch of the exterior jaw-feet is elongated, and almost linear. In these the claws are slender, with the hands cylindrical, and the fingers long.

Sometimes the testa is almost globular, and either very spiny, as the genus *arcania*, or even as in that of *ilia*.

Sometimes the testa is suborbicular and depressed, as in the genus *persephona*; or it is ovoïd, as in that of *myra*.

In some the claws are thick, with the hands ovoïd, and short fingers. These are the true *Leucosiæ* of this naturalist.

In the others, the superior extremity of the buccal cavity passes the front or forehead. The external branch of the exterior jaw-feet is short and arched; the testa is rounded and depressed.

This last division comprehends his genus *phylira*.

Other considerations, taken from the proportions of the feet, and the form of the external jaw-feet, support these characters. *Ilia nucleus*, Leach.

The fifth section, that of TRIGONA, is composed of species whose testa is generally triangular or subovoïd, narrowed into a point, or in the manner of a beak in front, usually very unequal and rough, and the eyes are lateral.

The epistoma, or interval comprised between the antennæ and the buccal cavity, is almost always square, as long, or almost as long, as broad. The claws, or at least those of the males, are always broad and elongated. The following feet are very long in a great number, and sometimes even the last two have a different form from the preceding. The third articulation of the external jaw-feet is always square or hexagonal, in those at least, whose feet are of the usual length.

The apparent number of the segments of the tail varies. In many of them it is seven in both sexes; but in others, or at least in the males, it is less.

Many of these crustacea have been vulgarly designated under the collective name of *Sea-spiders*.

Although the species of this tribe are very numerous, but two have as yet been discovered in the fossil state, and one of which *(Maia squinado)* is still existing in the same localities.

A first division will comprehend those, whose second and following feet are similar, and the size of which diminishes progressively.

Among those we shall form a first group, of all the species whose tail, either in both sexes, or in the females, consists of seven articulations. The third articulation of the external jaw-feet is always square, and truncated or emarginated at the upper internal angle.

Very large claws, especially when compared with the other feet, which are very short, directed horizontally, and perpendicularly to the axis of the body, as far as the carpus or articulation preceding the hand, afterwards folded in front upon themselves, with the fingers abruptly bent, and forming an angle; ocular pedicles very short, and not at all, or very little projecting from their cavities, and a shelly testa, very unequal, or very spiny, characterize

PARTHENOPE, *Fab.*

Some of them have the lateral antennæ very short, of the length of the eyes at most. Their first articulation is entirely situated below the ocular cavities.

If the tail presents in both sexes seven segments, these species will compose *Parthenope*, properly so called, of Dr. Leach. If those of the males present but five, we shall have his genus *Lambrus*.

The others have the lateral antennæ very sensibly longer than the eyes. Their first articulation is prolonged as far as the internal upper extremity of the cavities proper to these last organs, and appears to be confounded with the testa. Here the post-abdomen always consists of seven segments. The claws of the females are much shorter than those of the other sex. The same naturalist distinguishes generically these crustacea under the denomination of *Eurynoma.* (*Cancer asper* Pennant.)

In the following the claws are always advanced, and their length is at most double that of the body. Their fingers are not abruptly and angularly inclined.

Here the length of the longest feet (the second pair) little exceeds that of the testa, measured from the eyes to the origin of the tail. The under part of the tarsi is generally either denticulated or spiny, or garnished with a fringe of hairs, terminating like a knob.

We shall present, in the first place, those whose ocular pedicles are very short, and of middle length, capable of

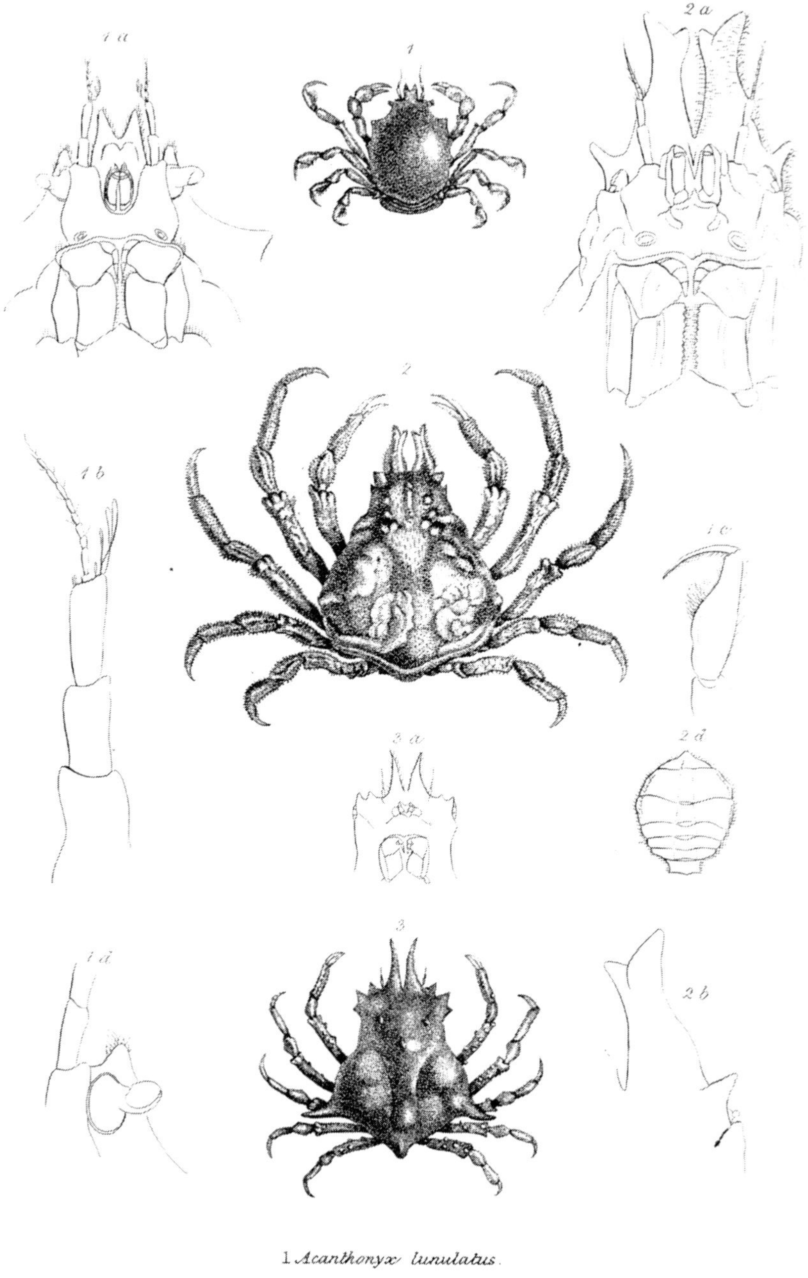

1 *Acanthonyx lunulatus.*

2 *Pisa serpulifera.*

3 *Pericera trispinosa.*

London, Published by Whittaker & C.º Ave Maria Lane, 1832.

being withdrawn altogether within their cavities, and whose claws, in the males at least, are notably thicker than the other feet.

MITHRAX, *Leach*.

Their claws are very robust, with the fingers hollowed like a spoon at the end. The stem of the lateral antennæ is sensibly shorter than the pedicle. The tail is composed of seven articulations in both sexes. (*Mithrax spinicinctus*, Latr.)

ACANTHONYX, *Latr.*,

Have an advancement in the form of a tooth, or spine, at the inferior side of the legs. The under part of the tarsi is hairy, and, as it were, pectinated, and the upper part of the testa is even. The tail of the males presents, at most, six complete segments. (*Meria glabra*, Collect. du Mus. d'Hist. Nat.)

PISA, *Leach*,

Whose claws are of a middle size, with the toes pointed. The legs have no spine underneath, and the tail consists of seven segments in both sexes. As well as in the preceding subgenera, the lateral antennæ are inserted at an equal distance from the fossets, which receive the intermediate, and from the ocular cavities, or more approximated to the latter.

Some, as in the genus *naxia* of Dr. Leach, have two ranges of denticulations under the tarsi, others have but a single range of denticulations, or a simple fringe of thick hairs in a knob, under the same articulation. Those which are in this last case form the genus *Lissa* of the same.

Among those which have but a single range of denticulations, sometimes, as in his *Pisa* properly so called, the length of the feet diminishes gradually; sometimes the third feet are abruptly shorter than the preceding in the males. This takes place in his *Chorinus.* Of these four genera of Dr. Leach, the prototype species are, *Pisa aurita,* Latr.; *Pisa chiragra,* ejusd.; *Pisa Xyphias,* ejusd.; and *Pisa heros,* ejusd.

PERICERA, *Latr.,*

Approximating to pisa in the form and proportions of the claws and the number of the segments of the tail, are remote from them as well as from the anterior subgenera, inasmuch as the lateral antennæ are inserted under the muzzle, and more sensibly approaching the fossets which lodge the intermediate ones, than those which receive the ocular pedicles. (*Maia taurus,* Lam.)

In the two following subgenera the ocular pedicles are short or middle-sized, as well as in the preceding. But the claws, even those of the males, are scarcely thicker than the following feet. The tail is always composed of seven segments.

MAIA, *Leach.*

In these the second articulation of the lateral antennæ appears to spring from the internal canthus of the ocular cavities. The hand, and the articulation which precedes it, are almost of the same length. The testa is ovoïd. (*Inachus Cornutus,* Fab., the *Cancer squinado,* of Herbst.)

MICIPPUS, *Leach,*

Have the first articulation of the lateral antennæ bent, dilated

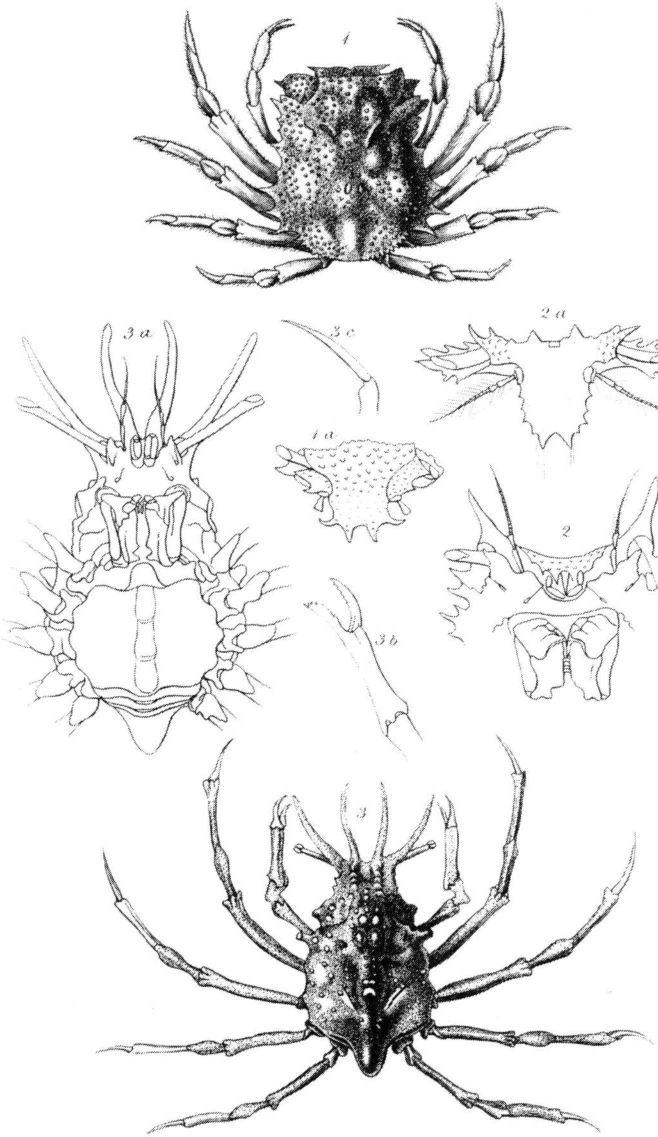

1 *Micippe phylira.*

2 *Detials of Mic. cristata.*

3 *Stenocionops cervicornis.*

London, Published by Whittaker & C.º Ave Maria Lane, 1833.

at its upper extremity, in the manner of a transverse and oblique lamina, closing the ocular cavities. The following articulation is inserted under its superior edge. The testa, seen above, appears, as it were, broadly truncated in front. Its anterior extremity is inclined, and is terminated by a sort of hood, or denticulated bill. (*Cancer cristatus*, Linn.)

STENOCIONOPS, *Leach*,

These are distinguished from all the subgenera of this tribe, by their long, narrow, ocular pedicles, projecting very much out of their fossets. (*Cancer cervicornis*, Herbst.)

Sometimes the under part of the feet presents neither ranges of denticulations nor fringes of hair, like a knob; those of the first pair at least exceed the testa in length by one half, and often by a much greater proportion. The body is generally shorter than in the preceding, either globular, or in the form of a contracted egg.

A crustaceum of this tribe (*Maia retuja*, Coll. du Jardin du Roi), whose testa is an ovoïd, truncated, or blunted in front, and woolly; whose ocular pedicles, elongated, and very much bent, proceed to lodge themselves behind, in fossets situated under the lateral edges of the testa; whose carpus, as well as in the maiæ, is elongated, offers another character which distinguishes it exclusively: the length of the feet, beginning from the second, appears to augment progressively, or at least to vary but little. Dr. Leach has formed of it the genus

CAMPOSCIA.

In the others, as usual, the length of the feet diminishes progressively, from the second pair to the last.

We are acquainted with some whose ocular pedicles, though much shorter than those of stenocionops, are always projecting; whose lateral antennæ have the third articulation of their peduncle as long or even longer than the preceding, and terminated by a long and setaceous stem. They approach micippus. Such are

KALIMUS, *Latr.*,

Two species, one very near the *Cancer superciliosus* of Linnæus.

Those which form the two following subgenera have the ocular pedicles susceptible of complete retraction into their fossets, and protected posteriorly by a projection in the form of a tooth or angle of the lateral edges of the testa; the second articulation of the peduncle of the lateral antennæ is much larger than the following; they are terminated by a very short stem, in the form of an elongated stylet.

HYAS, *Leach*,

Have the lateral edges of the testa very much dilated, in the manner of an auricle, behind the ocular cavities, which are oval, and tolerably large; the external side of the second articulation of their lateral antennæ compressed and carinated, and the ocular pedicles capable of being entirely discovered, when the animal raises them up. The body is subovoid. (*Cancer araneus*, Linn.)

In

LIBINIA, *Leach*,

The ocular fossets are very small, and almost orbicular; the ocular pedicles are very short, and admit but of little protrusion. The second articulation of the lateral antennæ is cylindrical, and but little or not at all compressed. The body is almost globular or triangular.

We shall unite to these the *Doclæa* and *Egeria* of Dr. Leach.

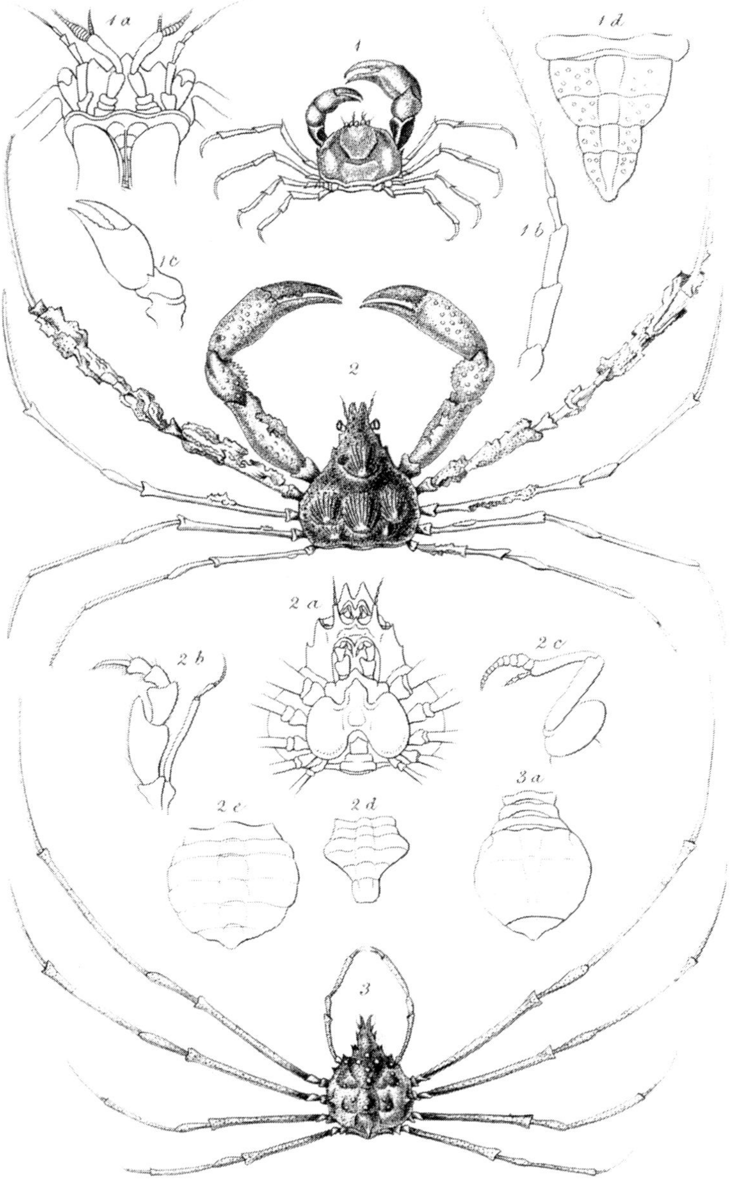

1 *Hymenosoma Leachii.*

2 *Inachus thoracicus.*

3 *Leptopus longipes.*

London, Published by Whittaker & C.º Ave Maria Lane 1833.

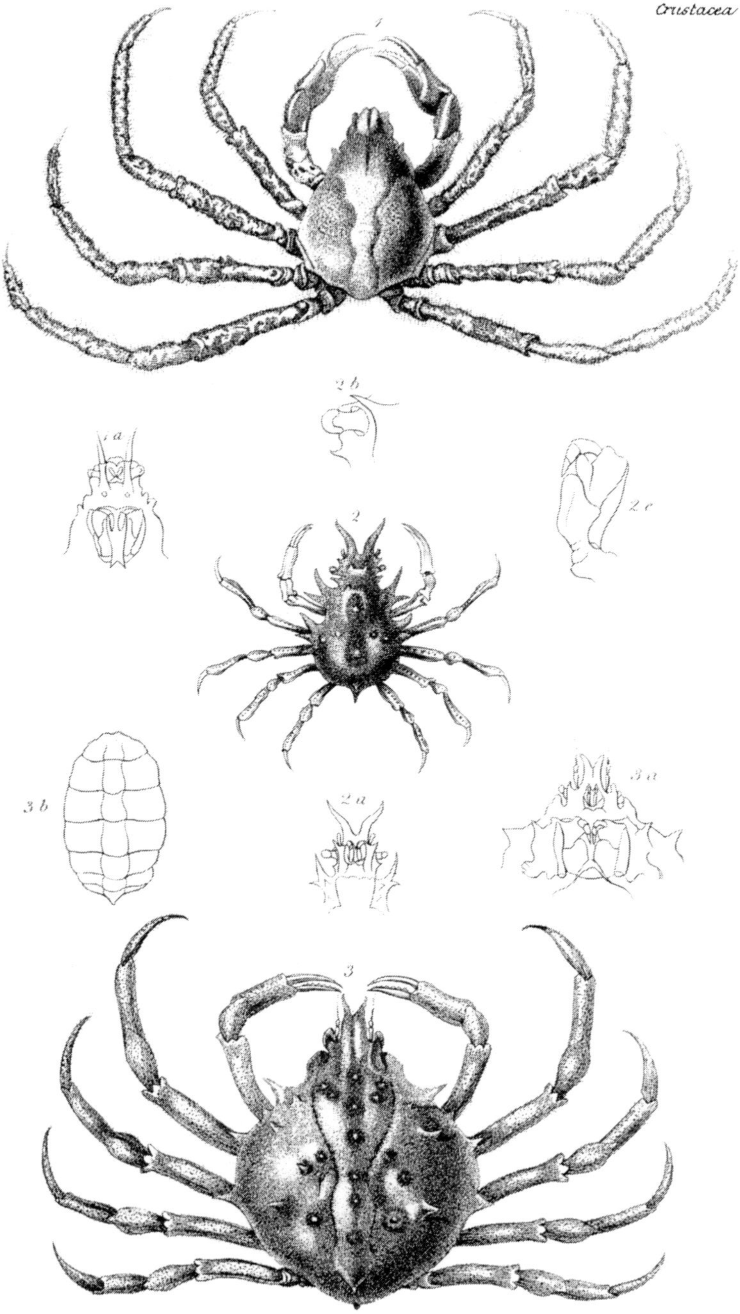

1 *Camposcia retusa* 2 *Halimus aries.*

3 *Libinia spinosa.*

London, Published by Whittaker & C.º Ave Maria Lane: 1833.

In his *Libinia*, properly so called, the claws of the males are thicker than the two following feet, and almost as long; the length of those which are the longest is not altogether double that of the testa. (*Libinia canaliculata*, Say. *emarginata*, Leach.)

The claws of the males of *Doclæa* are notably shorter than the two following feet; the length of these feet scarcely exceeds more than once and a half that of the testa, which is almost globular, and always covered with a brown or blackish down. (*Doclæa Rissonii*, Leach.)

In *Egeria* the claws are filiform, with the hands very much elongated, and almost linear. The following feet are five or six times longer than the testa. The body is triangular. (*Egeria Indica*, Leach.)

After having reviewed the subgenera of this tribe, whose feet coming after the claws are of an identical form, and whose tail is composed, in the females at least, and most frequently in both sexes, of seven articulations or complete segments, we shall pass to those in which it presents but six at most. The feet are generally long and filiform, as in the last subgenera. If we except *leptopus*, these crustacea are again remote from the preceding, in the relation of the form of the third articulation of the external jaw-feet; it is proportionally more narrow, contracted at its base, and the following articulation appears to be inserted at the middle of its upper edge, or more externally. The following subgenus differs from those which succeed it, by having but three segments in the tail of the males. The form of the third articulation of the external jaw-feet appears to me to be in other respects the same as in the preceding subgenera.

LEPTOPUS, *Lam.*

The tail of the females is formed of five segments. The

body is convex, and the feet are very long. But a single species is known. *Maia longipes*, Coll. du Mus.

If we except some species of *hymenosoma*, in which the tail presents distinctly but four or five articulations, in all the following subgenera this part of the body has six, either in both sexes or in the males. The third articulation of the external jaw-feet is sometimes in the form of a reversed triangle, or of an oval, narrowed inferiorly, sometimes in the form of a heart. The following articulation is inserted at the middle of its superior edge, or more externally than internally.

Some, such as the three following subgenera, approach to those which we have just described, in the isometrical, or at least transverse form of the epistoma. The basis of the intermediate antennæ is but little remote from the upper edge of the buccal cavity.

One of these subgenera is distinguished from the two others by the flatness of its testa, and by the upper extremity of the first articulation of its lateral antennæ (free in many), not exceeding that of the ocular pedicles. Such are

HYMENOSOMA, *Leach.*

The testa is triangular or orbicular. (*Hymenosoma orbicularia*, Desm.)

In the two following subgenera the testa is more or less convex, always triangular, and terminated in front in the manner of a bill. The first articulation of the lateral antennæ, always fixed, forms a crest, or projecting line, between the fossets of the middle antennæ and those of the eyes, and which is prolonged beyond the end of the ocular pedicles.

INACHUS, *Fab.*,

Have six segments in the tail; all the tarsi almost straight, or

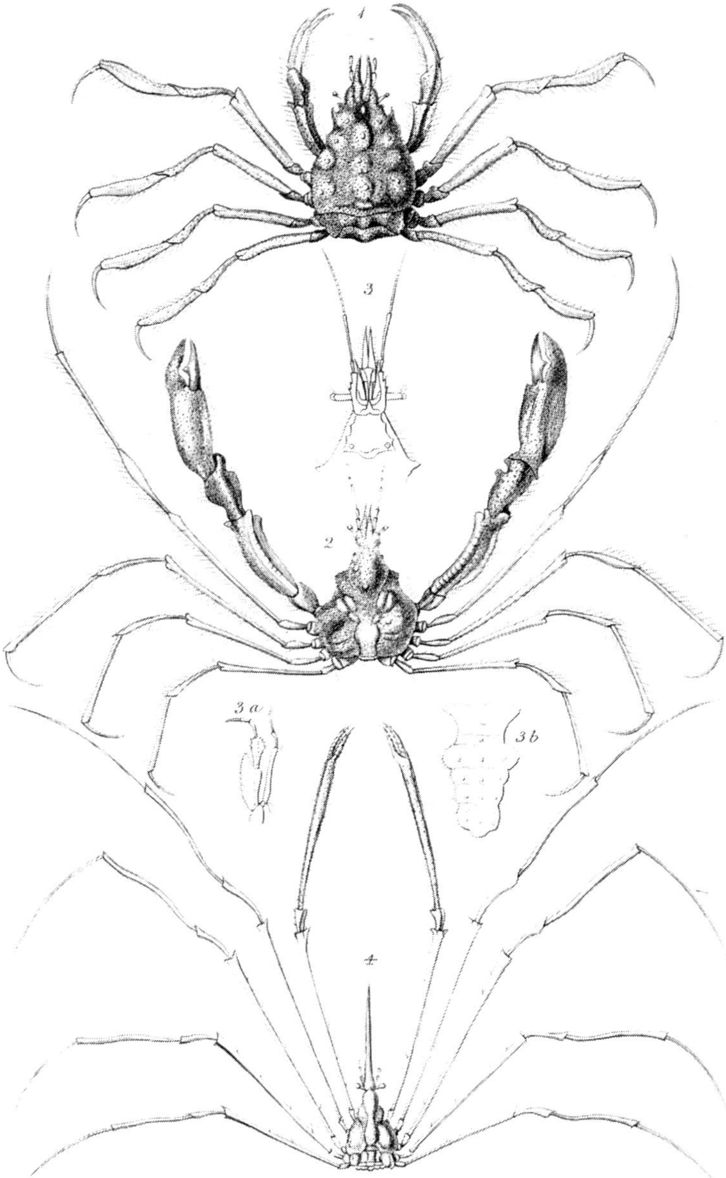

1 Eurypodius Latreillii. 3 Details of Sten. tenuirostris.

2 Stenorhynchus phalangium. 4 Leptopodia sagittaria

London, Published by Whittaker & C.º Ave Maria Lane: 1833.

but little arched; the ocular pedicles even susceptible of being concealed in their fossets, and a tooth or spine, in the males at least, at the posterior edge of these cavities. Dr. Leach has greatly contracted the primitive extent of this group. (*Cancer dodecos?* Linn. *Inachus dorsettensis,* Leach.)

ACHÆUS, *Leach,*

Have all equally six segments in the tail; but their four posterior tarsi are very much arched, or like a reaping hook; their ocular pedicles are always projecting, and present a tubercle in front. (*Achæus cranchii,* Leach.)

Next come those whose epistoma is longer than broad, in the form of a triangle, elongated and truncated at the summit, and in which the origin of the middle antennæ is removed, by a notable space, from the superior edge of the buccal cavity; the ocular pedicles are always projecting, when the testa is triangular and terminated in a point, more or less bifid, or entire.

STENORHYNCHUS, Lam., *Macropodia,* Leach,

Have six segments to the tail in both sexes. The anterior extremity of the testa is bifid. (*Macropodia tenuirostris,* Leach.)

LEPTOPODIA, *Leach.*

The tail of the males is composed of five segments; that of the females has one more. The testa is prolonged anteriorly into a long, entire, and denticulated point. (*Inachus sagittarius, Fab.*)

The last trigona differ from the preceding, in the dissimilarity of the posterior feet.

PACTOLUS, *Leach*,

Have the four or six anterior feet simple, or without forceps; the internal extremity of the penult articulation of the four hinder ones is prolonged into a tooth, forming with the last articulation a forceps or didactylous hand. The testa has the form of that of leptopodia, and the tail presents the same number of segments, but the feet are much shorter; those of the third pair are wanting in the individual which has served for the establishment of this section. (*Pactolus Boscii*, Leach.)

LITHODES, *Latr.*,

Resemble, as to the form of the first eight pair of feet, the other trigona. Their length, however, appears to augment progressively from the second to the fourth, but the last two are very small, folded, but little apparent, imperfect, and as it were, useless. The tail is membranaceous, with three crustaceous and transverse spaces on the sides, and another at the end, representing the segmentary divisions. The eyes are approximate inferiorly; the external jaw-feet are elongated, and projecting; the testa is triangular, very spiny, and terminated anteriorly in a denticulated point. These crustacea are proper to the northern seas. (*Cancer maia*, Linn.)

Our sixth section, that of CRYPTOPODA, is composed of brachyurous crustacea, singular in this, that the feet, with the exception of the two anterior, or claws, can be entirely withdrawn, and concealed, under an advancement, in the form of a vault, of the posterior extremities of their testa. This testa is almost semi-circular or triangular; the upper blade of the forceps is more or less raised, and denticulated like a ridge. In the species in which those are largest, they cover the front of the body.

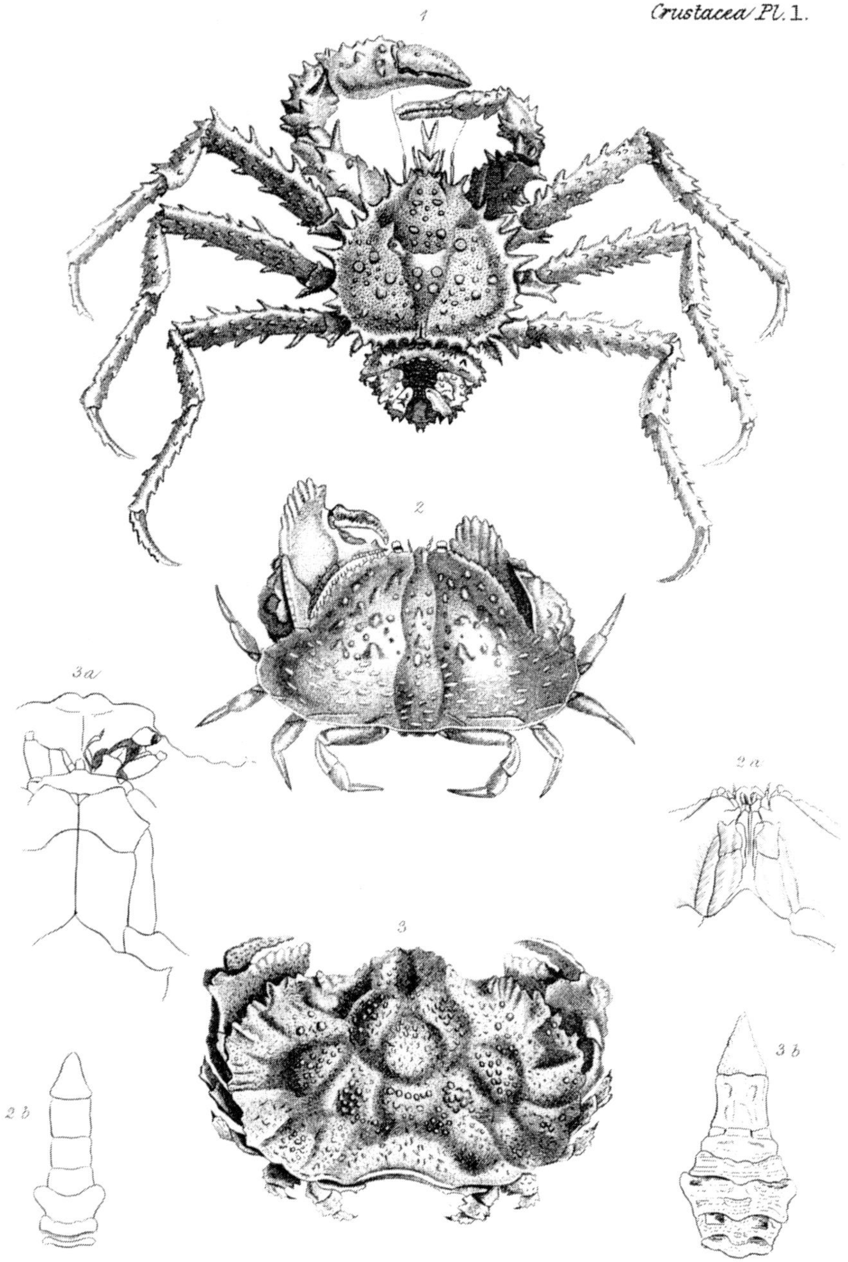

1 Lithodes artica.

2 Calappa tuberculosa.

3 Ethra depressa.

London: Published by Whittaker & C.º Ave Maria Lane 1832.

CALAPPA, *Fab.*,

Have the testa very gibbous, the forceps triangular, very much compressed, denticulated above, like a ridge or crest, and covering perpendicularly the front of the body, when the feet are contracted. The third articulation of the external jaw-feet is terminated like a hook; the superior extremity of the buccal cavity is narrowed, and divided longitudinally into two lodges by a partition.

Some, and the most numerous, have the two posterior and lateral dilatations of the testa incised, and denticulated. (*Cancer granulatus*, Linn., *Calappa granulata*, Fab.)

Others, such as *Calappa fornicata*, Fab., have the edges of the dilatations of the testa entire.

ÆTHRA, *Leach*,

Differ from calappa, by their very flatted testa, their forceps not raised perpendicularly, nor shading the front of the body, and the almost square form of the third articulation of the external jaw-feet.

Sometimes, (*Æthra depressa*, Lam.) the testa is in a transverse oval; sometimes, (*Parthenope fornicata*, Fab.) in the form of a short triangle, very broad, dilated, and rounded laterally. The claws are but little elongated, and tolerably thick; but sometimes they are longer, angular, and remind us, as does also the form of the testa, of the parthenopes. These last species might form a subgenus proper.

Finally, a seventh and last division, NOTOPODA, is formed of brachyuri, whose last four, or last two feet, are inserted above the level of the others, or seem to be dorsal and turn upwards. In those in which they are terminated by a sharp hook, the animal usually employs it to retain divers marine bodies, such as the valves of shells, of alcyones, with which it covers itself: the tail has seven segments in both sexes.

Some of them have, as well as the other brachyuri, the tail folded underneath; their feet are terminated by a sharp hook, and are not adapted for swimming.

In this division the testa is almost square, and terminated anteriorly by an advanced and denticulated point, or it is subovoid, or truncated in front.

HOMOLA, *Leach*,

Have the eyes supported by long pedicles, very much approximated at their base, and inserted below the middle of the front; the two posterior feet alone are raised; the claws are larger in the males than in the females.

The head is very spiny, with an advanced and denticulated projection at the middle of the forehead. The upper jaw-feet are elongated and projecting. (*Homola Spinifrons*, Leach.)

DORIPPE, *Fab.*,

Have the eyes very much apart, and situated at the lateral and anterior angles of the testa, the four posterior feet raised; the claws short in both sexes; the testa ovoid, broadly truncated, without any projection, in the form of a bill, and flatted.

As M. Desmarest has observed, we see on each side above the origin of the claws, an oblique cleft, cut longitudinally by a diaphragm, ciliated, like it, upon its edges, communicating with the gills, and serving as an issue for the water. (*Dorippe lanata, Cancer lanatus*, Linn., Desmarest, &c.)

In the others, the testa is sometimes almost orbicular or globular; sometimes arched in front, and contracted pos-

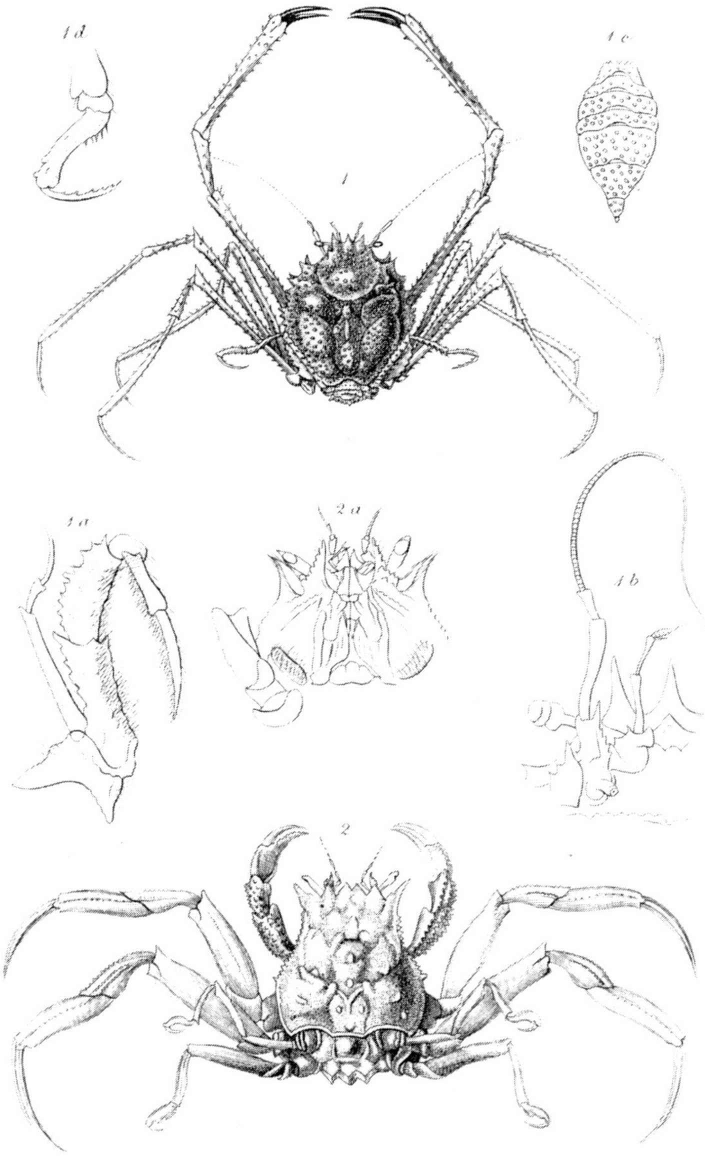

1 *Homola Cuvieri.*

2 *Dorippe nodulosa.*

London, Published by Whittaker & C°. Ave Maria Lane, 1833.

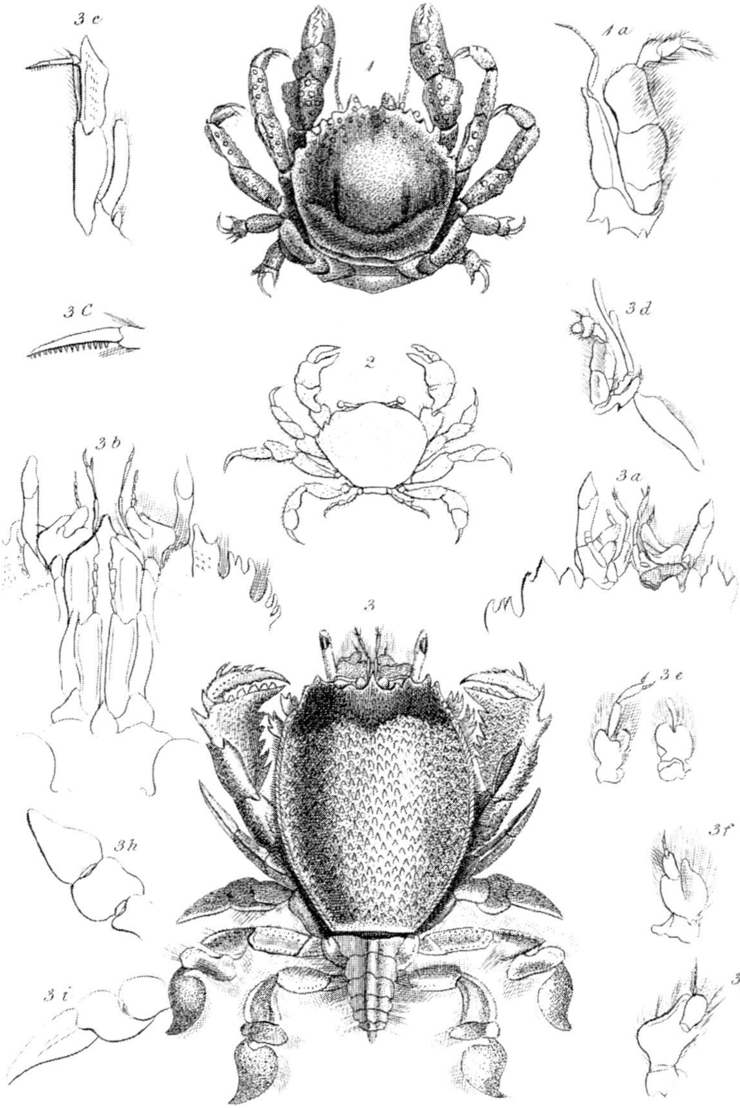

1 *Dromia nodipes*

2 *Dynomene hispida.*

3 *Ranina serrata.*

London, Published by Whittaker & Co. Ave Maria Lane, 1833.

teriorly, denticulated or spiny on the sides. The eyes are situated near the middle of the forehead, and carried on short pedicles.

DROMIA, *Fab.*,

Have the four posterior feet inserted on the back, and terminated by a double hook; the testa suborbicular, or almost globular, gibbous, and woolly, or well furnished with hairs. (*Cancer dormia*, Linn.)

DYNOMENE, *Latr.*,

In which the two posterior feet, much smaller than the others, are alone dorsal, and as it appears to us, imperfect. The testa is widened almost into the form of an inverted heart, and truncated posteriorly, like that of the last quadrilatera, and simply furnished with hairs. The ocular pedicles are longer than those of Dromia. (*Dynomene hispida*, Desmarest.)

The last notopeda differ from the preceding, in that all the feet, with the exception of the claws, are terminated like fins, and from all the other brachyura, in having the tail extended. Such are,

RANINA, *Lam.*

Their testa is elongated, proceeds narrowing from front to rear, and has generally the form of an inverted triangle, with the base denticulated. The ocular pedicles are elongated. The lateral antennæ are long and advanced. The external jaw-feet are equally elongated, narrow, and with the third articulation contracted into a point towards its extremity. All the feet are very much approximate, or almost contiguous at their origin, and beginning from the fourth pair, they ascend upon the back, but the last two alone are upper. The forceps are compressed, almost in the form of an inverted triangle, denticulated, and with the fingers abruptly bent. These crustacea have the greatest relations with the albumea of Fabricius, the first subgenus of the next family, and thus make the passage from the brachyura to the macroura. From the approximation of the feet it is even probable that the genital apertures of the female are situated as in macroura.

The second family, or

The Macrourous Decapods—Exochnata, *Fab.*,

Have at the end of the tail some appendages, forming most frequently on each side a fin, and the tail as long at least as the body, extended and discovered, and simply curved towards its posterior extremity. Its under part most frequently presents in both sexes five pair of false feet, each terminated by two laminæ, or two threads. This tail is always composed of seven distinct segments. The genital apertures of the females are situated on the first articulation of the feet of the third pair. The gills are formed of vesicular, barbed, and hairy pyramids, and disposed, in many, either on two ranges or by bundles. The antennæ are generally elongated and projecting. The ocular pedicles are usually short. The external jaw-feet are most frequently narrow, elongated, in the form of palpi, and do not completely cover the other parts of the mouth. The testa is more narrow, and more elongated than that of the brachyura, and usually terminating in a point at the middle of the forehead. For more ample details we must refer to the memoir before mentioned, of MM. Audouin and Milne Edwards. A character observed by them on the lobster (*Astacus marinus*, Fab.) and which would be decisive, if it applied to the other macroura, is, that besides the two venous sinuses of which we have spoken in the generalities of the order, there exists a third, lodged in the sternal canal, and extending between the two preceding, from one end of the thorax to the other. This very curious disposition would establish, according to them, a connexion between the venous system of the macroura, and that of the stomapod crustacea.

The macroura never quit the water, and with the exception of a small number are all marine.

After the example of Degeer and Gronovius, but a single genus has been formed of them, that of ASTACUS, which we shall thus divide.

Some, from the proportions, form, and uses of their feet, of which the first or second at least, are in the form of claws, and from the sub-caudal situation of their eggs, evidently approximate to the preceding crustacea, and still more to those which are vulgarly known under the names of *craw-fish*, *lobster*, and *prawn* or *shrimp*.

The others have very slender feet in the form of a thread or lash, and accompanied with an appendage or external and elongated branch, which seems to double their number. They are proper for swimming, and none of them terminates in a forceps. The eggs are situated between them, and not under the tail.

The first shall be subdivided into four sections, the ANOMALA, LOCUSTÆ, ASTACINI, and CARIDES.

The second shall compose the fifth and last section of this family and of the decapods, that of SCHIZOPODA.

In the first, or that of ANOMALA, the two or four last feet are always much smaller than the preceding. The under part of the tail never presents more than four pair of appendages, or false feet. The lateral fins of the end of the tail, or the pieces which represent them, are thrown on the sides, and do not form with the last segment a fin, in the form of a fan.

The ocular pedicles are generally longer than those of the macroura of the following sections.

Here (the *Hippides,* Latr.) all the upper segments are solid. The two anterior feet sometimes terminate by a monodactylous hand, or one without fingers, in the manner of a palette, and sometimes they go into a point. The six or four following finish with a fin. The last two are filiform, folded, and situated at the inferior origin of the tail. This tail grows narrow abruptly, immediately after its first segment, which is short and broad, and the last is in the form of an elongated triangle. The lateral appendages of the last but one are in the form of curved fins. The subcaudal appendages are in number four pair, and composed of a very slender and filiform stem. The antennæ are very hairy, or ciliate. The lateral ones at first approach the intermediate, and are afterwards arched, or turned outwards.

ALBUNEA, *Fab.*,

Have the two anterior feet terminated by a very compressed hand, triangular and monodactylous. The last articulation of the following feet is like a reaping-hook. The lateral antennæ are short. The intermediate are terminated by a single, long, and setaceous thread. The ocular pedicles occupy the middle of the forehead, and form, when united, a sort of muzzle, flat, triangular, with the external sides arched. The testa is almost plane, nearly square, but rounded at the posterior angles, and finely denticulate at the anterior edge.

The only species well known is (*Cancer Symnista,* Linn.)

HIPPA, *Fab.*—EMERITA, *Gronov.*,

Have the two anterior feet terminated by a very compressed hand, almost ovoïd, and without fingers. The lateral antennæ are much shorter than the intermediate, and rounded. These last are terminated by two short obtuse threads, placed one upon the other. The ocular pedicles are long and filiform, and the third articulation of the jaw-feet is very large, in the form of a lamina, emarginated at the end, and covering the following articulations. The testa is almost ovoïd, truncated at the two ends, and convex.

The last articulation of the second feet, and of the two following pair, is triangular, but approaches, in the last at least, to the form of a crescent. The last two of the fourth pair are straightened up, and applied on the preceding two. The first one of the tail has two impressed and transverse lines. (*Hippa adactyla*, Fab.)

REMIPES, *Latr.*,

Have the two anterior feet elongated, with the last articulation conical, compressed, and hairy. The four antennæ are very much approximated, very short, and nearly of the same length; the intermediate are terminated by two threads. The ocular pedicles are very short and cylindrical. The external jaw-feet are in the form of small claws, slender, arched at the end, and terminated by a strong hook. The testa is conformed like that of hippa.

The last articulation of the second and third feet forms a triangular plate, with an emargination at the external side; the same of the fourth, is triangular, narrow, and elongated. As well as in hippa, the first segment of the tail presents two impressed and transverse lines. (*Remipes testudinarius*, Latr.)

Sometimes (*Pagurini*, Latr.) the teguments are slightly crustaceous, and the tail most frequently soft, in the form of a sac, and rounded. The two anterior feet terminate in a didactylous hand; the four following go into a point, and the posterior four, shorter, finish by a sort of forceps or small didactylous hand. The first articulation of the peduncle of the lateral antennæ presents an appendage or projection proceeding into a point, or in the form of a spine.

These crustacea, which the Greeks name *carcinion*, and the Latins *cancelli*, live for the most part in univalve and empty shells. Their tail (*Birgus* excepted) presents, and in the females only, but three false feet, situated on one of the sides, and divided, each, into two filiform and hairy branches. The three last segments are suddenly more narrow.

In some, such as

Birgus, *Leach*,

The tail is tolerably solid, suborbicular, with two ranks of appendages, in the form of laminæ, underneath. The fourth feet are only a little smaller than the two preceding; the two last are folded and concealed, their extremity lodging in a depression of the base of the thorax. The fingers of the end, as well as those of the last pair but one, are simply hairy or spinous. With the exception of the claws, all the feet are separated at their origin by a very sensible interval. The thorax is in the form of an inverted heart, and pointed in front. (*Cancer latro*, Linn.)

Pagurus, *Fab.*

The last four feet are much shorter than the preceding, with the forceps charged with small grains. The tail is soft, long, cylindrical, narrowed towards the end, and usually presents but one rank of oviferous appendages, and which are in the form of a thread. The thorax is ovoïd or oblong.

Some species, Cœnobita, *Latr.*, are distinguished from the others by their advanced antennæ, and the middle ones are almost as long as the exterior or lateral, and have elongated threads. The thorax is ovoïdo-conical, narrow, elongated, very much compressed laterally, and with the anterior

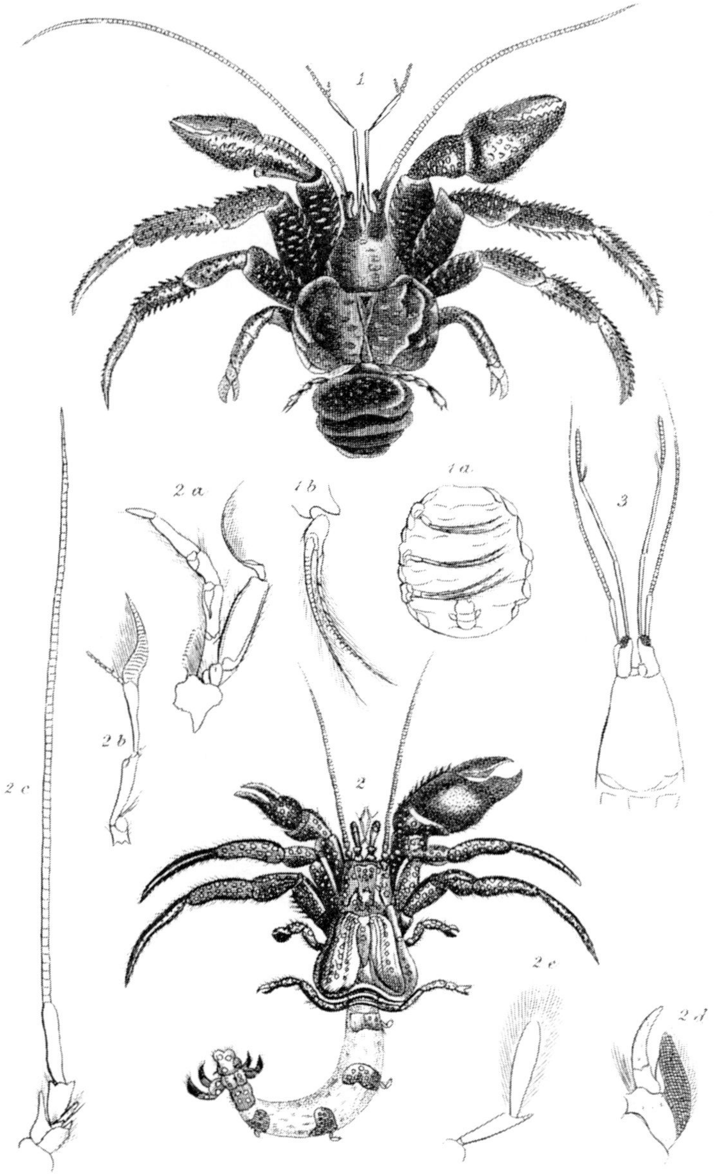

1. *Birgus latro.*

2. *Pagurus guttatus*

3. *Antennæ of Pagurus clypeatus (Cænobita Lat.)*

London, Published by Whittaker & Co. Ave Maria Lane, 1833.

or cephalic division in the form of a heart. (*Pagurus clypeatus*, Fab.)

Those which form the most numerous division, PAGURUS, proper, *Latr.*, have, on the contrary, the middle antennæ curved, notably shorter than the lateral, with the two threads short, and the upper one in an elongated or subulated cone. The anterior division of the thorax is squared, or in the form of a triangle inverted and curvilinear. (*Cancer Bernhardus*, Linn.)

A species inhabiting the Mediterranean should, from its peculiar characters, form a subgenus proper (PROPHYLAX, *Latr.*) In these the tail, instead of being, with the exception of the upper of the three last segments, soft and arched, and having but a single rank of oviferous threads, is covered with coriaceous teguments, is directed in a straight line, and is not curved underneath, except at its extremity. Its inferior surface presents a furrow, and two ranks of false feet. The body besides is linear, with two lateral appendages at the end of the tail almost equal, and whose'division is larger, foliaceous, and ciliated. The last four feet are slightly granular at their extremity, and appear to be terminated only by a single finger, or at least are not very distinctly bifid. Perhaps we must refer to this division, the paguri living in serpulæ, alcyones, &c.

In all the following macroura, the two posterior feet at most are alone smaller than the preceding. Most frequently the false subcaudal feet are five pair in number. The teguments are always crustaceous. The lateral fins of the penult segment of the tail, and its last one, form one in common, disposed like a fan.

The two following sections have a common character, which separates them from the fourth. The antennæ are inserted at the same height, or on a level. The peduncle of the lateral ones, when it is accompanied with a shell, is never entirely covered by it. The false sub-caudal feet are often but four in number. The two middle antennæ are never terminated but by two threads, and generally shorter than their peduncle, or scarcely longer. The external leaf of the natatory appendages of the last segment of the tail but one, is never divided by a transverse suture.

Our second section, LOCUSTA, thus designated from the word *locusta*, applied by the Latins to the most remarkable crustacea of this division, from which comes the French word *langouste*, have always but four pair of false feet. The posterior extremity of the fin terminating the tail is always almost membranaceous, or less solid than the rest. The peduncle of the middle antennæ is always longer than the two threads of the end, and more or less folded or bent; the lateral ones are never accompanied with scales: sometimes they are reduced to a single pedicle, which is dilated, very much flatted, and in the form of a crest; sometimes they are large, long, going into a point, and entirely bristling with prickles. All the feet are almost similar, and go into a point at the end; the first two are simply a little stronger. Their penult articulation, and that of the two last, is at most unidenticulated, but without forming with the last a hand perfectly didactylous. The pectoral space comprised between the feet is triangular. The thorax is almost square or subcylindrical, without any frontal prolongation, in the manner of a pointed bill or lance.

SCYLLARUS, *Fab.*,

Present, in the form of their lateral antennæ, a character altogether unusual; the stem is wanting, and the articulations of the peduncle, very much dilated transversely, form a large crest, flatted, horizontal, and more or less denticulated.

The external branch of the subcaudal appendages is terminated by a leaf, but the internal, in some males, only shows itself in the form of a tooth.

According to the proportions and form of the thorax, the

position of the eyes, and some other parts, Dr. Leach has established three genera: 1st. SCYLLARUS with the thorax as long or longer than broad, without lateral incisions, and the eyes always situated near its anterior angles. The last articulation but one of the two posterior feet is unidenticulated in the females. (*Cancer arctus*, Linn.)

2nd, His THENUS has the thorax measured in front broader than long, with a deep incision at each lateral edge, and the eyes situated at its anterior angles. (*Thenus Indicus*, Leach.)

3rd, His IBACUS does not differ from thenus but in the position of the eyes, which are approximated from the origin of the intermediate antennæ, or much more interior. (*Ibacus Peronii*, Leach.)

PALINURUS, *Fab.*,

Have the lateral antennæ large, setaceous, and bristling with prickles. Some of these crustacea arrive to a very considerable size. In the species of our coasts, and probably in others, the females have at the extremity of the last articulation but one of the two posterior feet, a projection in the form of a spur or tooth, exclusively proper to this sex. Scyllarus presents the same difference. (*Palinurus quadricornis*, Fab.)

The third section, that of ASTACINI, *Latr.*, is distinguished from the preceding, by the form of the two anterior feet, and often by that of the two following pair, which terminate by a forceps with two biters, or a didactylous hand. In some the last two or four are much smaller than the preceding, which approximates them to the anomala. But the fan-like fin of the extremity of the tail, and other characters remove them from these last. The thorax is narrowed in front, and the forehead advances more or less, in the manner of a beak or pointed muzzle.

Some GALATHADEÆ, *Leach*, have, as well as the preceding macroura, four pair of false feet; and the middle antennæ elbowed, and with two threads representing the stem, are manifestly shorter than their peduncle. That of the lateral antenna is never accompanied with a plate in the form of a scale. The two anterior feet only terminate in a didactylous hand, which is often extremely flatted. The last segment of the tail is bilobate, at least in the majority.

At the head of this division will come those whose two posterior feet are much more slender than the preceding, filiform, folded, and useless for running.

GALATHEA, *Fab.*,

Have the tail extended, the thorax almost ovoïd, or oblong, the middle antennæ projecting, and the forceps elongated. The upper part of the body is usually very much incised or striated, spiny, and ciliate. (*Galathea rugosa*, Fab.)

Dr. Leach forms with the *galathea gregaria* of Fabricius, a proper genus under the name of GRIMOTEA. The second articulation of the intermediate antenna is terminated in a knob, and the last three of the external jaw-feet are foliaceous.

The ÆGLEA of the same is not distinguished from the preceding and from galathea, but by having the mandibles denticulated, the second articulation of the external jaw-feet shorter than the first, and the upper part of the body generally even.

PORCELLANA, *Lam.*,

Form in the macroura, with reference to the tail, a very singular exception; it is folded underneath as in the brachyura. They are otherwise remote from galathea, in the more shortened, suborbicular, or almost squared form of the thorax; in having the middle antennæ withdrawn within their fossets; in their forceps being triangular; and finally, by reason of

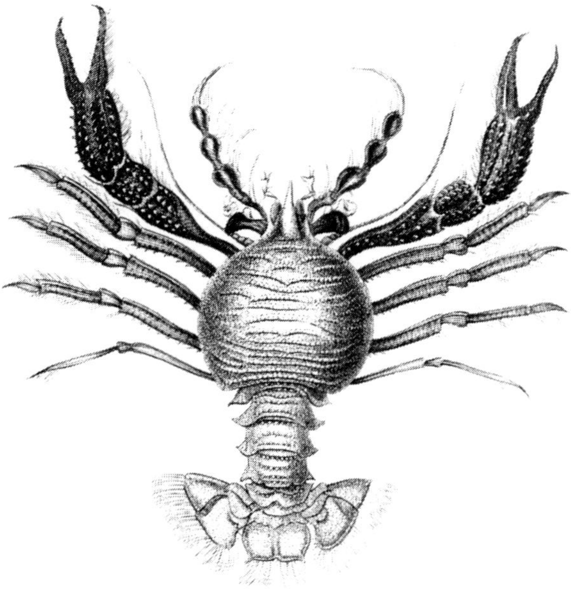

C. M. Curtis del[t]

Galathea gregaria.

London, Published by Whittaker & C.o Ave Maria Lane. 1833.

the internal dilatation of the lower articulations of their external jaw-feet. Their body is very much flatted.

Some are remarkable for their very large forceps, furnished with hairs, or very much ciliated. Such are 1st, *Cancer platychiles*, Penn. whose forceps are hairy only at the external edge, and thorax almost naked and rounded. 2nd, *P. hirta*, Lam., in which all the upper part of the forceps and thorax is hairy, and in which this part is almost oval, and narrowed in front. The others have the pincers without hair; such is *Cancer hexapus*, Linn.

The genus MONOLEPIS of M. Say appears to constitute the passage from porcellana to megalopus. It approaches the first in the relation of the two posterior feet and of the direction of the tail. But this tail should have but six segments, and the eyes should be very large, as in the second. It would also appear that the lateral fins of the end of the tail resemble those of the last.

The other crustacea of the same division differ from the preceding by having their posterior feet similar in form, proportions, and uses, to the preceding, or equally ambulatory. They are again remote from them, by reason of their body, which is more thick and more raised, and of their lateral antennæ, which are much shorter, also by reason of their claws being smaller, the bulk of the eyes and the lateral fins of the tail, which are composed but of a single lamina. This tail is extended, narrow, and simply curved underneath, towards its extremity.

MEGALOPUS, *Leach*—MACROPA, *Latr.*

We shall comprehend in our second division (*Astacini*, Latr.) those which have five pair of false feet, the middle antennæ straight, or almost straight, projecting, advanced, and terminated by two threads, as long or longer than their peduncle, and which a single subgenus excepted *(gebia)* have the four or six anterior feet terminated by a didactylous hand.

Their tail is always extended; their two posterior feet are never much more slender than the preceding, nor folded. The peduncle of the lateral antennæ is often accompanied by a shell.

Those whose first four feet at most are terminated by two fingers, whose lateral antennæ never have a shell or scale at the base, and in which the exterior leaf of the lateral fins of the end of the tail presents no transverse suture, will form a first subdivision. Their feet for the most part are ciliate or hairy.

Sometimes the index, or immoveable finger (formed by a projection of the last articulation but one,) of the claws, is very sensibly shorter than the thumb, or mobile finger, and forms only a simple tooth.

GEBIA, *Leach*,

Approach the preceding subgenera, in that the two anterior feet alone are didactylous. The leaves of the lateral fins of the end of the tail, proceed widening from their base to their extremity, and have longitudinal crests. The intermediate piece, or the last segment of the tail, is almost square. (*Gebia stellata*, Leach.)

THALASSINA, *Latr.*,

Have the four anterior feet terminated by two toes or fingers, the leaves of the lateral fins of the end of the tail narrow and elongated, without crests; and the last segment of this tail, or the intermediate piece, in an elongated triangle. (*Thalassina scorpionides*, Latr.)

Sometimes the four anterior feet, or the two first and one of the second, are terminated by two elongated fingers, forming the forceps completely.

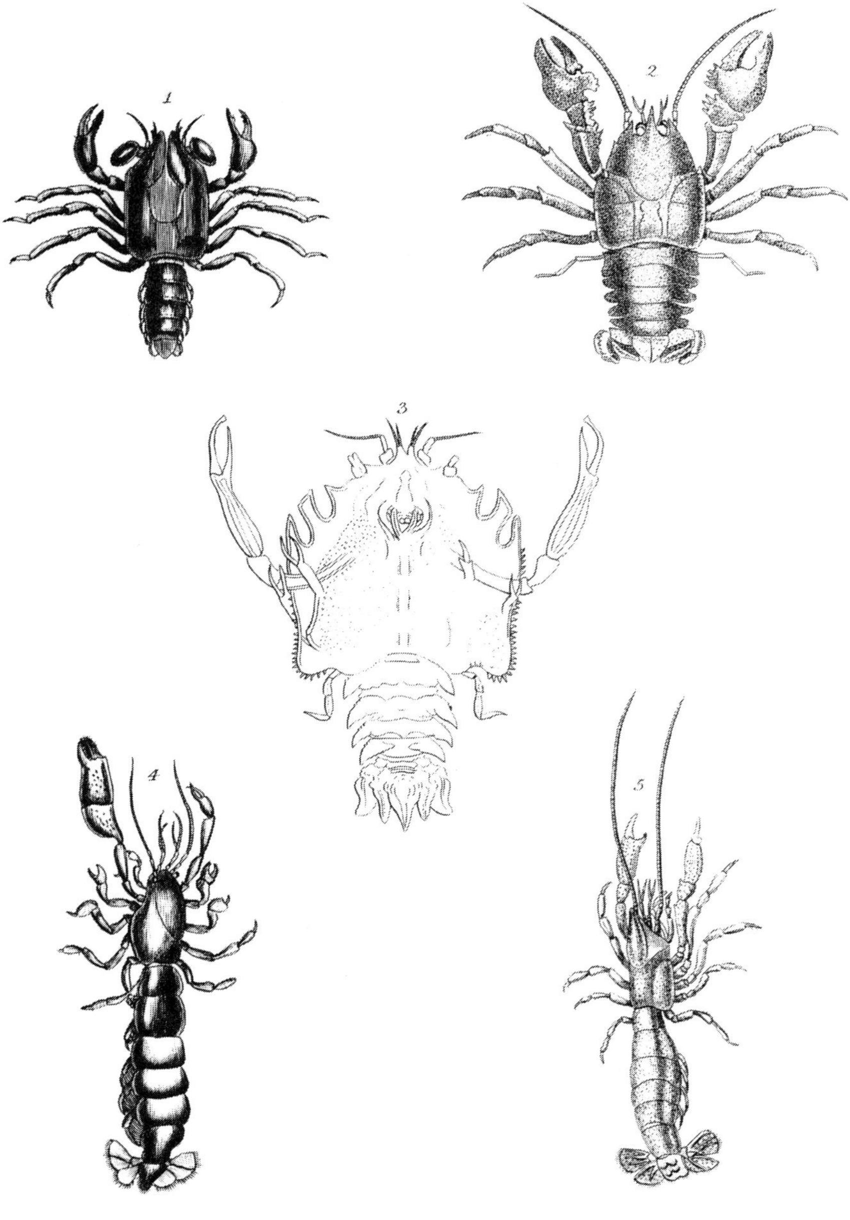

1 *Megalopus muticus* 3 *Eryon Cuvieri.*

2 *Æglea lævis* 4 *Callianassa subterranea.*

5 *Gebia stellata.*

London, Published by Whittaker & C.o Ave-Maria Lane 1833.

The two anterior claws are larger; the lateral leaves of the fin terminating the tail are in the form of an inverted triangle or broader at the posterior edge. The intermediate, on the contrary, grow narrow from the base to the end, and proceed into a point.

CALLIANASSA, *Leach*,

Have the claws very unequal, as well both in form as proportions; the carpus of the largest of the two anterior is transverse, and forms with the forceps a common body. The same articulation of the other claw is elongated. The two posterior feet are almost didactylous. The external leaf of the lateral fins of the end of the tail is larger than the internal, with a crest or ridge; this last is even.

The ocular pedicles are in the form of a scale or shell, and the cornea is situated near the middle of their external edge. The threads of the middle antennæ are but little longer than their peduncle. (*Callianassa subterranea*, Leach.)

AXIUS, *Leach*,

Differ from them in their claws, which are almost equal, and in which the carpus makes no part of the forceps; the posterior feet are similar to the preceding. The leaves of the lateral fins are almost of the same size, and have each a longitudinal crest; the threads of the middle antennæ are evidently longer than their peduncle. (*Axius stirhynchus*, Leach.)

Our second, and last subdivision, presents crustacea, in which the six anterior feet form so many claws, terminating in a forceps perfectly didactylous, a character which distinguishes them from all the preceding decapods, and which approximates them to the first of the following section; but here the claws of the third pair are the largest, instead of which, in the others, it is the first two, and their thickness, moreover, is much more considerable. The peduncle of the lateral antenna is accompanied with a scale, or spines; the external leaf of the lateral fins of the end of the tail is, in all the living species, as it were, divided into two by a transverse suture.

ERYON, *Desm.*,

Have all the leaflets of the caudal fin, narrowed at their extremity, and terminating in a point; the exterior one presents no transverse suture; the two threads of the middle antennæ are very short, and scarcely longer than their peduncle; the sides of the testa have very deep notches; the forceps of the two anterior claws are narrow and elongated. (*Eryon Cuvierii*, Desm., a fossil species.)

ASTACUS, *Gronov.*, *Fab.*,

Have the leaflets of the lateral fins of the end of the tail widened and rounded at their extremity; the external one is divided transversely into two by a transverse suture, the posterior extremity of that of the middle is obtuse or rounded; the two threads of the middle antennæ are notably longer than their peduncle; the sides of the testa are not incised.

In some, and all marine, the last segment of the tail, or that which occupies the middle of the terminal fin, presents no transverse suture.

Those whose lateral antennæ have a large scale on the peduncle, whose eyes are very large, and kidney-formed, and in which the pincers of the two anterior claws are narrow, elongated, prismatic, and equal, form the genus NEPHROPS of Dr. Leach. (*Cancer Norwegicus*, Linn.)

Those in which the peduncle of the lateral antennæ presents simply but two short projections, in the form of teeth or spines, whose eyes are neither very large nor reniform, and which have the pincers or forceps more or less oval, com-

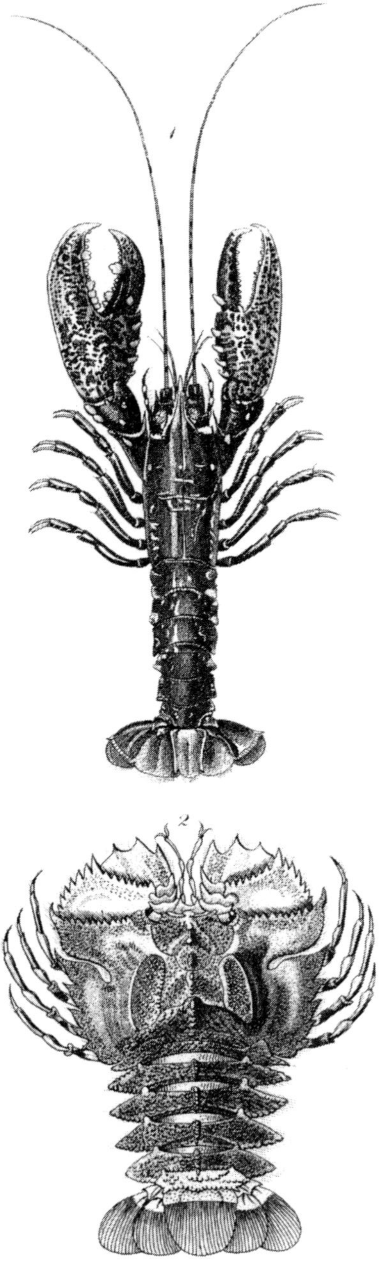

1 *Astacus Marinus.*

2 *Ibacus Peronii.*

London Published by Whittaker & C.º Ave Maria Lane 1833.

pose, with the fresh-water species, the genus ASTACUS, properly so called, of the same naturalist. (*Cancer gammarus*, Linn.)

In the fresh-water species, which, in their antennæ, eyes, and the form of the claws, otherwise resemble the preceding, the last segment of the tail, or the middle one of its terminal fin, is cut transversely in two by a suture. (*Cancer astacus*, Linn.)

In the fourth section, that of CARIDES, the middle antennæ are superior, or inserted above the lateral, the peduncle of the latter is entirely covered by a large scale.

Their body is arched, as it were humped, and of a consistence less solid than that of the preceding crustacea; the forehead is always prolonged in front into a point, and most frequently in the manner of a bill, or pointed lamina, compressed and denticulated on its two edges. The antennæ are always advanced; the lateral ones are usually very long, and in the form of a very attenuated thread; the intermediate, in a great number, terminate in three threads; the eyes are very much approximated; the external jaw-feet, more narrow, and more elongated than usual, resemble palpi or antennæ; the mandibles of the majority are narrowed, and arched at their extremity; one of the first two pair of feet is often folded on itself, or doubled; the segments of the tail are dilated or widened laterally; the external leaflet of the terminal fin is always divided in two by a suture, a character which is observed only in the last crustacea of the preceding section. The odd piece of the middle, or the seventh and last segment, is elongated, narrowed towards the end, and presents, above, some ranges of small spines; the false feet, four pair in number, are elongated, and usually foliaceous.

Some have the first three pair of feet in the form of a didactylous claw, and their length augments progressively, so that the third pair is the longest. Such are

PENÆUS, *Fab.*,

In which no articulation of the feet presents any annular division; their mandible-palpi are raised and foliaceous. We see a small appendage, in the form of an elliptical plate, at the base of the feet, a character which seems to approach these crustacea to pasiphaë, the last subgenus of this section, and to those of the following, (*Palæmon sulcatus*, Oliv.)

Other penæi have the intermediate antennæ terminated by long threads; these are those of our second division. (*Penæus monodon*, Fab.)

STENOPUS, *Latr.*,

Are distinguished from the last by the transverse and annular divisions of the two articulations before the last, of the four posterior feet; all the body is soft; the antennæ and feet are long and slender; those of the third pair are the broadest. (*Cancer setiferus*, Linn.)

The other Carides, many of which have the intermediate antennæ terminated by three threads, present at most but two pair of didactylous claws, formed by the four anterior feet.

A subgenus established on a single species proper to North America, that of

ATYA, *Leach*,

Is removed from all the analogous crustacea, by an anomalous character; the forceps terminating the four claws is cleft as far as its base, or appears to be composed of two fingers in the form of lashes, united at their origin; the articulation which precedes, is in the form of a crescent; the second pair is the largest; the middle antennæ have but two threads. (*Atya scabra*, Leach.)

In all the following subgenera, the fingers of the forceps arise only at a certain distance from the origin of the last articulation but one, or that which is in the form of a hand, and the body, or articulation which precedes it, is not annulated.

First come the Carides, whose feet are robust, and not filiform, and without any appendage at their external base; their body is never very soft nor very much elongated.

Among these subgenera with feet without appendages, the three following again present unusual forms as regards the claws.

In that of

CRANGON, *Fab.*,

The two anterior claws, larger than the following feet, have but a single tooth at the place of the index or fixed finger, and that which is mobile is in the form of a hook, and bent.

The upper and middle antennæ have but two threads; the second feet are folded, more or less distinctly bifid or didactylous at their extremity; none of their articulations are annulated; the anterior bill of the testa is very short.

We shall not separate from Crangon the EGEON of *M. Risso*, or the PONTOPHILUS of *Dr. Leach*. In the latter, the last articulation of the external jaw-feet is as long again as the preceding, while they are of equal length in the first; the second feet of Egeon are shorter than the third, and the smallest of all, whereas their length is the same in crangon. But the number of species being very limited, this generic distinction becomes so much the less necessary. (*Crangon vulgaris.*)

PROCESSA, *Leach*, NIKA, *Risso*,

Have one of the anterior feet simply terminating in a point, and the other in a didactylous forceps; the two following are unequal, slender, and also terminated by two fingers; one of these second feet is very long, with the carpus and the preceding articulation annulated; this character is proper to the other foot only at the first of these articulations; the feet of the fourth pair are longer than the preceding and the two following; the upper antennæ have but two threads. (*Nika edulis*, Risso.)

HYMENOCERA, *Latr.*,

Have the two anterior feet terminated by a long hook, bifid at the end, and with very short divisions; the two following are very large: the hands, the fixed finger, and the upper thread of the middle antennæ are dilated, membranaceous, and, as it were, foliaceous; the external jaw-feet are likewise foliaceous, and cover the mouth.

We shall now pass to some subgenera in which the claws present no remarkable or unusual particularity.

Sometimes the upper or middle antennæ are terminated but by two threads; the bill is generally short.

GNATHOPHYLLUM, *Latr.*,

The only ones, which, in the relation of the form and amplitude of the lower jaw-feet, approach hymenocera; the four anterior feet are in the form of didactylous claws; the second pair is longer and thicker than the first; none of the articulations of the four are annulated. (*Alpheus elegans*, Risso.)

PONTONIA, *Latr.*,

Have, like the two following subgenera, the four anterior feet in the form of claws, and didactylous, but the carpus is not annulated. (*Alpheus thyrenus*, Risso.)

ALPHEUS, *Fab.*,

Which have also the four anterior feet terminated by a didactylous forceps, but the carpus of the second is articulated; these last are shorter than the first. (*Alpheus malabaricus*, Fab.)

HIPPOLYTE, *Leach*,

Are removed from Alpheus only by the respective proportions of the claws; the second are longer than the first. (*Palæmon diversimana*, Oliv.)

The last two subgenera following, have this peculiarity, that one pair only of their feet is terminated in a didactylous forceps. In

ANTONOMEA, *Risso*,

It is the two anterior, which are distinguished from the others, by their size, thickness, and disproportion. (*Antonomea Olivii*, Risso.)

PANDALUS, *Leach*.

The two anterior feet are simple, or scarcely bifid; the two following are longer, of an unequal length, didactylous, with the carpus and the preceding articulation annulated.

The external jaw-feet are slender and very long, at least in some; the anterior projection of the testa is very long and much denticulated. (*Pandalus annulicornis*, Leach.)

Sometimes the upper antennæ have three threads; these crustacea have four didactylous claws, the smallest of which are folded and the bill elongated.

PALÆMON, *Fab.*,

Are distinguished from the two following subgenera, by their inarticulated carpus; the second feet are larger than the first; these last are folded. (*P. serratus*, Leach.)

The carpus is articulated, or presents some annular divisions in the two following subgenera:

LYSMATA, *Risso*,

Which have the second pair of claws larger than the first. (*Lysmata seticauda*, Risso.)

ATHANAS, *Leach*,

In which the first pair of claws, on the contrary, is more bulky than the second. (*Athanas nitescens*, Leach.)

The last subgenus of this section, that of

PASIPHÆA, *Sav.*,

Although very much approximating to many of the preceding by the superior antennæ being terminated by two threads; by the form of the four anterior feet terminating in a didactylous forceps, preceded by an articulation without annular divisions, and by the shortness of the muzzle or frontal horn, yet they differ from them in some relations. An appendage, in the form of a thread, or hair, is very distinctly visible at the external base of their feet; these feet, with the exception of the claws, which are larger and almost equal, are very narrow and filiform; the body is very much elongated, very much compressed, and very soft. (*Alpheus sivado*, Risso.)

Our fifth and last section of macroura, that of SCHIZOPODA, appears to connect the macroura with the following order: the feet, none of which are terminated in a forceps, are very slender, in the form of lashes, provided with an appendage more or less long, proceeding from their exterior side, near their base, and exclusively adapted for swimming; the eggs are situated between them, and not under the tail; the ocular pedicles are very short, as in the majority of the macroura, the forehead advances into a point, or presents the appearance of a sort of beak. The testa is slender; the tail terminates as usual in a fin. These crustacea are small and marine.

Here the eyes are very apparent; the lateral antennæ are accompanied with a scale; the middle ones are terminated by two threads, and composed of many small articulations, as well as the preceding.

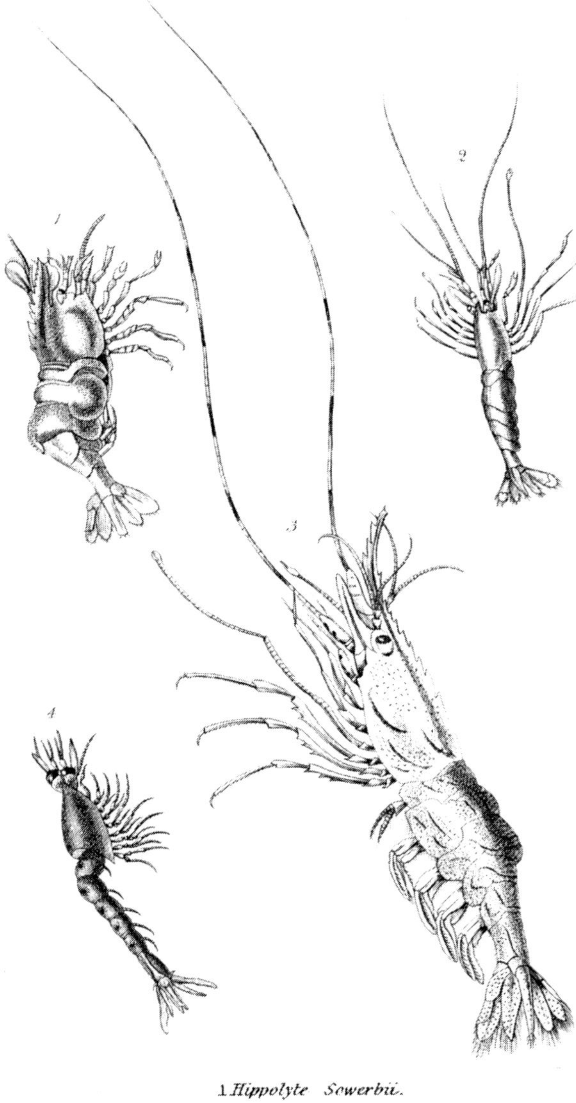

1 *Hippolyte Sowerbii.*

2 *Nika canaliculata.*

3 *Pandalus annulicornis.*

4 *Mysis Fabricii.*

London: Published by Whittaker & Co Ave Maria Lane 1833.

MYSIS, *Latr.*,

Have the antennæ and feet exposed, the testa elongated, almost square or cylindrical, the eyes very much approximate, and the feet capillary, as it were, formed of two threads. (*Mysis Fabricii*, Leach.)

CRYPTOPUS, *Latr.*,

Have a subovoïd testa, enlarged or swollen at intervals, folded inferiorly on the sides, enveloping the body as well as the antennæ and feet, and leaving discovered underneath nothing but a longitudinal cleft. The eyes are wide apart. The feet are in the form of lashes, with a lateral appendage. (*Cryptopus Defrancii*, Lat.)

Sometimes the eyes are concealed. The intermediate antennæ are conical, inarticulate, and very short. The lateral ones are composed of a peduncle and a thread, without distinct articulations. Their base presents no scale, at least projecting.

MULCION, *Latr.*

The body is very soft, with the thorax ovoïd. The feet are in the form of a lash, and the majority at least have an appendage at their basis. The fourth pair is the longest of all. (*Mulcion Lesueurii*, Latr.)

Nebalia, which we had at first placed in this section, not having natatory appendages under the last segments of their body, and their feet being pretty similar to those of cyclops, will pass with *condylura* into the order of branchiopods, of which they will make the beginning. *Nebalia*, from their very projecting eyes, which appear to be pedicled, and from some other characters, seem, with *zoë*, to link the schizopoda with the branchiopods.

THE SECOND ORDER OF CRUSTACEA.

STOMAPODA

Have their gills exposed, and adherent to the five pairs of appendages situated under the abdomen, (the tail) which this part presented us in the decapods, and which here, as in the majority of the macrouri, serve for swimming, or are fin feet. Their testa is divided into two parts, the anterior of which carries the eyes and the intermediate antennæ, or forms the head, without bearing the fin-feet. These organs, as well as the four anterior feet, are often approximated to the mouth, on two lines converging inferiorly, and from thence the denomination of Stomapods given to this order. The heart, to judge from the squillæ, the most remarkable genus of this order, and the only one which has been studied, is elongated, and similar to a thick vessel. It extends all along the back, reposes on the liver and the intestinal canal, and terminates posteriorly and near the anus in a point. Its parietes are slender, transparent, and almost membranaceous. Its anterior extremity, placed immediately behind the stomach, gives birth to three principal arteries, the middle one of which (the ophthamic) throwing out from both sides many branches, is more especially carried to the eyes and the middle antennæ, and the two lateral (the antennary) pass over the sides of the stomach, and lose themselves in the muscles of the mouth, and of the external antennæ. The superior surface of the heart produces no artery, but a great number are seen to issue from its two sides; and each pair, as it appears to us, corresponds with each segment of the body, commencing at the jaw-feet, whether these segments be exterior, or concealed by the testa, and even very small, as is the case with the anterior ones. At the level of the first five rings of the abdomen, or of those which carry the natatory appendages, and the gills, this upper surface of the heart receives near the medial line five pairs of vessels (one pair for each segment,) coming from these last organs, and which, according to MM. Audouin and Milne Edwards, are analogous to the branchio-cardiac canals of the decapods. A central canal situated below the liver and intestine, receives the venous blood, which flows in from all parts of the body. At the level of each segment, bearing the fin-feet and gills, it throws out on each side a lateral branch, repairing to the gill, situated at the base of the corresponding fin-foot. The parietes of these conduits have appeared to the same observers smooth and continuous, but formed rather by a stratum of cellular lamellary tissue, cemented to the neighbouring muscles, than by a proper membrane. It has seemed to them that these conduits communicate together towards the lateral edge of the rings; but of this they are not certain. The *afferential* or internal vessels of the gills, which in these squillæ form plumose tufts, are continued with the *branchio-cardiac* canals, are no longer lodged in little cells, pass between the muscles, turn obliquely round the lateral portion of the abdomen, gain the anterior edge of the preceding ring, and proceed to terminate at the upper surface of the heart, near the median line, riding slightly one upon the other. The medullary cordon presents, besides the brain, but ten ganglia, the anterior of which furnishes the nerves of the parts of the mouth; the three following those of the six natatory feet, and the last six those of the tail. Thus the last four jaw-feet, although representing the four anterior feet of the decapods, nevertheless form a part of the organs of mastication. The stomach of the same crustacea (squillæ) is small, and presents some very small teeth, towards the pylorus. It is followed by a thin and straight intestine, which occupies the whole length of the abdomen, accompanied on the right and left by glandular lobes, appearing to hold the place of the liver. An appendage, in the form of a branch, adhering to the internal base of the last pair of feet, appears to characterize the male individuals.

The teguments of the stomapods are slender, and even almost membranaceous or diaphanous in many. The testa or carapace is sometimes formed of two bucklers, the anterior of which corresponds to the head, and the other to the thorax, sometimes of a single piece, but free behind, leaving usually uncovered the thoracic segments, bearing the last three pairs of feet, and having an articulation in front, serving as a base to the eyes, and to the intermediate antennæ. These last organs are always extended, and terminated by two or three threads. The eyes are always interapproximate. The composition of the mouth is essentially the same as that of the decapods ; but the palpi of the mandibles, instead of being couched upon them, are always raised. The jaw-feet are without any whip-like appendage, such as they present in the decapods ; they have the form of claws or small feet, and in many at least (the squillæ) their external base, as well as that of the two anterior feet, properly so called, presents a vesicular body ; those of the second pair, in the same stomapods, are much larger than the others, and than the feet themselves ; accordingly, they have been considered as genuine feet, and fourteen have been reckoned. The four anterior feet have also the form of claws, but terminate, as well as the jaw-feet, in a talon, or by a hook, which is bent at the side of the head, over the inferior and anterior edge of the preceding articulation, or the hand. But in some others, such as phyllosoma, all these organs are filiform, and without forceps ; some of them, like the six last, and equally simple ones of the stomapods, provided with claws, have an appendage or lateral branch. The last seven segments of the body, enclosing a good part of the heart, and serving as an attachment for the respiratory organs, cannot, considering this relation, be assimilated to that portion of the body termed *tail* in the decapods; it is an abdomen, properly so called. Its last segment but one has, on each side, a fin, similarly composed as that of the tail in the macrouri, but often armed, as well as the last segment or intermediate piece, with spines or teeth.

We shall divide the stomapods into two families; in the first, that of

UNIPELTATA,

The testa forms but a single buckler, in the shape of an elongated quadrilateral figure, usually widened, and free behind, covering the head, with the exception of the eyes and antennæ (which are borne on a common and anterior articulation), and the first segments at least of the thorax. Its anterior extremity is terminated in a point, or is preceded by a small plate, finishing in a similar manner. All the jaw-feet, the second of which are very large, and the four anterior feet, are very much approximated to the mouth, on two lines, converging, inferiorly, and are formed like claws, with a single finger or hook, mobile and folded. If we except the second feet, all these organs have, externally, at their origin, a small pedicled bladder. The other feet, six in number, and the third articulation of which carries laterally and at its base an appendage, are linear, terminated by a brush, and simply natatory. The lateral antennæ have a scale at their base, and the stem of the intermediate is formed of three threads. The body is narrow and elongated, and the ocular peduncles are always short.

This family is composed of a single genus, that of

SQUILLA, *Fab.*,

Which we thus divide:

In some the crustaceous buckler is preceded by a small plate, more or less triangular, situated above the articulation, bearing the middle antennæ and the eyes, covers only the anterior portion of the thorax, and does not fold underneath laterally. The articulation serves as a peduncle to the middle antennæ, as well as to the ocular pedicles, and the external sides of the end of the abdomen are uncovered.

Sometimes the body is almost semi-cylindrical, with the last segment rounded, denticulated or spiny at the posterior edge. The lateral appendages of the last six feet are in the form of a stylet.

SQUILLA, proper, *Latr.*,

Have, along the internal side of the last articulation but one of the two large claws, a very narrow groove, denticulated on one of its edges, spiny on the other, and the following articulation, or talon, in the form of a reaping-hook, and most frequently denticulated. (*Cancer mantis*, Linn.)

GONODACTYLUS, *Latr.*

The groove of the last articulation but one of the large claws is widened at its extremity, and presents neither denticulations nor spines. The talon is bellied out into the form of a knot towards its base, and afterwards terminates in a compressed point, straight, or but little curved. All the species are exotic. (*Squilla scyllarus*, Fab.)

Sometimes the body is very narrow and depressed, with the last segment, almost square, entire, without denticulations or spines. The lateral appendage of the last six feet is in the form of a palette, almost orbicular, and jutting out a little. The antennæ and feet are shorter than in the preceding. The last articulation but one of the large claws is furnished at the internal edge with very numerous hairs, in the form of little spines. The talon is like a reaping-hook.

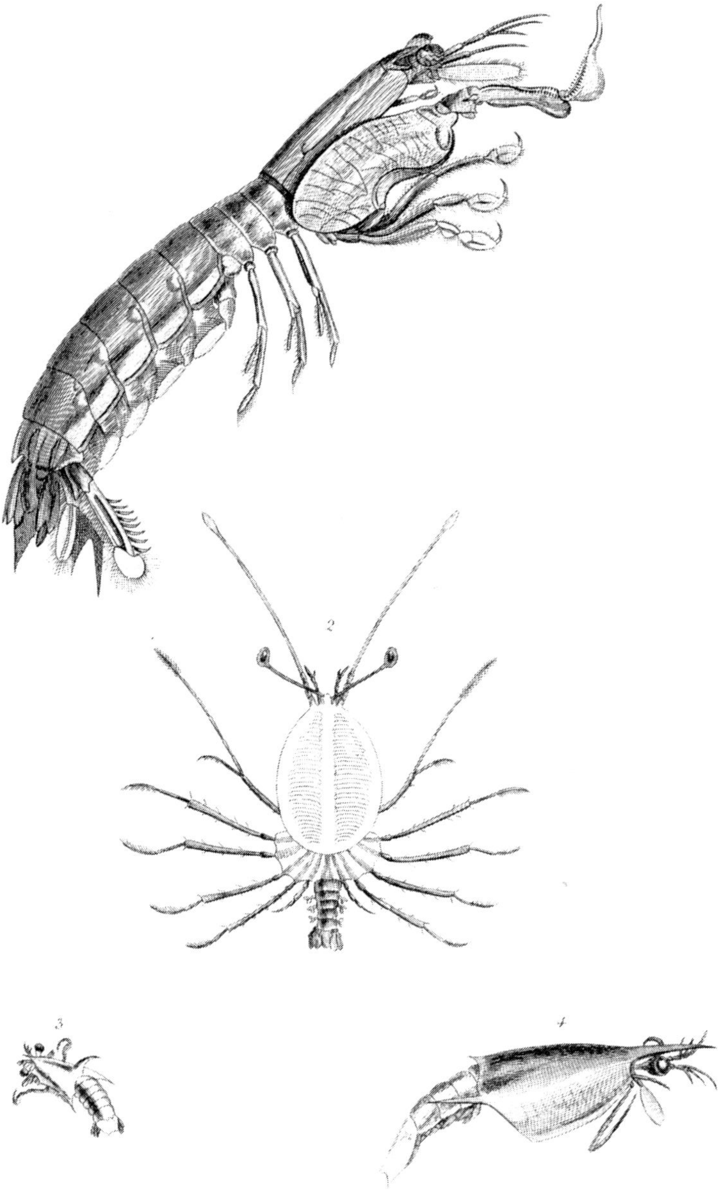

1 *Squilla chiragra Fab.* 3 *Erichthus armatus.*

2 *Phyllosoma clavicornis.* 4 *Erichthus vitreus.*

London, Published by Whittaker & C.º Ave Maria Lane, 1833.

Coronis, *Latr.*

But a single species is known, *Squilla eusebia?* Risso.

The other stomapods of this family have the testa, as it were membranaceous, diaphanous, covering the whole thorax, folded laterally underneath, prolonged anteriorly like a sword, or spine, and advancing above the support of the middle antennæ and the eyes. This support is capable of being bent underneath, and of being enclosed in the case formed by the curve of the buckler. The posterior fins are concealed under the last segment.

The talons of the great claws have no teeth. The second articulation of the ocular pedicles is much thicker than the first, in the form of an inverted cone. The eyes, properly so called, are thick, almost globular. The fin-like appendage of the feet resembles that of the squillæ and gonodactyli. In

Erichthus, *Latr.*, Smerdis, *Leach*,

The first articulation of the ocular pedicles is much shorter than the second. The middle of the lateral edges of the buckler is much protruded in the form of an angle, and their posterior extremity presents two teeth. (*Erichthus vitreus*, Latr.)

Alima, *Leach.*

The first articulation of the ocular pedicles is much longer than the following, slender and cylindrical. The body is more narrow and elongated than in Erichthus, the lateral edges of the buckler are almost straight, or but little dilated. Its middle is carinated longitudinally. Each of its angles forms a spine, of which the two posterior are the strongest. (*Alima hyalina*, Lat.)

The second family, that of

BIPELTATA,

Has the testa divided into two bucklers, of which the anterior, very large, more or less oval, forms the head, and the second, corresponding to the thorax, transverse, and angular in its outline, carries the jaw and ordinary feet. These feet, with the exception at most of the posterior two, and the last two jaw-feet, are slender, filiform, and for the most part very long, and accompanied with a lateral ciliated appendage. The four other jaw-feet are very small, and conical. The base of the lateral antennæ presents no scales. The middle are terminated by two threads. The ocular pedicles are long. The body is very much flatted, membranous, and transparent, with the abdomen small, and without spines at the posterior fin.

This family comprehends but a single genus, that of

PHYLLOSOMA, *Leach.*

All the species belong to the Atlantic Ocean and the Eastern Seas. With reference to their nervous system, they seem to be intermediate between the antecedent crustacea and those which follow.

OF THE MALACOSTRACA, WITH SESSILE AND IMMOVEABLE EYES.

The branchipes will be, from this point, the only crustacea which present us with eyes carried on pedicles ; but besides that these pedicles are not articulated, nor lodged in special cavities, these crustacea have no carapace, and are also remote from the preceding by many other characters. All the malacostraca of this division are alike destitute of carapace. Their body, from the head, is composed of a series of articulations, of which generally the first seven have each one pair of feet, and the following, and last, to the number of seven at most, form a sort of tail, terminated by fins, or appendages in the shape of stylets. The head presents four antennæ, of which the two middle are superior; two eyes and a mouth composed of two mandibles, of a tongue, of two pairs of jaws, and of a sort of lip, formed by two jaw-feet, corresponding to the two upper ones of the decapods. As in the stomapods, there exists no flagrum. The last four jaw-feet are transformed into feet proper, sometimes simple, sometimes terminating in a forceps, but almost always in a single finger or hook.

According to the observations of MM. Audouin and Milne Edwards, the two ganglionic cords of the spinal marrow should be perfectly symmetrical, and distinct in their whole length; and according to the observations of M. le Baron Cuvier, the onisci would not be remote from them, only that these cords do not present in all the segments of the body the same uniformity, and have some ganglia less. Thus, according to them, the nervous system of these crustacea must be the most simple of all. In cymothoë and idotea the two chains of ganglia would be no longer distinct. Those which come immediately after the two cephalic ones would form as many small circular masses, situated on the median line of the body; but the cords of communication which serve to unite them together to form a continuous chain, would remain isolated, and cemented one to the other. It would seem, according to these facts, that these last crustacea should be, in this point of view, more elevated in the animal scale than the preceding; but other considerations appear to us strongly to separate the talitri from the onisci, and to place in an intermediate rank cymothoë and idotea.

The sexual organs are situated inferiorly towards the origin of the tail. The first two appendages with which it is furnished underneath, and which are analogous to those which this part presents in the preceding crustacea, but more diversified here, and always, as it would appear, bearing gills, differ in these respects, according to the sexes. The coupling is the same as that of the insects.

These animals are divided into three orders: those whose mandibles are provided with a palpus, appear naturally to be connected with the preceding crustacea; such are the amphipodes. Those in which these organs are without palpi, will compose the following orders, the læmodipods and isopods. Yamus, a genus of the second, being parasites, will conduct us naturally to bopyrus and cymothoë, with which we shall commence the isopods.

THE THIRD ORDER OF CRUSTACEA.

AMPHIPODA

Are the only malacostraca with sessile and immoveable eyes, whose mandibles, like those of the preceding crustacea, are provided with a palpus; the only ones also whose subcaudal appendages, always very apparent, resemble in their narrow and elongated form, in their articulations, bifurcations, &c. as well as in the hairs with which they are furnished, false feet, or rather fin-feet. In the malacostraca of the following orders these appendages have the form of laminæ, or scales; these lashes or hairs appear here to constitute the gills. Many exhibit, as well as the stomapods and læmodipods, vesicular pouches placed between the feet, or at their external base, of which the use is unknown.

The first pair of feet, or that which corresponds to the

second jaw-feet, is always annexed to a particular segment, the first after the head. The antennæ, whose number, with but a single exception *(phronima)* is four, are advanced, gradually become slender, terminating in a point, and are composed, as in the preceding crustacea, of a peduncle, and of a single stem, or accompanied at most by a small lateral branch, and most frequently pluriarticulate. The body is usually compressed, and curved underneath posteriorly. The appendages of the end of the tail most frequently resemble small articulated stylets. These crustacea may be comprized in a single genus, that of

GAMMARUS, *Fab.*,

Which may be divided at first, according to the form and number of the feet, into three sections.

1. Those which have fourteen feet, all terminated by a hook, or the same number terminating in a point.

2. Those whose number of feet is also fourteen, but in which these organs, or the last four at least, are imperfect, and simply natatory.

3. Those which have but ten apparent feet.

The first section will be divided into two—

The first, UROPTERA, *Latr.*, have the head thick in general, the antennæ often short, and simply two in number in some, and the body soft. All the feet, the fifth pair at most excepted, are simple. The anterior are short or small, and the tail is either accompanied at the end with lateral fins, or terminated by appendages or points, widened and bidenticulated, or forked at their posterior extremity.

In some, as in

PHRONIMA, *Latr.*

There are but two antennæ, very short, and biarticulate. The fifth pair of feet is the largest of all, and terminated in a didactylous forceps. The appendages of the tail are six in number, and in the form of stylets, elongated, forked, or bidenticulate at their extremity. Six vesicular sacs are visible between the final feet. It appears that there are many species, but they have not been described in a comparative and rigorous manner. The type is *Cancer Sedentarius*, Forsk.

In others the number of the antennæ is four; all the feet are simple. The tail has, at each side of its extremity, a lamellary or foliaceous fin, the laminæ of which are pointed or unidenticulate at the end.

HYPERIA, *Latr.*,

Whose body is thicker in front, whose head is occupied for the most part by oblong eyes, a little emarginated at the internal edge, two of whose antennæ are as long at least as one half the body, and terminated by a setaceous stem, long, and composed of many small articulations. (*Cancer monoculoides*, Montague.)

PHROSINE, *Risso.*

Similar, in the form of the body, and that of the head, to hyperia, but the antennæ are at most only of the length of this part, with but few articulations, in the form of stylet, or terminated by a stem, in an elongated cone. (*Phrosine macrophthalma*, Risso.)

DACTYLOCERA, *Latr.*,

Whose body is not thickened in front, whose head is of middle size, depressed, almost squared, with the eyes small, and whose antennæ, very short, and with few articulations, are of diverse forms; the inferior being slender, and in the form of a stylet, the superior being terminated by a small lamina, concave at the internal side, and representing a sort of spoon or pincer. (*Phrosina Semilunata*, Risso.)

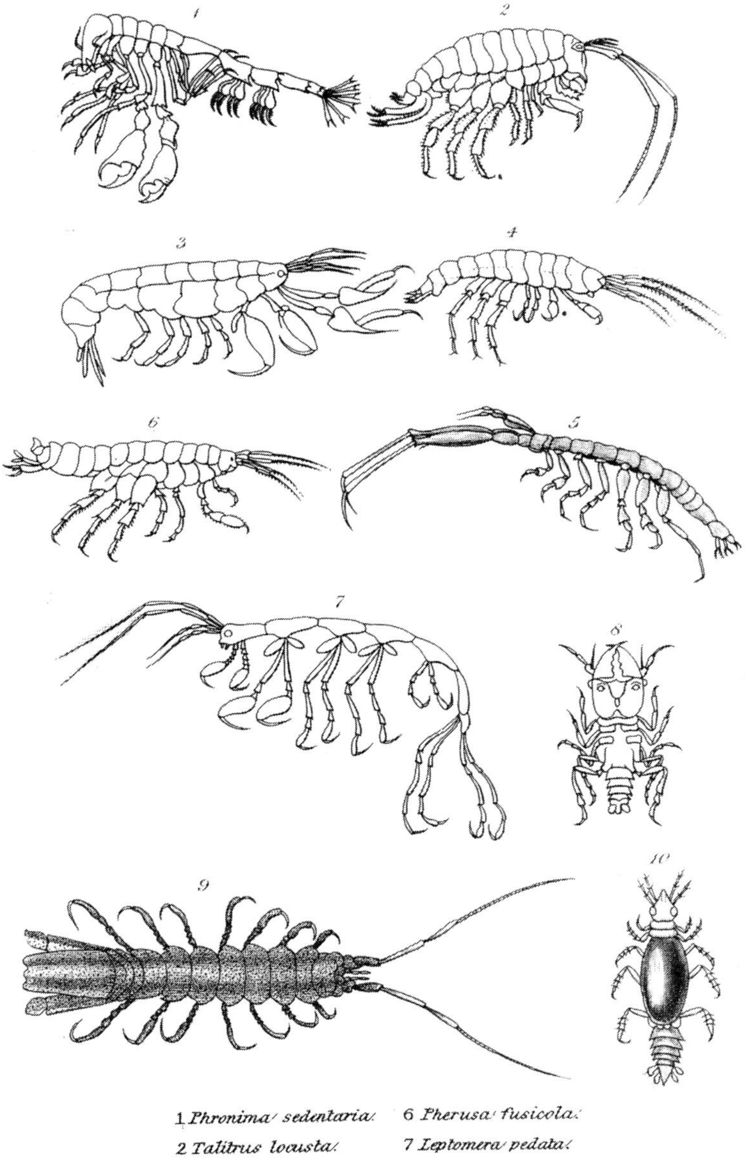

1 *Phronima sedentaria.* 6 *Pherusa fusicola.*

2 *Talitrus locusta.* 7 *Leptomera pedata.*

3 *Leucothoe articulata.* 8 *Anceus maxillaris.*

4 *Orchestia littoralis.* 9 *Stenosoma linearis.*

5 *Corophium longicornis.* 10 *Praniza Cæruleatus.*

London, Published by Whittaker & C.º Ave Maria Lane 1833.

The second, GAMMARINÆ, *Latr.*, have always four antennæ; the body is invested with coriaceous elastic teguments, and generally compressed and arched; the posterior extremity of the tail is without fins; its appendages are in the form of cylindrical or conical styles; two at least of their four anterior feet are most frequently terminated in a pincer.

The vesicular pouches, in those in which they have been observed, are situated at the external base of the feet, beginning from the second pair, and accompanied by a small lamina. The pectoral scales enclosing the eggs are six in number.

Sometimes the four antennæ, although of different proportions in many, have essentially the same form and the same uses. The inferior do not resemble feet, nor perform their functions.

One subgenus, which we have established under the denomination of

IONE,

After a figure of Montagu's (*Oniscus thoracicus*, Linn. Trans. IX. iii. 3, 4.) presents us with very peculiar characters, and which remove it from all the others of the same order. The body is composed of about fifteen articulations, but distinguished only by lateral incisions, in the form of teeth; the four antennæ are very short; the external, longer than the two others, are alone visible, when the animal is observed at the back; the first two segments of the body are each provided, in the female, with two elongated cirri, fleshy, flatted, and similar to oars; the feet are very short, concealed under the body, and hooked; the last six segments are provided with lateral appendages, fleshy, elongated, fasciculated, simple in the male, branchy in the other sex. We also see at the posterior extremity of the body, six other simple, curved appendages, two of which are larger than the others; the abdominal valves are very large, cover the whole inferior part of the body, and form a sort of receptacle for the eggs.

All the following Amphipoda have the segments of the body perfectly distinct in their whole extent, and none of either sex exhibit the long cirri in the form of oars, which are seen in the first two of Ione.

In these, the talon or mobile finger, when it exists, of the feet terminating in pincers, is formed only of a single articulation.

Among these last there are some whose upper antennæ are much shorter than the under, and even than their peduncle. The stem of these is composed of a great number of articulations.

ORCHESTIA, *Leach*,

Have the second feet terminated in the males by a large pincer, with the talon, or long moveable finger, a little curved, and by two fingers in the females; the third articulation of the lower antennæ is, at most, of the length of the preceding two united. (*Oniscus gammarellus*, Pallas.)

TALITRUS, *Latr.*,

Have no feet in the form of a claw; the third articulation of the inferior antennæ is longer than the preceding two united; these antennæ are large and spiny. (*Oniscus locusta*, Pall.)

In the following, the upper antennæ are never shorter than the lower.

Some, having, moreover, their antennæ elongated, setaceous, and terminated by a pluri-articulate stem, and without remarkable claws, approach the preceding, in that the upper antennæ are a little shorter than the lower, and are again removed from the following, by the form of their head, contracting in front, like a muzzle; such are

ATYLUS, *Leach.* (*Atylus carinatus*, Leach.)

All those which succeed, have the upper antennæ as long, or longer than the lower, and the head does not advance in the manner of a muzzle.

Here, as in the five following genera of Dr. Leach, the peduncle of the antennæ is formed of three articulations.

Some present, in their superior antennæ, a character unique in this order; the interior extremity of the third articulation of their peduncle has a small articulated thread. It distinguishes

GAMMARUS, (proper) *Latr.*

The four anterior feet are in the form of little claws, with the talon, or mobile finger, folded underneath. (*Cancer pulex*, Linn.)

The antennæ of the following, are, as well as in all other amphipods, simple, or without appendages.

MELITA, *Leach*,

Have the second feet terminated in the males, by a large compressed pincer, with the talon folded under its internal face; the antennæ are almost of equal length; the posterior extremity of the body presents, on each side, a small foliaceous plate. (*Cancer palmatus*, Montagu.)

MÆRA, *Leach*,

Whose second feet are likewise terminated in the males, by a large compressed pincer, but the talon is folded back on its inferior edge, and is not concealed; the upper antennæ are longer than the lower, and the posterior extremity of the body presents no foliaceous lamina. (*Cancer gammarus grossimanus*, Montagu.)

AMPITHOE, *Leach*,

In which the four anterior feet are nearly identical in both sexes, and the last articulation but one, or the hand, is ovoïd. (*Cancer rubricatus*, Montagu.)

PHERUSA, *Leach*,

Which do not differ from Ampithoe, except that the hands of the claws are filiform. (*Pherusa fucicola*, Leach.)

Sometimes the peduncle of the antennæ is composed of only two articulations—(the third, from its smallness, being confounded with those of the stem, or forming that of its base);—the superior are longer than the inferior; all the feet are simple, or without pincers. Such are

DEXAMINE, *Leach*.

In those, the talon, or mobile finger of the two pincers, is biarticulate; the antennæ are of equal length. (*Cancer gammarus spinosus*, Montag.)

LEUCOTHOE, *Leach*,

Have the antennæ short, and their peduncle with two articulations; the four anterior feet strongly terminated in a pincer, the talons of the two anterior biarticulate; those of the second pair with a single articulation and long. (*Cancer articulosus*, Montagu.)

CERAPUS, *Say.*,

Whose antennæ are large, with the peduncle in the upper of three, in the lower of four articulations; the two anterior feet are small, with a talon of a single articulation; and the two following terminate in a large triangular hand, smooth, denticulated, with the talon biarticulate. (*Cerapus tubularis*, Say.)

Sometimes the lower antennæ, much larger than the upper, and whose stem is composed at most of four articulations, have the form of feet, and appear to serve, at least occasionally, as organs of prehension.

Here the second feet are terminated by a large pincer.

PODOCERUS, *Leach*,

With projecting eyes. (*Podocerus variegatus*, Leach.)

JASSA, *Leach*.

Eyes not projecting, (*Jassa pulchella*, Leach.)

Sometimes none of the feet are terminated by a large pincer.

COROPHIUM, *Latr*. *Cancer grossipes*, Lin.

The second section, HETEROPA, *Lat.*, is composed of those which have fourteen feet, the last four of which, at least, are imperfect at the end, and exclusively proper for natation. It comprehends two subgenera,

PTERYGOCERA, *Lat.*,

Which have the thorax divided into several segments; four antennæ furnished with hairs forming plumes; all the feet natatory, and the last large and pinnated; and cylindrical articulated appendages at the posterior extremity of the body. (*Oniscus arenarius*, Encyc. Method.)

APSEUDES, *Leach*. EUPHEUS, *Risso*.

Which have also the thorax divided into many segments, but the two anterior feet are terminated by didactylous pincers; the following two are widened into a knob, terminating in a point, and denticulated on the edges; the six following are slender and unguiculated at the end; the last four are natatory; the antennæ are simple; the body is narrow, elongated, with two long appendages, in the form of a thread, at their posterior extremity. (*Eupheus ligioïdes*, Risso.)

The third and last section, DECEMPEDES, *Lat.*, is composed of Amphipodes presenting but ten distinct feet.

TYPHIS, *Risso*,

Have but two very small antennæ. The head is thick, and the eyes not projecting. Each pair of feet is annexed to a proper segment: the anterior four are terminated by a didactylous pincer. On each side of the thorax are two mobile laminæ, forming sorts of shutters, or valves, which, united, when the animal folds its feet and tail underneath, close the body inferiorly, and give it the form of a spheroïd. The posterior extremity of the tail is without appendages. (*Typhis ovoïdes*, Risso.)

ANCEUS, *Risso*, GNATHIA, *Leach*,

Which have also the thorax divided into as many segments, as pairs of feet, but in which all those organs are simple and monodactylous. They have, besides, four antennæ (setaceous.) The head is strong, squared, with two large projections in the form of mandibles. The extremity of the tail has some foliaceous appendages in the form of fins. (*Anceus forficularis*, Risso.)

PRANIZA, *Leach*,

Have four setaceous antennæ, as well as Anceus; but the thorax seen above presents only three segments, the first two of which, very short and transverse, have each one pair of feet, and the third, much larger and longitudinal, bear the others. All the feet are simple. The head is triangular, pointed in front, with the eyes projecting. The posterior extremity of the body also presents a fin on each side. (*Oriscus cæruleatus*, Montagu.)

THE FOURTH ORDER OF CRUSTACEA.

LÆMODIPODA,

Are among the malacostraca with sessile eyes; the only ones in which the posterior extremity of the body presents no distinct gills; which have almost no tail, the last two feet being inserted at this end, or the segment serving as an attachment for them, being followed only by one or two other articulations very small. They are, again, the only ones in which the two anterior feet, which correspond to the second jaw-feet, form a portion of the head.

They have all four setaceous antennæ, and carried on a peduncle of three articulations, mandibles without palpi, a vesicular body, at the base of four pair of feet at least, commencing from the second or third pair, those of the head comprized. The body, most frequently filiform or linear, is composed, reckoning the head, of from eight to nine articulations, with some small appendages, in the form of tubercles, at its posterior and inferior extremity. The feet are terminated by a strong hook; the anterior four, the second of which are largest, are always terminated in a monodactylous pincer or talon. In many the four following are shortened, less articulated, without hook at the end, or rudimentary, and by no means proper for the ordinary uses.

The females carry their eggs under the second and third segments of the body, in a pouch formed of approximated scales.

These crustacea are all marine. M. Savigny considers them as neighbouring the pycnogonides, and making with them the passage to the arachnida from the crustacea.

But a single genus can be found of them, to which, from ancient usage, we shall preserve the name of

CYAMUS, *Latr.*

Some, FILIFORMA, *Latr.*, have the body long and very slender, and linear, with the segments longitudinal. The feet equally elongated and slender, and the stem of the antennæ composed of many small articulations.

LEPTOMERA, *Lat.* PROTO, *Leach*,

Have fourteen feet (the two annexed to the head comprised) complete, and in a continued series.

In some, as in our LEPTOMERA proper (*Gammarus pedatus*, Mull.) all the feet, with the exception of the anterior two, have a vesiculary body at their base. In others, as the PROTO of Leach (*Cancer pedatus*, Montagu), these appendages are proper only to the second feet and the four following.

NAUPREDIA, *Latr.*,

Have but ten feet, all in a continued series, the second and the two following pair have at their base a vesicular body.

CAPRELLA, *Lamarck*,

Have likewise but ten feet, but in an interrupted series, commencing inclusively at the second segment, not counting the head. This segment and the following each present the vesicular bodies, and are totally deprived of feet. (*Squilla lobata*, Muller.)

The other læmodipods, OVALIA, *Lat.*, have the body oval, with the segments transverse. The stem of the antennæ appears to be inarticulate. The feet are short, or but little elongated; those of the second and third segments are imperfect, and terminated by a long cylindrical articulation, and without hooks. They have at their base an elongated vesicular body. These læmodipods form the subgenus of CYAMUS (proper), *Lat.* LARUNDA, *Leach.* (*Oniscus ceti*, Lin.)

THE FIFTH ORDER OF CRUSTACEA.

ISOPODA,

Polygonata, Fab., the genus *Monoculus* excepted,

Approach the læmodipods by the want of palpi to the mandibles, but they are remote from them in many respects. The two anterior feet are not annexed to the head, but depend, as well as the following, on a peculiar segment; they are always fourteen in number, unguiculated, and without vesicular appendage at the base. The under part of the tail is furnished with very apparent appendages, in the form of leaflets or vesicular pouches, and the first or the exterior two usually cover totally, or in a great part, the others. The body is generally flatted, or more wide than thick. The mouth is composed of the same pieces as in the preceding crustacea; but here the parts which correspond to the two upper jaw-feet of the decapods present still more than in these last the appearance of a lower lip, terminated by two palpi. Two of the antennæ, the middle, are almost obliterated in the last crustacea of this order, which are all terrestrial, and differ from the last in their respiratory organs. The male sexual organs are most frequently announced by the presence of linear or filiform appendages, and sometimes of hooks, placed at the internal origin of the first subcaudal laminæ. The females carry their eggs under the breast, &c.

This order in Linnæus embraces the genus

ONISCUS,

Which we shall divide into six sections.

The first, EPICARIDES, *Lat.*, is composed of parasite isopods, without eyes or antennæ, the body of which is very flat, very small, and oblong in the males, much larger in the females, in the form of a narrowed oval, and a little curved posteriorly, hollow underneath, with a thoracic border, divided on each side into five membranaceous lobes. The feet are situated on this border, very small, curled up, and of no use either for walking or swimming. The under part of the tail is furnished with five pair of small ciliated imbricated leaflets, answering to as many segments, and disposed on two longitudinal ranges; but the posterior extremity is deprived of appendages. The mouth presents distinctly but two membranaceous leaflets, applied on another of the same consistence, in the form of a large quadrilateral figure. The lower concavity, forming a sort of flat basket, is filled by the eggs.

These crustacea form but a single subgenus, that of

BOPYRUS, *Latr.* (*Bopyrus crangorum*, Lat.)

The second section, CYMOTHOADA, *Lat.* comprehends the isopods with four very apparent antennæ, setaceous, and almost always terminated by a pluri-articulated stem; having eyes, a mouth composed as usual, vesicular gills, disposed longitudinally in pairs; the tarsi formed of from four to six segments, with a fin on each side near the end, and the anterior feet most usually terminated by a strong claw or hook. These crustacea are all parasites.

Sometimes the eyes are carried upon tubercles, at the summit of the head. The tail is composed of only four segments.

SEROLIS, *Leach.*

But one known species, *Cymothoa paradoxa*, Fab. The antennæ are placed on two lines, and terminated by a pluri-articulate stem. Under the three first segments of the tail, between the ordinary appendages, there are three others, transverse, and terminated posteriorly in a point.

Sometimes the eyes are lateral, and not carried upon tubercles. The tail is composed of five or six segments.

Here the eyes are not composed of simple eyes approximated, and in the form of little grains. The antennæ are on two lines, and of seven articulations, at the least. The six anterior feet are commonly terminated by a strong claw.

In some, whose tail is always of six segments, the length of the lower antennæ never exceeds half that of the body.

We shall commence with those whose mandibles, as is most usual, project but little, or not at all. Here come the

CYMOTHOA, *Fab.*,

Whose antennæ are almost of equal length, the eyes but little apparent, with the last segment of the tail in a transverse square, and the two pieces terminating the lateral fins, linear and equal, in the form of a stylet. (*Cymothoa œstrum*, Fab.)

ICHTHYOPHILUS, *Latr.* (*Nerocila Livoneca*, Leach.)

Having also the antennæ of equal length, and the eyes but little visible; but the last segment of the body is almost triangular, with the two pieces terminating the lateral fins, in the form of leaflets or laminæ, the exterior of which is larger in *Nerocila*, and of the same size as the other in *Livoneca*.

In the four following subgenera the upper antennæ are manifestly shorter than the lower.

Many have, like Cymothoa, all the feet terminated by a

strong and very arched claw. The last eight are not spiny. The eyes are always apart, and convex. They form their genera in the method of Dr. Leach, but may be united in a single subgenus, under the common denomination of one of them, that of

CANOLIRA, *Leach.* (*Anilocra, Olencira, ejusd.*)

Olencira have the laminæ of their fins narrow, and armed with prickles. In anilocra the external lamina of these fins is longer than the internal. The inverse of this is the case with canolira; besides the eyes are but little granulated, while they are very sensibly so in the preceding.

In the three following subgenera the second, third, and fourth feet only are terminated by a claw strongly curved, and the last eight are spiny. The eyes are usually but little convex, large, and converge anteriorly.

ÆGA, *Leach,*

Have the first two articulations of their upper antennæ very broad and compressed, while in the two subgenera which succeed, these articulations are almost cylindrical.

ROCINELA, *Leach,*

Differ from Æga, as we have just said, in the form of the first two articulations of their upper antennæ, and approach them in their large eyes, which are approximated anteriorly.

CONILIRA, *Leach,*

Resemble rocinela in their antennæ, but the eyes are small, apart, and the edges of the segments are almost straight, and not in the form of scythes, and prominent.

The last subgenus, among those of this section whose antennæ are on two lines, whose tail is of six segments, and whose inferior antennæ are always short, is distinguished from all the preceding by its strong and prominent mandibles. It is that of

SYNODUS, *Latr.*

In those which follow the tail is most frequently composed of but five segments. The length of the lower antennæ exceeds the half of that of the body.

CIROLANA, *Leach,*

Have six segments to the tail.

NELOCIRA, *Leach,*

Have but five, the cornea of the eye is smooth.

EURYDICE, *Leach,*

Similar to Nelocira in the number of the caudal segments, remote from them in the character of their granular eyes.

This subgenus conducts us to those in which these organs are formed of small grains, or simple eyes approximated, and which have besides the four antennæ inserted on one and the same horizontal line, of four articulations at most, and all the feet ambulatory. The tail is composed of six segments, the last of which is large and suborbicular. Such are

LIMNORIA, *Leach.* (*Limnoria Terebrans,* Leach.)

The third section, SPHÆROMIDES, *Latr.*, presents four very distinct antennæ, setaceous or conical, and a single subgenus excepted, *(anthura)* always terminated by a stem, divided into many small and short articulations; the lower always longer, are inserted beneath the under part of the first articulation of the upper, which is thick and broad. The mouth is composed as usual. The gills are vesicular or soft, naked, and disposed longitudinally in pairs. The tail presents but two complete and mobile segments, but has often on the first some impressed and transverse lines, indicating the vestiges of other segments. On each side of its posterior extremity is a fin terminated by two leaflets, the inferior of which alone is mobile, and the superior is formed by an internal prolongation of the common support. The branchial appendages are curved interiorly. The internal side of the first is accompanied in the males by a small linear and elongated piece. The anterior part of the head, situated under the antennæ, is triangular, or in the form of an inverted heart.

Some have the body oval or oblong, in a state of contraction assuming the form of a bell. The antennæ are terminated by a pluri-articulate articulation, and the inferior at least sensibly longer than the head. The lateral and posterior fins are formed of a peduncle and two laminæ, composing with the last segment a common fan-like fin.

In these the impressed and transverse lines of the anterior segment of the tail, always shorter than the following or the last, do not reach the lateral edges. The first articulation of the upper antennæ is in the form of a triangular palette.

The head, seen from above, forms a transverse square. The leaflets of the fins are very much flatted, and the intermediate piece, or last segment, is widened, and rounded laterally.

ZUZARA, *Leach*,

In which the leaflets of the fins are very large, and the upper, being shorter, is removed from the other to form a border to the last segment.

SPHÆROMA, *Latr.*,

In which the leaflets are of middle size, equal, and applied one upon the other. *Sph. dentata*, Desmarest.

In these the impressed lines, or transverse sutures of the anterior segment of the tail, reach its lateral edges and divide it. The first articulation of the upper antennæ forms an elongated palette, square or linear.

The leaflets of the fins are usually narrower and more thick than in the preceding. The exterior one sometimes emboxes the other. The latter is prismatic. Their point of union presents the appearance of a knot or articulation.

Sometimes the sixth segment of the body is sensibly of greater breadth than the preceding and the following. One of the two leaflets of the fins alone is projecting.

NÆSA, *Campecopea*, Leach.

Sometimes the sixth segment is of the length of the preceding and the following. (*Næsa bidentata*, Desm.)

CILICÆA, *Leach*,

In which one of the leaflets of the fins alone is projecting, the other being laid back against the posterior edge of the last segment. (*Cilicæa, Latreillii*, Desm.)

CYMODOCEA, *Leach*,

In which the two leaflets of the fins are projecting, and equally directed backwards. The sixth segment is not prolonged posteriorly, and the last presents but a small lamina in an emargination.

DYNAMENE, *Leach*,

Like cymodocea in the projection and direction of the leaflets of the fins, but the sixth segment is prolonged behind, and the last presents but a simple cleft, without lamina.

The others, such as

ANTHURA, *Leach*,

Have the body vermiform, and the antennæ scarcely as long as the head, with four articulations. The leaflets of the posterior fins form by their disposition and approximation a sort of capsule.

The anterior feet are terminated by a monodactylous pincer.

In the fourth section, IDOTEIDES, *Leach*, the antennæ are also four in number, but on the same horizontal and transverse line. The lateral are terminated by a stem, finishing in a point, gradually growing more slender, and pluri-articulated. The intermediate are short, filiform, or a little thicker towards the end, of four articulations, none of which is divided. The composition of the mouth is the same as in the preceding sections. The gills are in the form of bladders, (white in the majority) susceptible of being swelled to serve for swimming, and covered by two laminæ, or valvules, of the last segment, adhering laterally to its sides, longitudinal, biarticulate, and opening in the middle by a straight line, like two shutters. The tail is formed of three segments, the last of which is much larger, without appendages at the end, or lateral fins.

IDOTEA, *Fab.*,

Have all the feet strongly unguiculated, identical; the body oval, or simply oblong, and the lateral antennæ shorter than one half of the body. *Oniscus entomon*, Lin.

STENOSOMA, *Leach*,

Differ only in the linear form of the body, and the length of the antennæ exceeding the half of that of the body. (*Stenosoma lineare*, Leach.)

ARCTURUS, *Latr.*,

Are very remarkable for the form of the second and third feet, which are directed forward, and terminate by a long barbed articulation, and imperfect, or but slightly unguiculated. The anterior two are applied on the mouth, and unguiculated. The last six are strong, ambulatory, thrown backwards, and bidenticulated at their extremity. In the relation of the length of their antennæ, and the form of the body, they approach Stenosoma. (*Arcturus tuberculatus,* Lat.)

The fifth section (*Asellota,* Lat.) presents us with isopods with four very apparent antennæ, disposed on two lines, setaceous, terminated by a pluri-articulate stem; two mandibles, four jaws, usually covered by a sort of lip, formed by the first jaw-feet; vesicular gills disposed in pairs, covered by two longitudinal and biarticulate leaflets, but free; a tail formed of a single segment, without lateral fins, but with two bifid stylets, or two very short appendages, in the form of tubercles, at the middle of its posterior edge. Some other appendages, in the form of laminæ, situated at its inferior base, more numerous in the males, distinguish the sexes.

ASELLUS, *Geoff.*,

Have two bifid stylets at the posterior extremity of the body, the eyes apart, the upper antennæ of the length, at least, of the peduncle of the lower, and the hooks of the end of the feet entire. (*Aselle d'eau douce,* Geoff. *Idotea aquatica,* Fab.)

ONISCODA, *Latr.*,

Or janira of Dr. Leach, differ from asellus in the approximation of their eyes, their upper antennæ being shorter than the peduncle of the lower, and in the knobs of the tarsi, which are bifid. (*Janira maculosa,* Leach.)

JÆRA, *Leach*,

Instead of the stylets at the end of the tail, have but two tubercles. (*Jæra albifrons*, Leach.)

Finally, the isopods of the sixth and last section, ONISCIDES, *Lat.*, have four antennæ, but the two intermediate very small, but little apparent, and of two articulations at most. The lateral are setaceous. The tail is composed of six segments, with two or four appendages, in the form of stylets, at the posterior edge of the last, and without lateral fins. Some are aquatic and others terrestrial. In the latter the first leaflets of the end of the tail present a range of little holes, through which the air penetrates, and is carried to the organs of respiration, which are inclosed there.

Some have the sixth articulation of their antennæ or their stem composed in such a manner, that, counting the little articulations of this part, the sum total of all the articulations is nine at least. These isopods are marine, and form two subgenera.

TYLOS, *Latr.*,

Appear to have the faculty of rolling themselves into a ball. The last segment of the body is semicircular, and fills exactly the notch formed by the preceding. The posterior appendages are very small, and entirely inferior. The antennæ have but nine articulations, of which the last four compose the stem. On each side is a sunken tubercle, representing each, one of the intermediate antennæ. The intermediate space is raised. The gills are vesicular, imbricated, and covered by laminæ.

LIGIA, *Fab.*,

Have the stem of the lateral antennæ composed of a great number of small articulations, and two very projecting stylets, divided at the end into two branches, at the posterior extremity of the body. (*Oniscus Oceanicus*, Lin.)

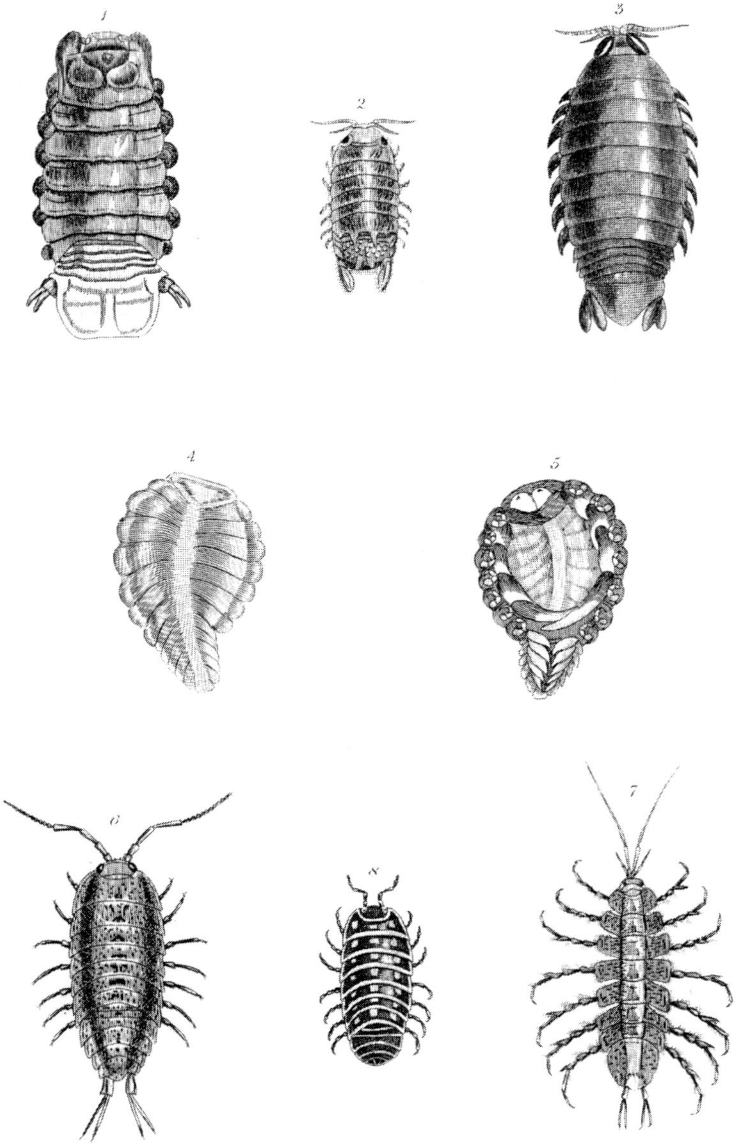

1 *Cymothoa œstrum.* 4 e 5 *Bopyrus crangorum.*

2 *Næsa bidentata.* 6 *Ligia oceanicus.*

3. *Æga emarginata.* 7 *Asellus aquaticus.*

8. *Armadillo pustulatus.*

London, Published by Whittaker & Co. Ave Maria Lane 1833.

In the others and all terrestrial, the lateral antennæ present at most but eight articulations, whose proportions towards the extremity diminish gradually, without any of them appearing to be divided or composed.

Here the appendages, or posterior stylets, advance beyond the last segment. The body does not contract itself, or at least very imperfectly, into a ball.

PHILOSCIA, *Lat.*,

Have the lateral antennæ divided into eight articulations, and uncovered at their base. The four posterior appendages are almost equal. (*Oniscus Sylvestris*, Fab.)

ONISCUS *(proper) Lin.*, *Wood-lice*, *Vulg.*

Have also eight articulations to the lateral antennæ, but their base is curved, and the two exterior appendages of the end of the tail are much larger than the two interior. (*Oniscus Murarius*, Fab.)

PORCELLIO, *Lat.*,

Are distinguished from oniscus by the number of articulations of the lateral antennæ, which is only seven. (*Oniscus asellus*, Cuv.)

ARMADILLO.

The posterior appendages of the body make no projection. The last segment is triangular; a small lamina, in the form of an inverted triangle, or more wide, and truncated at the end, formed by the last articulation of the lateral appendages, fills, on each side, the vacancy comprized between this segment and the preceding. The lateral antennæ have but seven articulations. The upper subcaudal scales have a range of small holes. (*Oniscus Armadillo*, Lin)

SUPPLEMENT

ON

THE CRUSTACEA.

We shall insert here, after the text of Latreille on the MALACOSTRACA, our supplementary observations on the class in general, and also on that division in particular.

All the animals of this class, as their name indicates, are covered with integuments of a crustaceous substance, more calcareous than that which envelopes the *myriapods*, the *arachnida*, and the *insects*. Most of them feed on bodies in a state of putrefaction, and in all the sexes are distinct.

The ancients were very well acquainted with the division which we call *malacostraca*, which they placed between the mollusca and the fish. Aristotle has devoted a particular chapter to the species which were known to him. Athenæus has given an enumeration of such as are edible, and Hippocrates has noticed some which, in his opinion, may be usefully employed for medicinal purposes.

Pliny has scarcely added any thing to the observations of Aristotle, and those who have since spoken of the crustacea, such as Rondelet, Belon, Gesner, Aldrovandus, and Jonston, who also place them between the mollusca and the fish, have produced nothing which could throw any additional light on the natural history, or on the structure of this class of animals.

The ancient naturalists, however, and even the modern, to the time of Linnæus, perceived, as we have seen, the necessity of classing these animals separately. That great naturalist, however, united them to the insects, and having taken as the basis of his first partitions, the presence or absence of wings, he placed them at the end of this class, in the order aptera, of which they compose, with the palpous arachnida, the second division. His opinion was generally followed. But Brisson, in his "*Regne Animal*," continued to distinguish the crustacea from the insects, placed them immediately after the fish, and associated with them the arachnida and myriapoda. His crustacea, therefore, are all the insects of Linnæus which have more than six feet. The apterous order of the Swedish naturalist subsequently underwent some modifications, but it preserved the same rank in all the methods which were established on the same principles.

Fabricius at first composed with the crustacea his fourth order of insects, that of *agonata*, which he thus named because they have no under lip. In this order he placed the scorpions, and removed the onisci and monoculi of Linnæus.

No regard was paid, in these and other methodical distributions, to the essential differences presented by the internal structure of these animals, although some great naturalists, such as Swammerdam, Rœsel, and Degeer, had already observed a circulation and gills in the crustacea; and though it was easy to conclude from hence, that their organization differed from that of insects, and approximated more to that of superior animals, it was reserved for the first naturalist of onr age, whom it is unnecessary to name, to awaken our attention to this most important point, and to direct our steps into the true road to a natural classification. At first, in his *Elementary View of the Natural History of Animals*, founded upon such considerations, he transported the crustacea to the head of the insect class, and formed with them a special and well-circumscribed division. Soon after, in his *Comparative Anatomy*, he made a peculiar class of the crustacea; and his example was followed at the same time by M. de Lamarck, in his public lectures on the invertebrated animals.

As we have thus touched, however slightly, on the history of the classifications of crustacea, it would be highly improper to pass over in silence the name of our distinguished countryman Dr. Leach. His labours on certain parts of this order are highly valuable, and at once excite our regret at their unhappy interruption, and our earnest hope of their auspicious recommencement and completion. The nature of our work will not permit us to follow all his details, or give a minute analysis of his distribution. His generic divisions are generally noticed in our text, and nothing important in his researches shall be omitted in these supplementary additions.

Although the name of *crustacea* has become one of general usage, we may yet consider that of *malacostraca*, as in some measure, if not altogether synonymous, although the latter has been used by M. Latreille and other recent writers, to indicate a single division of the class, in which those beings are comprehended. The term μαλακοστρακος (molli crustâ obtectus) designated, among the Greeks, those marine animals without blood, whose external envelope, much less solid than the testa of the shelled mollusca, is yet considerably more so than the skin of the naked mollusca.

The crustacea, considered under the various relations which their organization presents, should incontestably occupy a very elevated rank among invertebrated animals, and those which are provided with articulated limbs. They cannot be placed at a remote distance from the arachnida and insects, whose body is like their's, symmetrical, encompassed with a corneous, solid, and resisting skin, which performs the functions of the skeleton in the animals of the superior classes; whose members are, like their's, composed of several distinct pieces; whose eyes are always apparent, and whose generation is bisexual.

They are more distant from the animals of the class annelides of Lamarck, whose body is destitute of true limbs, in which the eyes are usually wanting, and the generation is frequently hermaphrodite.

It would seem, that the crustacea ought in strictness to be placed after certain of the mollusca, such as the cephalopoda, and before others, as the gasteropoda, and more especially the acephala, which by certain shades present evident passages to the animals of the lowest classes. Nevertheless, as the mollusca of the different orders have well-established relations to each other, it would not be right to cut their series into two parts, for the purpose of intercalating between them the articulated animals, and consequently the crustacea. We must therefore decide either to place, after these last, the entire class of the mollusca, as the ancients did, or to leave this class before them, as has been judged expedient by the more recent zoologists. Of all the moderns, M. de Blainville alone has inclined to the notions of the ancients on this subject: he has proposed that the crustacea should be followed by the mollusca and the worms, and placed after the insects and arachnida, which should themselves immediately follow the fish. But the other mode of arrangement is justified by the consideration of those characters which connect the fish with the cephalopod mollusca, and which have been luminously exposed by M. Latreille in a memoir addressed some years ago to the Society of Natural History in Paris.

In spite, however, of all the pains which can possibly be taken, it will ever remain impracticable to allocate the crustacea, so as not to injure any of their affinities with the animals of the other classes. This would alone be feasible, if the animated productions of nature, as was long pretended, composed but a single series, unbroken by interruptions, and undeviating into digression. But modern science cannot recognize this continuous chain: she finds that Being in its wonderful varieties of organization is distributed into different groups, connected together by more or less complicated branches, thus forming a kind of reticulated web, whose tissue will perhaps never be thoroughly unravelled by the art of man.

There exist, in fact, intermediate groups between the class of crustacea and the other classes, especially those of insects and arachnida; and it is more especially the genera of the families of *oniscides*, of *asellota*, of *myriapoda*, and *pycnogonides*, which form these links. These genera have been alternately placed by authors in one or other of these classes of invertebrated animals. They form their true points of contact. Nevertheless, these classes are in general very distinct, as may easily be seen by a reference to our text for their comparative characters.

We shall enlarge a little here on the brief review in the text, of the general form and structure of the crustacea, endeavouring, as far as the duty of perspicuity will permit us, to avoid repetition.

The body of all insects, as we have seen (myriapoda excepted) is constantly divided into three very apparent parts. Such, however, is not the case with the crustacea. The head is most frequently indistinct, and its position not to be recognized but by the existence of the antennæ, and the aperture of the mouth. It is intimately confounded with the most considerable part of the body, that which encloses the principal viscera, and affords points of attachment for the feet. The posterior portion of the body, divided into isolated segments, merely contains the posterior extremity of the intestinal canal, and is not provided with genuine feet. Such is the organization of *cancer* and *astacus*, or, to speak more generally, of the brachyurous and macrourous decapods.

In other crustacea, the head is decidedly detached, but there is no thorax, and the whole of the body is divided into intersimilar segments. This is the case with *squillæ*, *asellæ*, &c.

In some others the head is distinct, but the first rings of the body are united above, so as to form a sort of buckler of no great extent.

In some others *(limulæ)*, the segmentary division of the body is only apparent underneath, while the head, above, presents a vast buckler, and the trunk and abdomen are confounded, and covered by a second large plate, terminating in an ensiform appendage.

Sometimes we find the head more or less distinct, and the body not divided neatly into trunk and abdomen, but exhibiting scarcely any trace of segments, and comprised in a bivalve testa, formed by a hardened expansion of the dorsal skin.

The general form of the antennæ is that of a thread or lash, that is, they are longitudinally conical, or diminished insensibly in thickness, from a round base to a very attenuated extremity. They are composed of small hollow cylinders of corneo-calcareous substance, or of articulations superadded one to the other, and whose cavity encloses muscles, nerves, and without doubt ramifications of the circulating system. Each antenna has its peduncle and thread. The *peduncle* (a term borrowed from botany) is a sort of stem or stalk, composed of three or four articulations much thicker than the rest, and frequently affording an attachment to certain appendatory leaflets. The thread is single, double, or triple, varying in the number of its articulations, but often composed of a multitude of small ones.

The antennæ, in certain genera, assume anomalous forms, which assimilate them to organs of locomotion. At other times their peduncle alone exists, and is transformed into very broad and crenulated plates. In the decapod crustacea, the base of the external antennæ presents a little rounded subtriangular body, strong in the short-tailed, a little membranaceous in the long-tailed species, which closes the external issue of a cavity traversing the testa or shell, and which is considered to be the auricular organ.

The simple eyes of the crustacea, when they do exist, are always sessile, *i. e.* fixed. The complex eyes are often *pedunculated*, and mobile, and this character is exclusively peculiar to the present class. The peduncle of these eyes is usually formed of a single cylindrical piece, varies in its dimensions, and is lodged in a fosset, sometimes very deep. The branchipes have pedunculated eyes, but not placed in a particular fosset.

The principal parts of the mouth, most frequently destined for the operations of grinding and lacerating, are in pairs, and placed laterally, as in the masticating insects. But sometimes, united to other parts, which might be termed lips, they are modified so as to form a sort of bill, or sucker.

In the ordinary crustacea, or malacostraca, the parts of the mouth present variations pretty frequent as to dimension and form, so that the most exterior among them are often similar to feet, and perform the functions of those organs. In the entomostraca, those pieces which are less numerous, present such varied modifications that it is impossible to describe them in a general manner. A clear understanding, however, of this subject is of so much importance, that even at the hazard of repeating some of the substance of the text, we must venture on a few details respecting the composition of the mouth in the different orders of the crustacea.

In general, the pieces which compose it, are attached to the edges of an emargination, presented by the testa underneath, which has received the name of *buccal aperture*, and is sometimes regularly quadrilateral, sometimes in the figure of a trapezium or triangle. This aperture is not distinct, except in the species which are provided with a calcareous testa of greater or less solidity.

The decapod and short-tailed crustacea are provided, 1st, with a transverse upper lip, articulated with the anterior edge of the buccal aperture: 2d, with a pair of mandibles, or thick, solid, lateral pieces, compressed, and trenchant internally, carrying on their back, and near their point of articulation, an appendage or palpus formed of three articulations; these mandibles being placed anteriorly, and underneath all the other even pieces: 3rd. with a thin, lamellate, and bifid tongue, placed against the posterior basis of the mandibles: 4th. with a first pair of jaws, membranaceous, deeply lobate, and ciliated on their edges, without palpi, and applied on the lower face of the mandibles: 5th. with a second pair of jaws, without palpi, applied on the first, equally membranaceous, lobate, and ciliated: 6th, with a third pair of membranaceous jaws, provided externally with a palpus, formed of a long peduncle, which carries at its extremity a small arched stalk, setaceous and multi-articulate: 7th, with a fourth pair of jaws, formed by a stem rather narrow, compressed, not membranaceous, divided, like the feet, into six articulations, and by an external flagelliform palpus, analogous to that of the preceding jaws, but more distinct: 8th. with a last pair of pieces, composed like the preceding, of two parts or stems; the interior, crustaceous and compressed, is divided into six articulations, of which the second and third are much larger than the others, and the last small; the exterior is in the form of a palpus similar to those of the two pair of jaws which are situated before these last.

M. Savigny considers these three pair of external jaws as nothing but feet, so modified as to serve for manducation, and his opinion is founded upon this, that the palpus with which they are provided is analogous to the threads remarked in the anterior feet of many entomostraca, that the two external ones are articulated like feet, properly so called, and generally composed of the same number of pieces, and that at the base they serve as a point of attachment for the gills like ordinary feet. According to this naturalist, all the true crustacea should have sixteen feet, and not differ among themselves, but in the number of those feet which are thus converted into auxiliary jaws. These are what are designated in our text, by the appellation, somewhat startling to the uninitiated, of *jaw-feet.*

In the crabs, the external jaw-feet, or third auxiliary jaws of M. Savigny, are always very apparent. They close the mouth above, and cover all the space comprised by the buccal cavity. The second piece of their internal stem, the largest of all, is pretty generally applied by its interior edge, against the corresponding edge of the same piece, in the opposite jaw-feet, but sometimes these pieces are separated, and leave a triangular space between them. The third piece is smaller and of a form sometimes square, sometimes triangular, trapezoïdal, or oblong, and its point, or internal edge, presents an emargination, for the juncture of the fourth articulation, which, itself, affords an attachment to the two last.

The second, and especially the third, articulation of the external jaw-feet, present the most numerous modifications in their forms, and most usually serve to characterize the genera of the brachyurous decapods. In the long-tailed decapods, (*astacini*) the mandibles and the two genuine pair of membranaceous and lobate jaws, differ but little from the same parts in crabs. But the jaw-feet, and particularly those of the external pair, are elongated, prismatic, and strong. The final articulations are nearly as thick as the second and third, and these pieces have an incontestable analogy with the ambulatory feet. In pasiphaë and mysis they are visibly employed in locomotion.

The squillæ of the order stomapods, crustacea, very anomalous in their organization, are provided with a large conical upper lip; with two very strong mandibles, denticulated and palpigerous; with a tongue formed of two compressed pieces, placed one on each side, and performing the office of jaws; with a first pair of membranaceous jaws composed of two pieces, and bearing, externally, a small palpiform appendage; with a second pair of foliaceous jaws, triangular, formed of four pieces, and covering like a lip, but longitudinally, all the parts of the mouth which have been just mentioned. After these come eight pairs of appendages or members, to which it is difficult to assign precise names, and five of which surround the mouth. M. Savigny considers the first two as auxiliary jaws, and the other fourteen as feet.

The crustacea, with sessile eyes, amphipods, and isopods in general, besides the upper lip, palpigerous mandibles, cartilaginous bifid tongue, and two pair of jaws with two laminæ and without palpi, have also an under lip resulting from the union of the two jaw-feet, or auxiliary jaws. There are, farther on, fourteen feet, properly so called. In bopyrus, the principal parts of the mouth are indistinct, but its orifice is covered by two anterior membranaceous pieces, a little convex, under which are two appendages, soft, compressed, and placed on each side, like the jaws in the other crustacea.

In the entomostraca, the limulæ are equally anomalous, as the squillæ among the malacostraca. The pharynx is placed in the middle of ten appendages in the form of feet or claws. The haunches of these appendages, situated on the sides of the aperture of the œsophagus are spiny, and serve as jaws for the trituration of the aliments. In front, are two appendages, called by M. Savigny, succedaneous mandibles, and palpi by Baron Cuvier, also in the form of forceps, but much smaller than the others, and annexed to the sides of a lanceolate, flatted piece, which is composed of their haunches united, and which M. Savigny considers as performing the functions of an upper lip. The posterior edge of the pharynx presents a piece, also flatted, but bifid, and which may be regarded as the lower lip, formed by the union of the haunches of a pair of feet not developed. There are no true mandibles or antennæ.

In apus the mouth more resembles that of the crustacea

proper. There is an upper lip, two large mandibles, two pairs of jaws, and a tongue. The caligi and some entomostraca of the neighbouring genera, are provided with a bill or sucker, formed of the union of two lips, and two very small mandibles, and with many of these (cecrops) M. Latreille has recognized, besides the bill, three pairs of jaw-feet.

Independently of the antennæ, &c. the head of some crustacea is provided with certain prolongations, to which different names have been assigned. Thus, in many decapods, the portion of the carapace which is situated between the eyes, is more or less advanced, and takes the name of *rostrum*. Its dimensions vary, and it is sometimes bifurcated, sometimes denticulated, and sometimes spiny.

In *ancœus* the head of the males is provided with two large projections, which very much resemble mandibles, but do not perform their functions, and there are two on the head of branchipes, resembling the mandibles of the *lucanus cervus*, and intended to seize the female in the act of coupling. Concurrently with these are two soft productions, spiral, in the form of a proboscis, which are situated between the others, and a little underneath. The first of these appendages are also found with females, but are much more simple and less voluminous, and the others do not exist.

When the anterior edge of the head is not prolonged to form a rostrum, the interval separating the eyes assumes the name of *front* or *forehead*, and sometimes that of *hood* (Fr. *Chaperon.*) The forehead is remarkable in the crabs and other brachyurous decapods, sometimes straight, sometimes arched, sometimes entire, sometimes lobate, emarginated, or denticulated. It also varies in extent.

The body, at its lower face, is pretty constantly divided into transverse segments, but the upper is very often formed of a single piece called *testa* or *carapace*, like the tortoises. This vast buckler covers the body of the crabs altogethe:, under which the abdomen is fixed. It is fastened by two points of its middle to some appendages of the lower or sternal pieces, which support it like pillars. All its inferior and anterior part is articulated with the pieces of the mouth and the first segments of the inferior face of the body. But there is a breach of continuity on the sides, so as to allow the water to penetrate through two clefts into the cavities of the gills. Its general forms and dimensions, the smoothness or inequalities of its surface, &c. vary considerably, according to the genera. We may remark, however, that whatever be the irregularities of the surface of that of the crabs, the arrangement of it is always constant, and subjected to invariable laws. The masses which they form, or the projections which they constitute, are marked by deepened lines, more or less perceptible. M. Desmarest has given them the general name of *regions;* and, to distinguish between them, he has added to each a peculiar designation which is indicative of the organ which it covers. Thus we have the *gastric*, the *genital*, the *cardiac*, the *branchial*, and the *hepatic* regions. These vary in extent and distinctness. In the macrourous crustacea, with a very slender and flexible testa, they are nearly obliterated.

The carapace is wanting in all the isopod and amphipod crustacea, but we find it again in the sub-class of the entomostraca. In some genera of the latter, such as *daphnis*, *lyncœus*, *cypris*, &c. this buckler or mantle is large, and assumes a greater degree of solidity. It has a keel on the middle, as in *apus*, but here this keel becomes a sort of hinge, the sides of the carapace change into valves analogous in their use to those of the shells of the acephalous mollusca; and, by means of certain muscles which appertain to the dorsal region of the animal, these valves can be opened or closed at pleasure. Here we have a decided link between the crustacea and the testaceous mollusca.

The body of the crustacea which are provided with carapace, and particularly that of the decapods, is formed underneath this testa of very distinct segments, and these segments themselves are composed of several pieces. The under part of the body in the brachyurous decapods presents a surface more or less extensive, which may be compared to the breast-plate of the tortoises. Its middle is hollowed by a gutter or furrow, more or less broad, more or less prolonged in front, but, in general, of a greater extent in the females than in the males. This lower surface of the breast-plate is composed of two orders of pieces: the first, which are medial, and much larger than the others, are called *sternal* pieces; the others, being lateral, are termed *latero-sternal* pieces. Between these pieces and the lateral and inferior edges of the carapace, are situated the feet.

Among the entomostraca, some, as *apus* and *branchipes*, have the body annulated underneath, as well as above, and have no traces of latero-sternal pieces, while others have no indication of divisions whatsoever.

The name of *tail* or *abdomen* is reserved for the terminal part of the body, which contains only the posterior part of the intestine. The anus is at its inferior face, to which also the branchial feet are sometimes attached. In some crustacea, it contains the organs of generation; and in many it is furnished at its extremity with natatory appendages. In the decapod crustacea, this tail is always folded under the body, and closes the furrow or longitudinal gutter of the sternum. It forms with this furrow a sort of box, in which the eggs are received and lodged at the period of laying. The tail of the males is placed entirely in this furrow.

The macrourous crustacea have received this appellation in consequence of the extent of the tail; it is sometimes soft, and without distinct rings, as in pagurus, sometimes very solid and very muscular. A most remarkable circumstance is, that the *paguri* betake themselves to the cavities of univalve shells, and that the spiral form of these cavities takes away the symmetry of their tail, by causing it to assume the turbinated figure of the shell. In this case, the terminal appendages of the tail are transformed into hooks, for the purpose of fixing it in its dwelling. That of the other macroura, always twice as long as the body, is at first extended in the direction of the latter, and bent underneath, at its extremity, which is provided with five natatory laminæ, simple or double, which can be unfolded like a fan, and which, acting simultaneously, perform the office of a fin.

The abdominal or caudal segments are provided on each side with small appendages, which have been called *false feet*, and the use of which, in the females, is to serve as points of attachment for the eggs.

The feet of the crustacea are either proper for walking or swimming. Their number, disposition, and more especially their functions, differ considerably, for in certain cases some of these feet are changed into organs of manducation, and in others into respiratory organs. The feet, properly so called, are always larger, more solid, and less variable in their forms than the others, and especially than those called the branchial feet.

The feet which may be considered as the normal feet of crustacea are constantly formed of six pieces or articulations. Some are designated by the name of *claws* or *forceps*, the others are called *simple feet*. Their parts are described in the text, and the claws do not differ from the simple feet in their composition, but that their penult articulation is more swelled than the preceding, is prolonged underneath the last in front, and thus forms an immoveable finger; and that this last articulation, corresponding in its length to this appendage, is articulated above, so that it moves upon it from top to bottom, to form the forceps. In the brachyuri the forceps are always two in number, and belong to the anterior pair of feet, except in the genus *Pactolus*, where it is the last two which are terminated by small claws. They are in general larger, and more especially thicker than the feet properly so called; nevertheless, the latter are sometimes much the largest. In a great number of genera they are equal in size one to the other, but in some one is always thicker. Sometimes they are very long and slender; sometimes very short, and almost concealed. They also present considerable varieties of surface.

The proper feet vary only in length, position, and the form of the tarsus: in general they decrease in size regularly from the first pair, but in some genera the second and third are the largest. In the crabs which swim well they are all larger than in those which frequent the land, and in a direction more horizontal. The land-crabs, and those which frequent shores, all have the last articulation of their feet more robust, but little arched, and conical. In those which swim more than they walk, this articulation is very much depressed, ovoïd, and ciliated on its edges. In the macrouri the feet are very similar to those of the brachyuri, but in general more elongated. Some macrouri have no claws or forceps. Besides the true feet, these crustacea have five pair of what are called *false feet* under the tail, terminated either by two plates or two threads.

The Squillæ have received the denomination of *Stomapods*, from the disposition of the feet, or rather of the appendages so regarded, which surround the mouth. We have already seen, in describing the parts of the mouth, how embarrassing it is to give a suitable designation to those appendages, which many naturalists consider as feet, while others regard them as dependencies of the mouth. Be this, however, as it may, they present the same number of articulations as the ordinary feet of the decapods. The modifications of the feet in the other divisions are very various, but these modifications need not be described here. The caligi, however, we may observe, have very short feet, arched in the form of hooks, and serving, like those of Cymathoë, to fasten on the fleshy parts of those fishes on which they live. The arguli have three sorts of feet; the first two like cupping-glasses, round, and broad; the second with two hooks, proper for prehension, and the others, eight in number, soft, fleshy, and terminated by a fin formed of two leaflets.

Finally, the names of *branchiopods*, *phyllopa*, &c. have been appropriated to those *entomostraca*, whose feet are at once both organs of locomotion and of respiration. Apus, limnadia, and branchipus, which present this mode of conformation, have often a great number of these *gill-feet*; there are sixty pair at least in apus, eleven in branchipus, and two-and-twenty in limnadia. They are all composed of several thin and soft laminæ, diversely configurated, articulated together, and one, at least, of their edges furnished with numerous hairs. In apus, the first of these feet have four articulated threads, the two upper resembling antennæ; all the others have, underneath, near their base, an ovaliform, vesicular sac, and those of the eleventh pair, support a capsule, with two valves, which incloses the eggs.

These animals, like the insects, have their functions very distinct, and accordingly, like them, they should occupy an elevated rank in the series of beings. Being provided with articulated members, they are evidently in the relation of the locomotive faculty superior to the mollusca, and annelides, as well as to the radiated and infusory animals. All of them possess a nervous system, whose first centres, and first ramifications are easily to be observed. They are scarcely ever destitute of the organ of vision. In some of them the organ of hearing has been discovered, and every thing goes to prove, that the senses of taste and smell exist in the crustacea, as well as in the insects, although the peculiar seats of these senses have not yet been recognized. In these respects it is certain that the crustacea have the priority over very many of the mollusca, over the annelides, and over all the animals which have been placed subsequently to the articulated classes.

They have the greatest resemblance to the arachnida, since they possess, in the same degree of energy, the first two animal functions of which we have just spoken. The arachnida also exhibit an additional relation with the crustacea, namely, that which results from the presence of a heart, or centre of circulation, communicating with the assemblage of vessels, destined to carry the nutritive fluid or lymph, into the various parts of the body.

The organs of *locomotion* in the crustacea, consist, 1st. of passive organs performing the functions of the skeleton, in vertebrated animals, and principally composed of the external skin, which is hardened and divided into segments, or portions of segments more or less complicated, for the body and limbs, but always symmetrical; 2d. of active organs, soft and fibrous, or muscles which are contractile by the agency of the nervous system.

The solid pieces are articulated together, either with, or without motion. Those which are in the first predicament, such as the plates composing the breast-piece of crabs and astaci, are distinct only by straight sutures. Those which are in the second, ordinarily move, one upon the other, by ginglymus, or a hinge-like articulation. The mobile parts of the crustacea are those which we have already described, or which are described in the text, such as antennæ, the parts of the mouth, the peduncles of the eyes, &c.

The muscles in the crustacea, as in the insects, are formed of fibres not adherent to each other, not united by a cellular tissue, and not enveloped by aponeuroses, or tendinous expansions. These muscles are numerous, and always placed under or within the solid parts, and disposed so that each articulation in ginglymus, has its flexor, and its extensor muscle.

It forms no part of our plan to describe minutely the muscles of the crustacea. The reader who is desirous of further information on this subject, must be referred to special works on the anatomy of these animals, and more particularly to the Comparative Anatomy of Cuvier. We shall simply confine ourselves to stating that those of the feet of the brachyuri are very powerful, and placed in sorts of lodges, which are formed under the testa by certain vertical solid partitions which separate the different pieces of the breast-plate, that those of the tail of the macrourous decapods, when arrived at the maximum of development, are very complicated, and form a dorsal mass, which is rather thin, and a ventral mass very thick, both composed of three orders of well-marked fibres, finally, that in certain small entomostraca, particular muscles which do not exist in others are destined to fix the animal to its shell, and to enable it to open or shut the valves of the latter, according to inclination.

As to the function of *sensibility*, the crustacea have a nervous system very similar to that of insects and arachnida. It principally consists in a brain placed in front of, and above the intestinal tube, and in an elongated medulla, composed of a double knotty cord, placed at the lower face of the body, sometimes, as in the macrourous decapods, extending through the entire length of the body; and sometimes, as in the brachyuri, forming towards the middle of its lower face, a medullary circle from which the nerves issue in radiations.

" The brain," says M. Cuvier, in his Comparative Anatomy, " in the animals of these two families, is situated at the anterior extremity of the body. Its mass is more broad than long, and its superior face is divided into two rounded lobes. The middle lobes furnish, each from the anterior edge, an optic nerve, and which proceeds directly into the peduncle of the eye. This nerve is divided into a multitude of threads, each of which is carried to one of the particular eyes which form the assemblage of the composite eyes. From the lower face of the brain, originate four other nerves, which go to the antennæ, and which give out some threads to the neighbouring parts. From its posterior edge, spring two very elongated nervous cords, which comprehend the œsophagus between them, and unite underneath, in an enlargement or medial ganglion, and which give out towards the middle of their length, a thick nerve which repairs to the mandibles and to their muscles. The ganglion inferior to the œsophagus, furnishes the nerves which proceed to the jaws, and to the jaw-feet."

"In the *astaci*, and the other macrourous decapod crustacea, the two cords remain inter-approximated throughout the entire length of the body, and form there five successive ganglia, placed between the articulations of the five pairs of feet. Each foot receives a nerve from the ganglion, which corresponds to it, and this nerve penetrates as far as its extremity. That of the claw, or forceps, is the thickest. The medullary cords, when arrived at the tail, unite so intimately that it is no longer possible to distinguish them. They form then six ganglia, of which the first five furnish each two pairs of nerves. The last produces four, which are distributed in radii to the scaly fins that terminate the tail." In the crabs all the anterior part of the nervous system is the same, but the two œsophageal cords are united a little more backwards than in the astaci. "They are," continues the Baron, "in the middle of the thorax, and there commences a medulla, shaped like an oval ring, grooved in the middle, and eight times larger than the brain. From the circumference of this ring spring the nerves which proceed to the different parts. It furnishes six nerves on each side for the jaws, and the five feet, and there is an odd one which comes from the posterior part, and repairs to the tail. It represents, as it were, the ordinary knotty cord; but its ganglia, if it have any, are not visible. In the paguri, the nervous cord is longitudinal, as in the astaci, but the ganglia of the part corresponding to the tail, are less numerous. In the squillæ, there are ten ganglia, without reckoning the brain. That which is at the union of the two cords which have formed the collar, gives out nerves to the two large claws, and to the three pairs of feet which immediately follow them, and which in these animals are almost ranged on a transverse line; accordingly this ganglion is the largest of all. Each of the three following pairs has its particular ganglion. Then follow six in the length of the tail, which distribute their threads to the thick muscles of this part. The brain gives out immediately four trunks on each side, namely, the optic, those of the antennæ, and the cord which forms the collar, and as the antennæ are placed more behind than the brain, their nerves are directed backwards to repair to them."

" In *oniscus*, the two cords which compose the middle part of the nervous system, are not altogether inter-approximate. They can be perfectly distinguished through their whole extent. There are nine ganglia, without reckoning the brain, but the first two, and the last two, are so much approximated, that they may be reduced to seven."

In the entomostraca, the brain is often the only part which is visible. That of apus is a small transparent globule, situated under the interval of the eyes. The medullary cord is double, and has an enlargement at each of the numerous articulations of the body; but the whole is so slender, and so transparent, that it is difficult to ascertain the true nature of this organ. In daphnia, and branchipus, the brain is apparent, as well as the optic nerves, of which even the divisions can be observed.

Among the crustacea, without doubt, many degrees may be distinguished relative to the perfection of *vision*. Certain of them, as the crabs, and especially the land-crabs, appear to distinguish objects at a sufficiently great distance, while others seem to possess the power of vision only for those which are very close. Finally, some of them are absolutely destitute of eyes altogether.

We have already seen that the eyes of these animals are of two kinds, simple and compound. The former, which are sometimes called stemmata, are too small to be dissected in a satisfactory manner.

As to the compound, or complex eyes, they are better known. We need hardly repeat that they are divided, like those of the insects, into a number of hexagonal facettes, slightly convex, and which form so many small particular corneæ, whose substance is very transparent, and thicker at the middle than at the edges. M. de Blainville informs us that their internal surface in the Palinuri, which we may take as an example, is invested with a sort of black vascular membrane, which must be considered as a true choroid. In fact, it is evidently pierced in the middle of each little cornea, by a small orifice, which should be analogous to the pupil. From this orifice proceeds a small membranaceous production, in the form of an extremely short tube, which is applied on a corresponding nipple of a considerable, subgelatinous translucid mass, and which is indubitably the analogue of the crystalline, or vitreous humour. M. de Blainville has been unable to ascertain whether this mass is divided into as many parts as there are small tubes, by the prolongation of their very transparent envelope. But he has clearly recognized that this mass of vitreous humour, convex on one side, and concave on the other, is applied on a thick ganglion, or enlargement of the extremity of the optic nerve, which ganglion appeared to him also to present at its surface, as many small alveoli, as there are small ocular tubes.

M. Cuvier has not found in the eyes of Astacus all the details of organization which M. de Blainville states that he has observed in those of Palinurus. According to him, the optic nerve traverses the ocular peduncle by a cylindrical canal which occupies its axis. Arrived at the centre of the convexity of the eye, it forms a little button, from which proceed in all directions very fine threads, meeting at some distance, the choroïd membrane, which is nearly concentric to the cornea, and which envelopes this spherical tuft of the extremity of the nerve, like a hood. All the distance between this choroïd, and the cornea, is occupied, as in the insects, by compact whitish threads, which go perpendicularly from one to the other, and whose extremity, which touches the cornea, is equally invested with a black varnish. These threads are the continuation of those produced by the button, which terminates the optic nerve, and which have pierced the choroïd.

The eyes of *oniscus*, *gammarus*, and other isopoda, or amphipoda, have not been examined; but those of certain entomostraca, such as daphnia and branchipus, have. The daphnia, in the first moment of their development, appear to have two distinct eyes, but when they are more aged, these two eyes are confounded into a single one. Swammerdam and Leeuwenhoek regard as double the single eye of these animals in the adult state, while Geoffroy, Degeer, Jurine, and Straus, consider it as simple. "Placed at the most anterior part of the head," says this last naturalist, "this single eye is covered by the general envelope, which communicates no modification to this spot. Its form is that of a sphere, moveable on its centre in all directions. Its surface is furnished with about twenty crystallines (*areoles*, Jur.), perfectly limpid, placed at small distances one from the other, and rising in a hemisphere on a black ground, which forms the mass of the eye, but being isolated, these crystallines present themselves under the form of a pear, being in their natural situation encased by their lesser extremity in the globe of the eye, as far as beyond one-half their height. Their consistence is that of the cornea very much softened, and they are easily crushed under a feeble pressure. Their surface is perfectly smooth, and does not allow us to perceive any indication of adherence. The black part, when separated, looks like a heap of little blackish brown grains in a filamentous substance. All this assemblage is enveloped by a membrane, spheroïdal, perfectly transparent, applied immediately to the crystallines. The terminal ganglion of the optic nerve presents, like that of the decapod crustacea, a bundle of little nerves, the number of which appears equal to that of the crystallines. These crystallines being directed all ways, form by their union a composite eye nearly similar to that of the insects, and appear to form, each with the part of the globe of the eye which is related to it, a simple eye, independent of the others. The general spheroïdal envelope may be considered as being a cornea common to all these simple eyes." M. Straus presumes that each of these simple eyes is provided with a retina, or a choroïd.

This same system of organs is found again in lynceus, polyphemus and branchipus, but in these last the composite eye is pedunculated, and its general cornea is exterior, instead of being enclosed in the head.

The eyes of many entomostraca are moved by four muscles, which carry them in very various directions.

It is certain that many of the crustacea possess the sense of *hearing;* for noise produces an evident impression upon them. Nevertheless, it is probable, that this sense is very much obliterated in most of the entomostraca. It is only in the macrouri that the organ of hearing has been discovered with any approximation to certainty. Situated in the testa, at the lower part of the first articulation of the external antennæ, it consists, in Astacus and Squillæ, of a cavity pierced in the thickness of this testa, and enclosing a little sac or oval vestibule, formed by a slender membrane, of a white colour, and filled with an aqueous fluid, into which a nerve penetrates. Its external orifice is applied against a round, thick, white membrane, which closes an aperture of the same form, pierced at the posterior part by a tubercle of the crustaceous envelope, and which is a sort of tympanum.

In the crabs, and other brachyurous crustacea, we find at the base of the external antennæ, the same cavity of the testa; but its external projection is much less if it exist. When this projection is found, it is altogether strong, and has no posterior aperture provided with a membrane analogous to the tympanum.

The sense of *smell*, which seems to be very perfect in the decapod crustacea, appears also to be sufficiently delicate in many isopods. Its seat is not better known in these animals than in many insects; and from the same reasons, which we have already noticed in treating of the last mentioned class, it has been supposed to reside in the antennæ. It has been remarked, that the first pair of nerves proceed into these appendages in the same manner as the first pair of nerves in vertebrated animals is carried into those organs which are so indubitably known to be olfactory; the analogy of function has therefore been inferred from analogy of position.

This question, nevertheless, remains totally unresolved, for if the antennæ be the organs of smell in the insects and crustacea, where are those of the arachnida, which have no antennæ, and which, nevertheless, exhibit an equal perception of odorant emanations?

M. Dumeril, in adopting the conjecture of Baster, has endeavoured to demonstrate that the seat of smell in insects should be found in the points through which the air necessary for respiration was introduced into the body, that is to say, towards the entrance of the stigmata. But where should we place this seat in the crustacea which respire by gills?

M. Cuvier, who, in his Lessons on Comparative Anatomy,

appears to approve the system of Baster and Dumeril respecting the position of the organs of smell in insects, says nothing particular with regard to the crustacea. M. de Blainville, in his last work, adopts as most probable, the opinion that the antennæ are the seat of smell in all the articulated animals, because it is in accordance with many considerations *à priori.* He thinks that in the invertebrated animals, the apparatus of olfaction presents this difference to what takes place in vertebrated animals: that the skin, more or less modified, no longer lines a cavity or pouch, lodged in the head; but that it clothes the extremity of the appendages which may project more or less in front of the animal, such as the antennæ or tentacula.

Of the four antennæ, which exist in the crustacea, M. de Blainville seems to think that the seat of smell resides rather in the two intermediate, than in the two exterior.

There can be no sort of doubt that the sense of *taste* exists in the crustacea, and it appears probable that its seat is placed at the commencement of the intestinal canal, for we find that some of the nervous threads which are furnished by the two cords surrounding the œsophagus, proceed to this part. Nevertheless, one might also suppose it to be in the flagelliform palpi, which are annexed to the back of the jaw-feet, just as it was, for a long time, admitted to exist in the maxillary and labial palpi of the insects; but these palpi of the crustacea are by no means conformed for the perception of savours, and are not even organs of tact. They can be considered as nothing but true appendages of locomotion, a little modified, and which can serve at most only to direct the prey towards the jaws.

Touch would appear to be an extremely obtuse sense in the majority of the animals of this class. The very name of crustacea sufficiently indicates that their skin, the ordinary seat of this sense, is hardened, and changed into a truly solid crust. None of their appendages, that is to say the palpi, the antennæ, or the feet, appear to be modified for the exercise of tact.

Nevertheless, we may, in this respect, admit of some shades between the divers crustacea in proportion to the greater or less solidity of their testa. Thus the brachyurous decapods, and some of the macrouri, have their envelope generally thicker, more calcareous, and more solid than all the others. After them, come certain macrourous decapods, as Palæmon, Peneus, &c. and the stomapods, whose testa is flexible, corneous, semi-transparent; and, finally, the entomostraca of the genera Apus and Branchipus, the softest of all these animals, which have a skin so fine that in all parts of the body it may prove a sufficiently delicate organ of tact. The male branchipoda have at the head two soft organs capable of being rolled into a spiral form, like a sort of proboscis, and which may possibly be endued with a great degree of sensibility.

At a certain period of the year, indeed, crustacea, even the hardest, lose their old envelope, and are clothed with a new testa, extremely thin and very flexible. Then their sensibility is very great, and for fear of being wounded by the contact of external bodies, they remain concealed in the hollows of the rocks, until their new skin has acquired a sufficient consistence to protect them against accidents of this kind.

The skin of the crustacea is composed of many superposed layers, as has been ascertained by M. de Blainville. In Palinurus, we may distinguish, 1st. a first internal stratum, more fibrous than the others, translucid, evidently living, and forming the interior lamina of the parts which do not become crustaceous; 2d. a second stratum, more cartilaginous, of an opaline colour, a little thicker, and still appertaining to the membranaceous parts; 3d. a third stratum, still thicker, of a less compact texture, in which the calcareous molecules are deposited, which give solidity to the testa; 4th. a last stratum, altogether external, composed of colouring matter, or *pigmentum*, and an epidermic layer.

The name of *moulting* has been given to the renovation of the testa in crustacea. These moultings are more or less frequent according to the age of the animals, and the more or less rapid degrees in which their growth is developed.

In the decapod crustacea, the moulting takes place every year, towards the middle of the spring. Reaumur has studied that of the river crawfish, or astaci, and to him we are indebted for every thing we know concerning the mode in which this operation takes place. When these crustacea are desirous of changing skin, they rub their feet one against the other, and put themselves into very considerable motion. They afterwards swell out their body in a very sensible degree, and the first segment of the tail appears more separated than usual from the posterior edge of the carapace. The membrane which unites them breaks, and the body, with its new skin, appears. After a short term of repose, they begin to agitate themselves afresh, inflate and raise themselves more than they did at first; the carapace rises, is detached, and remains adherent in no place except towards the mouth. Soon after, the eyes are disengaged from their old skin, which remains fixed to the former testa, then the antennæ, as well as the parts of the mouth, and finally the carapace, are almost totally separated. At last, after divers reiterated movements, the astaci strip their claws and feet in an indeterminate order. Then they quit their carapace altogether, and suddenly extending their tail, they disengage themselves entirely from their old envelope.

After the moulting, the astaci are very soft, and remain in a state of exhaustion which lasts many days, until the most external part of the dermis is filled with calcareous molecules, which re-establish its solidity.

In the entomostraca, whose growth is much more rapid than that of the crustacea properly so called, and in which the duration of life is extremely short, the moultings succeed each other quickly. Thus, M. de Jurine, who observed the daphnia from the moment of their birth to that of their first ovideposition, counted, in an interval of seventeen days, eight moultings, which took place pretty nearly within two days' interval of each other. He did not pursue his observations on those changes of skin beyond this point, as they succeed each other in the same manner in summer up to the death of the animal. In winter the moultings are very much retarded, and it is not uncommon for eight or ten days to elapse between them.

In cypris, apus, branchipus, lynceus, limnadia, and polyphemus, the moultings are also very frequent.

In all the crustacea and entomostraca, it is remarked that the old skin is composed of all the principal or accessory parts which belong to the animal, and that often each spine or each hair there is hollow, and covers another spine or another hair. The chemical analysis of the old testa demonstrates that it is formed of carbonate and phosphate of lime, united to gelatine in various proportions, which are generally in relation to the solidity of the testa.

We proceed to a few observations on the function of NUTRITION in the crustacea. Most of these animals feed upon solid substances, and generally on animal matters more or less in a state of decomposition. There are, however, some among them that live on fluids, which they suck from the animals to which they are parasitically attached.

The former are all provided with a mouth more or less complicated, and composite, as we have already seen, and furnished with a variable number of organs for mastication or trituration. The latter have those parts combined, so as to form a sort of sucker or siphon. But as the buccal organs have been already described with some detail both here and in the text, we shall confine ourselves now to those of nutrition, properly speaking.

The intestinal canal of the crustacea is generally short and straight, and often in the course of its passage presents a remarkable dilatation, which is the stomach ; but sometimes the stomach is apparent only by a very slight enlargement of this canal. The œsophagus is short.

The stomach, as we have said, varies : that of the decapods, both brachyurous and macrourous, placed above, and a little in front of the mouth, occupies a very considerable space under the anterior portion of the carapace. It is very capacious, membranaceous, and its parietes are supported by complicated cartilaginous arches, which keep them apart, even when the stomach contains no food. Its figure is that of a trapezium, the angles of which are rounded in the form of lobes, and of which the two largest are anterior. In the middle of the upper paries, we are told by M. Cuvier, that there is a transverse cartilaginous ridge, which has within a first tooth, or oblong osseous plate, fixed to its external face, directed towards the pylorus, and terminating behind in a tubercle. On their posterior extremity is articulated a second ridge, directed backwards, bifurcated like a Y, and on each of the lateral branches of this another is articulated, which returns in front and externally, to arrive at the lateral extremity of the first ridge. It is on these two lateral ridges that the largest pyloric teeth are inserted : they are solid, oblong, and have a flat crown furrowed crosswise, and its inequalities and furrows vary according to the species. From the point of union of the transverse and lateral ridge on each side proceeds another lateral ridge, which goes lower than the first, and bears at its extremity a lateral tooth smaller than the preceding, placed a little in front of and below its anterior extremity, and bristling with three or four small sharp and curved points. These two small teeth, according to the Baron, take hold of the food which comes from the mouth, and carry it between the two large teeth with flat crowns, which grind it between them and against the first plate which we have mentioned. Near the pylorus, a fleshy and oval projection is found behind the large teeth, in the interval which separates them; and the pylorus itself is divided into two semi-canals by a middle ridge. The stomach has its peculiar muscles, and also extrinsic muscles, which are attached to the parts in the neighbourhood of the thorax, and which serve, with the first, to move the apparatus of five teeth that furnish the pylorus.

When the astaci, or crawfish, are ready to moult, we find, applied within the stomach, and on each side, a calcareous, round, flatted, white stone, with concentric strata. These stones appear destined to furnish the matter, or a part of the calcareous matter, of the new testa; for they diminish in size from the day after the moulting, and become totally dissolved in proportion as the new envelope acquires consistence. There is reason to believe that these bodies, which are vulgarly designated under the name of *crabs'-eyes*, and to which certain imaginary properties have been attributed, are found in all the crustacea properly so called, and especially in those which possess a very solid testa.

In the squillæ the stomach is small: its form is that of a triangular prism, membranaceous, and furnished on each side of its posterior extremity with a range of small pointed teeth. The onisci have the anterior part of their canal merely a little more bulky than the rest, and this increase of volume represents the stomach.

In Daphnia, the portion of the intestinal canal to which the name of stomach may be given, is in like manner merely larger than the rest of the tube. Its pylorus is not distinct, and the cardia alone is very apparent by the difference of volume in the œsophagus. Two blind vessels, rather short and thick, which lead to the stomach, have been considered by some naturalists as cœca, and by others as replacing the liver.

From the end of the stomach, the intestinal canal proceeds pretty directly to the anus. Its diameter is then nearly equal throughout, and sometimes very inconsiderable. Sometimes, towards the middle, it presents a sort of swelling within, which is a strong valvule, and from it proceeds a very long cœcum, as in the decapods. Sometimes, as in the entomostraca, there is no trace of these parts. Finally, its termination is always situated on the lower face of the last segment of the tail or abdomen.

The liver in the crabs, astaci, and other decapod crustacea, is a very voluminous organ, especially at certain periods of the year. It is situated at the lower face of the body, that is to say, underneath the stomach, heart, and preparatory organs of generation; and in the paguri it fills up more than the entire base of the tail. Its general form is indeterminate, for it is not comprised in a proper membranaceous envelope, such as the conglomerate glands of vertebrated animals possess. It is composed of an innumerable multitude of small secretory sacs, intermingled together, of a yellow colour, whose parietes appear spongy, and which contain a brown and bitter humour, which is the bile. Their communication with the intestinal canal by hepatic ducts has not been yet pointed out; but there is reason to believe that it exists not far from the stomach, if it be not in the stomach itself.

In the Squillæ, the liver, solid, and very similar to a conglomerate gland, is divided into lobes, and these lobes are ranged on the two sides of the whole length of the intestinal canal. In the limulæ, the liver pours the bile into the intestine, through two canals on each side. In oniscus we merely remark, close by the œsophagus, four voluminous and blind vessels, floating, undulated, of a yellow colour, altogether similar to the vessels considered as hepatic in the insects.

Finally, in the entomostraca, we can admit as organs analogous to the liver, only the two small vessels which lead to the anterior part of the stomach in daphnis.

We are not acquainted with any organ analogous to the pancreas in crustacea. It nevertheless is possible that this viscus may be replaced by the cœcum, of which we have spoken above, which admits no aliments in digestion into its interior, and which may be a gland intended to pour a peculiar fluid into the intestinal canal.

There is no peritoneum; the stomach, as we have seen, is supported by particular muscles. But the intestinal canal is supported only by the vessels, and by the compression of the surrounding parts.

The heart, in the decapod crustacea, is placed pretty nearly towards the middle of the body properly so called, in the rear of the stomach, and of a portion of the preparatory organs of generation, and between the gills. It is lodged in a sort of cavity, surrounded by solid partitions, to which are attached the muscles of the base of the feet, the assemblage of which forms two sorts of buttresses on each side, which sustain the upper part of the testa in the points where we see externally two small longitudinal impressions upon it. Its form is oval, a little depressed, its colour is whitish, and its parietes, which are semi-transparent, are yet tolerably thick. Its movements of dilatation and contraction are very apparent, and in general rather slow. It has no auricles, and no valvules are found in its interior.

This heart, by its contractions, distributes the lymph to the gills, by the assistance of as many vessels as there are packets of branchial plates, and these vessels all proceed from one or two principal trunks. The lymph which has received the influence of respiration, issues from the gills through an equal number of vessels, which proceed to unite in a ventral canal, situated underneath the intestine, and this canal distributes it to the entire body, from whence it returns to the heart, through a thick *vena cava.*

Thus the circulation is double, as the heart must be considered as the pulmonary ventricle, and the ventral canal as the aortic ventricle.

In the squillæ, the heart is elongated into a thick fibrous vessel, passing along the back and the upper part of the tail.

That of the small entomostraca, such as daphnis, lynceus, limnadia, is small, globular, situated near the back, above the intestinal canal, and its contractions are very perceptible.

In the limulæ it is a thick vessel, furnished within with fleshy columns, running along the back, and giving out, like that of the squillæ, branches from both its sides.

Finally, in branchipes, we see from the head as far as the last articulation but one of the tail, a brilliant organ, perfectly diaphanous, which is composed of a series of utricles or little cells, corresponding in number to that of the rings of the body (eighteen or nineteen), which contract and dilate successively with considerable quickness, with movements which may be compared to those of the systole and diastole. This organ is very analogous to the dorsal vessel in insects.

Respiration is a very active function in the crustacea. The respiratory organs are voluminous, and of two kinds, gills or branchial laminæ, and kinds of air sacs.

The gills are sometimes concealed, sometimes visible; they are frequently situated on the sides of the body, but often also on the posterior extremity of its inferior face. They are almost always annexed to the base of the ambulatory feet, or to that of the most exterior parts of the mouth; but also, in many cases, they of themselves alone constitute feet, which serve at the same time for locomotion and for respiration.

In the brachyurous decapod crustacea, they are placed at

the root of the feet, under the lateral and inferior jut of the carapace. They repose upon two solid, oblique tables, of the interior of the body, which serve to close the lodges above, where the first muscles of the feet are fixed. The water can penetrate to them through a cleft which is behind this edge of the carapace, and issue out through an anterior aperture situated near the mouth. In one genus, that of Dorippus, this anterior aperture, pierced in the body itself of the carapace, is very remarkable: their gills have each the form of a triangular pyramid, elongated, attached only by its base, and with the point directed upwards and inwards. They are composed of a stem of a cartilaginous nature, supporting numerous soft and membranaceous laminæ, separated into two longitudinal masses by a medial furrow, and piled one upon another, perpendicularly to the axis of the stem which sustains them. In the furrow are found two thick vessels, one venous, the other arterial, which distribute their branches *ad infinitum* over the surface of the membranaceous and double laminæ of the gills, so that the lymph there receives the impression of the respirable air mixed in the water.

These gills are seven in number on each side, five depending on the feet properly so called, and two on the first and second jaw-feet. They are continually rubbed by two long thin cartilaginous and flexible laminæ, attached near the base of the jaws, one above, the other underneath these organs; and the function of which appears to be, as M. Cuvier presumes, to express the water, which has served for respiration, from the intervals of the leaflets of the gills, so as to allow fresh supplies to enter.

The gills of the decapod macrourous crustacea differ from those of the brachyuri, in that the leaflets or respiratory laminæ are replaced by cylindrical filaments, disposed in tufts, which have each a vein and an artery. They are also much more numerous, being twenty-two on each side, divided into five principal groups, of four each, corresponding to the base of the first four feet, and of the external jaw-feet. Moreover, one isolated branch is placed altogether in front, and fixed to the second jaw-foot, while another equally isolated corresponds to the last or fifth foot. These gills are compressed by elongated, cartilaginous laminæ, mobile, and attached one to the basis of each foot, for the purpose of expressing the water. These laminæ separate the groups of gills, and in each group there is one of these gills, the most external, which is fixed to the basis of the lamina, and like it, moveable; the others, being adherent to the body itself, have no proper movement. Two similar laminæ, without gills at their base, are attached to the most anterior jaw-feet, and to the last jaw properly so called.

The testa of the crustacea also presents an anterior aperture below its edge and on each side of the mouth, for the issue of the water.

In the Squillæ the gills are visible, and can serve for locomotion: they are situated under the body and behind, to the number of five pairs, annexed to some short fins, divided into two lobes, and formed of membranaceous laminæ, ciliated on their edges. It is at the root of the external lobe of these fins, and at its internal edge, that the gill, which is very complicated, and which at first sight resembles a thick pencil, is attached M. Cuvier, who was the first to observe this organ, describes it thus: "The gill is at first formed of a conical peduncle, composed of two vessels. There proceeds from it a range of cylindrical tubes, which go on diminishing from the base of this peduncle to its point, similar to the arrangement of the pipes of an organ. Each of these is curved, and forms a long conical and flexible tail, which itself bears a very numerous range of long floating filaments, like lashes. Each filament contains two vessels, each tail and each tube two likewise, as does the general peduncle.

The amphipod crustacea *(Gammarus)* are provided with vesicular appendages, placed at the interior base of the feet, with the exception of those of the anterior pair, and which have been considered as gills. Among the isopods, some, such as *Leptomera* (which is placed by M. Latreille in his fourth order, *Læmodipoda*) and some others, have, as apparent or presumed, respiratory organs, only some very soft vesicular bodies, sometimes six in number, and situated at each side, on the second, third, and fourth segments, at the external base of the feet, which are attached there; sometimes there are but four of them, which are annexed to as many true or false feet of the second and third segment, or to their place, if these segments are absolutely destitute of locomotive organs. Others, such as *typhis*, &c. have gills under the tail, always naked, and in the form of stems, more or less complicated. Others, in fine, such as Cymothoë, Asellus, Oniscus, &c. have gills under the tail, either free, and in the form of vascular scales or membranaceous pouches, sometimes naked, sometimes covered by plates; or enclosed in scales. Among these last are those crustacea which can only respire the atmospheric air directly, and not through the watery medium.

The sub-class of entomostraca presents very numerous variations in the respiratory organs. The limulæ have, under the second part of their testa, five large transverse laminæ or jaw-feet united at their base, and bearing at their posterior face a great number of fine leaflets, piled up, which are the gills. Similar laminæ are remarked under the second part of the body of the caligi, and probably likewise cover some branchial leaflets. In daphnis, the ten feet are composed of many shortened articulations, and the last eight are provided, among these articulations, with a membranaceous lamina, ciliated on its edges, and which serves for respiration. It was for a long time believed, that in Cypris the organs of this function were the threads which terminate the antenna and feet; but M. Straus has proved that they are in the form of pectinated laminæ, annexed to the base of the two pairs of jaws. Finally, in apus, branchipes, and limnadia, they consist, in many species, of membranaceous leaflets, the assemblage of which composes the natatory feet of these animals. The name of branchiopods, which has been particularly applied to them, is derived from the alliance which has been remarked between the organs of motion and the organs of respiration.

The function of generation in most crustacea is well known, as is also the distinction of the sexes; but in some of them, naturalists have been as yet unable to distinguish the male sex. All the individuals among these last lay eggs, from which animals similar to themselves proceed, apparently without any previous sexual intercourse.

The decapods, stomapods, isopods, and amphipods, are those in which the sexes are well known, and in which, in consequence of their size, the organs have been better studied.

In the crabs, astaci, and the crustacea of the two families in which these animals are ranged, the exterior organs of generation are very distinguishable; and in the interior of the body the preparatory organs of this function are easily discovered. These last are visible when the carapace is raised up on the sides and in front of the heart, and they are more especially apparent at the period of coupling or that of laying. The sexual organs are double.

Certain individuals of the genus squillæ, presumed to be males, have near the internal origin of each of their last ambulatory feet, a small, crustaceous, filiform appendage, arched and not articulated, which is presumed to be a dependence of the sexual organ of the male.

The amphipods, whose organs of generation are not well known, couple after the manner of insects. Some isopods, in which the sexual organs of the male have been observed, have them double, and placed under the first leaflets of the tail, where they are indicated by threads and hooks.

The entomostraca are the only animals of this class, among which some are found whose sexes are not distinct. Nevertheless, they are still separate in the limulæ, for a considerable portion of the testa of these animals is filled in some of them by ovaries, and in others by organs which may be compared to the vasa deferentia, &c. of crabs and astaci.

The caligi have at the posterior part of their body two cylindrical threads, more or less long, divided into a multitude of small articulations, and which have been considered as external ovaries (but sometimes also as respiratory organs.) The branchipes have the sexes separate, and the males are easily distinguished from the females, by the claws in the form of forceps, and the prehensile tentacula with which their head is furnished.

All the individuals of the genus apus are conformed in a similar manner, and appear to be females, if they are not hermaphrodites. They have never been observed to couple. All have on each of the feet of the eleventh pair a capsule with two valves, enclosing the eggs, which are of a fine red. The limnadiæ appear to exhibit the same mode of generation.

In the genus daphnis there are both females and males; but the latter are infinitely more rare, and appear to exist like the males of the aphides, only at a certain period of the year. A single act of sexual intercourse suffices in these entomostraca, as in the aphides, to beget and impregnate seven or eight generations of females, which are developed in succession. The organ of generation in the female consists of two ovaries, whose form is that of vessels, and which extend on each side of the abdomen, from the first segment to the sixth, where they open separately on the back of the animal into an empty space left by the valves of the shell, which has been considered as a matrix, and the function of which is to preserve the eggs after the laying, until the entire development of the young. The male organs are not known. The males, however, are sufficiently distinguished by their large antennæ, and generation appears to take place something after the manner of that of the batracian reptiles.

The males of cypris are not known: all those which M. Straus submitted to the microscope were females. Their ovaries are very considerable, in the form of two thick, simple, conical vessels, terminating in a cul-de-sac, at their extremity, placed externally on the sides of the posterior part of the body, and opening one at the side of the other in the anterior part of the abdomen, where they communicate with the canal formed by the tail. In the present state of our knowledge, it is impossible to affirm whether these animals are hermaphrodites, and necessitated to a reciprocal fecundation, or whether the males are only to be found at a certain period of the year. If, however, these animals be hermaphrodites, M. Straus is of opinion that we might consider in them, as male preparatory organs, two very short blind vessels, filled with a gelatinous substance, and which are situated above the mandibles; but, on the other hand, these same vessels might be taken for salivary glands, if they communicated with the œsophagus, as M. Straus suspects.

Finally, in cyclops, the sexes are separate; and we see at the time of laying, in the females, two vesicular sacs, or external ovaries, situated at the base of the tail, and which are in all respects analogous to that which is found, but single, in the females of branchipes. In the interior of the body, on each side of the intestinal canal, is an ovary, in the form of a vessel, similar to those of daphnis, and which communicates with the external ovaries. In the males, the second ring of the tail bears underneath two oval bodies, sufficiently distant from each other, and which appear to give rise to two small organs, which the elder M. Jurine presumes to be those of generation. Each of them is composed of three rings, which diminish in size. The second furnishes two or three threads; and the third is terminated in a point.

With respect to the *products* of generation, the crustacea are either oviparous or ovo-viviparous. The eggs which they lay have a corneous envelope, solid, and usually transparent, through which the gum may sometimes be perceived. These eggs, developed in a liquid conduit, which in its bottom part receives the name of ovary, and in its most external portion that of oviduct, are small, often very numerous, of a spherical or oval form, and present, according to the species, very various colours.

After they issue from the body, they are usually carried for a longer or a shorter time by the females, sometimes under the tail, attached by filaments, resembling, from the desiccation of the viscous matter which invests them, two peculiar appendages, which have received the name of false feet, as is the case with crabs and astaci; sometimes between leaflets, at the base of which the gills are fixed, as in certain isopods; and sometimes, in fine, in an exterior membranaceous envelope, forming an ovary or external matrix, as in cyclops and branchipes, or in a dorsal cavity, as in daphnis and lynceus.

In certain genera they disclose the young while yet contained in the body of the animal, or in the dorsal cavity which we have just mentioned, as is remarked in Argulus and Daphnis, which, in consequence of this phenomenon, are distinguished from the other crustacea, as being ovo-viviparous.

The little ones which issue from the eggs are, in the generality of the crustacea, altogether similar to their parents; yet sometimes they differ so much from them, that they have at first been considered to belong to particular genera, as has been observed in the case of cyclops, whose young, at different ages, have been named *anyone* and *nauplia,* and also in argulus and branchipes.

These eggs, in one and the same species, are sometimes of two sorts, according to the seasons. Thus the ordinary eggs of daphnis are abundant and naked, while those which are to pass the winter at the bottom of the mud are expelled, only two in number, each enclosed in a capsule, with a double envelope, and surrounded, moreover, with the membranaceous lining of the dorsal cavity, where they were first deposited. The parietes of this cavity are then thickened and become opake; and it has appeared to some observers to be affected with a peculiar malady, to which they have given the name of *ephippium.*

The development of the eggs is more or less prompt, according to the duration of the life of the species to which they appertain, and the rapidity of their propagation. We have just seen, that in certain genera they are disclosed in the body of the mother. In others, they appear to grow thicker after being laid, before they give birth to the young, and remain in this state many days. There are some, in fine, such as those of apus, which seem capable of being preserved dry for many years, without the gum which they enclose undergoing any alteration; for, without this supposition, it would be impossible, unless we had recourse to the theory of spontaneous generation, to explain the sudden appearance, and in myriads, after heavy rains, of those aquatic flabby crustacea, which are destitute of all means of transportation, in places where, in the memory of man, they have never been remarked before.

One of the most remarkable characteristics of the crustacea is the *reproduction* of the lost members. The *astaci* and crabs are subject to lose their feet, which are detached with the greatest facility in the joints of the articulations. A little after a member is plucked away, a reddish pellicle is formed upon the flesh, which is left exposed. Some days later, this pellicle assumes a surface a little convex, it is elongated, becomes conical, still increases in size, and being cleft, leaves to view a soft body, which is composed exactly of the parts which are wanting to the member, but smaller in proportion than those which remain. These new parts soon acquire consistence; but it is not until after many moultings that they acquire the volume of the old ones. It has been remarked, that this reproduction does not take place when the member is broken between two articulations; and it has even been observed, that when this is the case, the crustacea themselves tear out the remaining stump, so as to have the rupture in the joint, where the new part can then be formed.

The animals of this class compose but a limited assemblage of species, but the individuals referrible to each of these species are very numerous. They are met with in all latitudes, but more abundantly, however, in warm and temperate climates than in the frigid zone; and their species are not indifferently proper to all countries. Thus the amphipod and isopod crustacea seem to be more peculiarly the inhabitants of the colder regions of the globe, while the decapods are more common between the tropics. In the medial zones we observe a mean proportion of the species of these different orders.

Certain genera, such as the ocypods, gecarcinus, gelasimus, uca, hippa, limulus, grapsus, &c. are found in more southern latitudes than the others, pretty nearly under the same parallels, on the American, Asiatic, and African shores; others, on the contrary, such as the crabs proper, portunus, and inachus, occupy more space, and are extended even to the polar circles.

As to the small entomostraca, they have as yet been observed only in temperate countries; but there is reason to believe that, from the degree of temperature necessary to their existence, they abound in the fresh waters of warmer climates; while, on the contrary, they are very rare, if even they exist at all, in the very northern regions.

The local habitations of the crustacea, considered generally, are very various: the most numerous of these animals are aquatic and marine, and some few genera, such as oniscus, armadillo, &c. are alone truly terrestrial. Certain brachyurous decapods penetrate very far into the land, but are forced to re-enter the sea at the period of coupling and laying. Some others, such as thelphusa, though they have forms very analogous to those of the marine crabs, do not quit the fresh water, and all the entomostraca, except limulus, caligus, and some animals approximating to the last, are also inhabitants of fresh water only.

Among the marine species, the majority do not quit the shores, while others live in the high seas, and have nothing to repose on but those floating banks of sea-weed so abundant between the tropics. Moreover, the littoral crustacea do not all sojourn in similar localities: some, as dorippus, and certain inachi, reside at depths of from two to four hundred feet, while others continually sport on the surface of the waters, and pass one half of their existence on the shore, which is continually washed by the waves. Many species are met with only in rocky situations, abounding in madrepores, and of difficult access; while others seek the bottom of the fine and shifting sand, into which they sink themselves.

Among those which come to land and make a tolerably long sojourn there, many brachyurous crustacea (the ocypods) excavate tolerably deep burrows, at the entrance of which they usually remain on guard. Some (the ranini) are said even to climb on elevated places, and often to mount on the roofs of the Indian cabins

The onisci, aselli, and ligiæ, are fond of humidity and shade, and most generally fix themselves under stones or in the breaches of rocks.

The macrourous decapods, such as craw-fish, lobsters, palinurus, palæmon, or carides, as well as the entomostraca, are the only crustacea which never come to land.

We may observe, that most of the animals of this class walk as well as swim, their feet being adapted to both these modes of motion.

The brachyurous decapods are evidently the crustacea most eminently adapted for walking. In such of them as run best, the eight posterior feet alone are employed, and they are all terminated by strong and pointed claws. They walk with the same facility forward and backward, on one side or the other, or in all oblique directions possible; they are seen to climb very inclined planes, and even a perpendicular surface, with the greatest celerity, providing that these planes be not altogether smooth. Many, such as the ocypods and gecarcini, are famous for the rapidity of their course, which is so great, that we are assured that a man would be unable to overtake them.

Many brachyurous decapods walk much less than others, and are more decidedly aquatic. These, provided with members, whose articulations, flatted and ciliated on the edges, are transformed into true oars, can execute in the water all the movements which the others perform on land, and in directions equally varied. Such are the portuni, the podaphthalmi, &c.

As for the macrouri, such as *astacus* and palæmon, if their feet serve for walking, it is only at the bottom of the water. Their swimming, which is usually backward, is performed by the motions of their strong tail, whose extremity, folded underneath, is widened by laminæ, which can be separated like a fan. Some, such as crangon, when swimming, have the back underneath, and the belly upwards.

Many amphipods swim by means of the contractions of their tail, aided by the movements of their feet, and some are forced, in consequence of the extreme compression of their body, and the very strong curvature of the tail, to remain continually couched on one side.

Although the squillæ possess feet adapted for motion, they appear to make no more use of them than the macrourous crustacea do of their's, and their natation appears to be effectuated principally by the assistance of the ten branchial feet, which are placed under a tail less robust and less curved than that of the macrouri, but equally terminated by flabelliform natatory laminæ.

In the sub-class of the entomostraca, all the animals which have numerous soft feet, and furnished with gills, as apus, limnadia, and branchipes, advance only by the action of these limbs, whose motions are soft and undulating. Daphnis and lynceus appear to jump in the water, which has caused the first to be sometimes termed *aquatic fleas*, because their swimming takes place by violent movements of their ramous antennæ, which are frequently repeated, leaving but few intervals of complete repose. In cypris, it is the feet, and especially the hinder feet, by which the animal advances.

Among the amphipods, some can leap with considerable vigour when they are on land, making use of their tail, folded underneath, as a spring.

The *instinct*, or mental faculty, of the crustacea in general, when compared with that of other classes, especially the insects, appears to be but very moderately developed. The crabs and their allied genera, seem to possess it in a higher degree than the rest. Those animals, in fact, exhibit great cunning, especially in avoiding their enemies: they are then observed to run along the ground with great rapidity, choosing for a retreat the places most difficult of access to their pursuers. Many among them, whose carapace is particularly tender, such as the pinnotheres, make their habitual residence in the valves of certain mollusca, such as the muscles and pinna marina; and others, which have a soft and vulnerable abdomen, place it in the cavities of deserted univalve shells, or in the hollows of rocks, so as to preserve it. These change their dwelling at certain periods as their body grows, and make choice of a more commodious retreat. Some macrourous crustacea (the thalassini) sink themselves into the sand or mud to escape from their enemies.

The cymothoës and isopods approximating to them, the caligi and bopyri, which live as parasites on the bodies of the crustacea, of fishes, or even under the testa of other crustacea, possess an instinctive quality which causes them to distinguish the beings on which they can fix, and the parts of those beings where they ought to place themselves in preference, for the purpose of finding the nutriment which is suitable to them.

The land-crabs have a constant habit of uniting, at a certain period of the year, in vast numbers, and of walking, by the most direct course, towards the sea, over almost every obstacle which may occur in their passage. After the deposition of their eggs, they re-assemble and return to their former abode.

Some species of different orders always live in numerous societies, and we may more particularly mention crangon, talitrus, and most of the small entomostraca, especially daphnis, whose colour sometimes gives to the water a tolerably deep red tint.

The crabs are courageous, and when their retreat is cut off, they advance their claws fiercely, and endeavour to pinch with their fingers, which they do very strongly, in proportion to their size. Some of them, closing these fingers with force and rapidity, produce a considerable noise.

The crustacea in general, as we have already remarked, live on animal matters, especially on such as are in a state of decomposition. The crabs, craw-fish, and prawns, come from all quarters to the dead bodies which float upon the water, or which are cast out by the sea upon the shore; and there is every reason to suspect that they are brought thither by the sense of smell.

It also appears that certain isopods live on the substance of the gelatinous animals, which compose the sponges; at least it is always on these marine bodies that some of them are to be found in very large numbers. Some others, such as the aselli, &c. are accused of destroying the nets of the fishermen, by gnawing the ligneous fibres of the cordages with which they are formed. The onisci, or wood-lice, live, as is well known, on rotten vegetable substances.

Finally, there is no doubt that the smallest entomostraca eat, along with the little animalculæ which abound in fresh waters, the remains of vegetables equally microscopic, for their alimentary canal, visible at the middle of their body in consequence of its transparence, is often of a fine green colour.

Among the carnivorous crustacea, there are some which seek a living prey, and fight to procure it. In these combats they often lose their forceps, but they shoot out again after a little time.

Those of the crustacea whose sexes are separate, never exhibit unions by pairs, which are observed in animals of the first two classes, namely, the mammifera and birds, and which are also to be found again among the insects. In general there is no relation between the sexes but at the period of reproduction.

The females, as we have said, preserve their eggs, after

deposition, for a greater or less time. In some they are fixed to the false feet by means of filaments, which result from the solidification of the mucus that surrounded them at the moment of their issuing forth; in others, they are placed in external membranaceous sacs, or in a dorsal cavity.

In most species of crustacea, when the little ones are disclosed, they remain some days near the mother, and place themselves under her tail, as has been observed in some crabs, and in the river craw-fish, or between the leaflets of the gills, as has been remarked in regard to the onisci.

Most crustacea are used by us as food. The larger and middle sized species, of which the individuals are numerous, are those which are in most request. Their flesh is nourishing, but difficult of digestion, and much of it cannot be eaten with impunity.

The decapod crustacea are the only animals of this class which are eaten in Europe. Among the brachyuri, the most esteemed are *cancer pagurus*, *cancer puber*, and the *maia squinado*. As to the *C. mœnas*, it is not so much in request, and its most frequent use is as a bait to catch fish and other crustacea. Among the macrouri, the palinurus and the lobster hold the first rank in consequence of their size, and then come *palæmon penæus*, many species of *nika*, the river craw-fish, and crangon. These last, which are eaten very much, though more on the continent than in England, are also employed as bait.

Many of these crustacea, such as *penæus* and *Palæmon*, are salted in some parts of the coast of the Mediterranean, and sent into the East, where the Greeks make an abundant use of them, especially in the season of Lent.

Formerly the trade in the stones which are found in the stomach of craw-fish, vulgarly called *crabs' eyes*, was sufficiently productive, when these bodies were made use of in medicine as absorbents, and it was particularly from Hungary, where these crustacea are very common, that they were derived.

It is more especially to the *decapods* that the name of crustacea has been given. Of all the animals of this class, the decapods are the most remarkable for their size, the complication of their organs, the solidity of their teguments, and their longevity. They all possess a very carnivorous instinct. Some travellers do, nevertheless, inform us, that certain species feed on fruits; but this appears to be doubtful. Some pass a very considerable portion of their lives out of the water, and repair thither only at the period of their amours, and for the purpose of laying their eggs. They go in very numerous bands, always pursuing the shortest and straightest direction, without embarrassing themselves with any obstacles in the way, and they return in the same manner, after having fulfilled the decrees of nature. The others scarcely ever, or at all events, but for a very short time, abandon the element in which they were born. Some among them live in the fresh waters; but the majority inhabit those which are salt or brackish, most frequently near the shore, at depths and localities which vary according to their means of subsistence, and the resources with which nature has provided them, to escape from the dangers by which they may be menaced. Thus the pinnotheres withdraw themselves into bivalve shells. The dromiæ form with alcyones a sort of mantle, which completely covers them. Many apply, upon their back, the valve of a shell, and on that account, these crustacea, like dorippus, have their four posterior feet recurved upwards, and adapted for retaining, with the strong hook that terminates them, and which in some species, is even accompanied by another, but a smaller one, the bodies which they think proper to appropriate.

Many, such as the matuti, portuni, orithyiæ, and the macrouri in general, swim with facility. But the other brachyuri are rather runners than swimmers, and suffer themselves to proceed at the will of the waves, assisting themselves in a trifling degree with their feet. They usually walk sideways, or backwards, and run with a velocity equalling or surpassing that of a horse, gaining, on the slightest appearance of danger, their habitual retreats, or any which chance may happen to present. They pinch strongly with their claws or forceps, and sometimes sacrifice them by leaving them within the hands of the person by whom they have been seized, that they may themselves escape. Nature repairs this loss the more speedily, if the breach be made at the joints, especially the second or third.

The growth of the craw-fish is slow, and according to the testimony of fishermen, it is hardly saleable at the end of seven or eight years from its birth. Some crustacea, such as those which remain habitually at great depths, in inaccessible situations, and which are better protected by the nature of their testa, may attain to a very considerable age. It was believed, in the time of Pliny, that some species could live for a period exceeding the duration of the life of man. M. Latreille has seen a great lobster *(palinurus)* of nearly six feet in length from one extremity of the body to the other, whence we may infer its great age.

The family of Brachyurous decapods is composed of the genus CANCER or CRAB. Under this name are commonly designated all the crustacea analogous to craw-fish, and lobsters, but whose body is proportionally shorter and broader, with a small tail refolded on the breast. In the method of Linnæus, the genus cancer has an acceptation much more general, since it comprehends the first three orders of crustacea, and even a part of the fourth.

The crabs remain in preference on coasts where there are rocks, in clefts where they are sheltered from the waves of the sea, and secured from the pursuit of their enemies.

When the waters rise, they approach the shore, and seize on marine animals, incapable of resisting, or which have perished. They are very voracious, and assemble in great numbers on the carcases on which they feed. It is principally during the night that they proceed to plunder. Nevertheless, as they do not always regain the sea with sufficient promptness, and they cannot swim, they are often exposed to be stranded in the low waters. If they do not find, in their neighbourhood, some hole to take refuge in, they contract their feet, squat down in some corner, and await in tranquillity the return of the tide, to arrive at the open sea. It is the individuals thus detained, that are most frequently collected by fishermen; for these crustacea will but seldom take the bait, and are not easily caught in nets. Around the islands of America and India, where the bottom of the sea, near the coasts, is visible in calm weather, they are harpooned with a long pole to which a fork is attached. In other places they are taken by divers.

The PORTUNI, called also *etrilles* by M. Cuvier, scarcely differ from certain crabs, and particularly from the *carcins* of Dr. Leach.

According to the report of M. Bosc, the portunus which he regards as the species named *pelagicus*, by Fabricius, swims almost continually, with facility, and even with a sort of grace. It can sustain itself upon the water, and for a considerable space of time, without giving itself any apparent motion. It has no other points of repose, excepting the varecs, and other plants of the Atlantic ocean, where it is found in great numbers; it lives on other marine animals. Another portunus, the *hastata* of M. Bosc, and which he has observed on the coasts of Carolina, also swims extremely well; but it walks as much as it swims. In general, it parades slowly, on the edge of the sea, or at the mouth of rivers, when the tide is flowing, for the purpose of seeking its food. But when the tide ebbs, it returns with it by swimming, because it is then afraid to be left upon the sand, and has no further prey to expect. Most generally it swims and walks forward; but should it be seized by any sudden fear, it escapes by swimming sideways, and even backwards. During the winter it disappears from the coast and retires into the depths of the sea. It returns in spring, and the female, in consequence of the eggs which she then carries, is very much esteemed as food. It is reported that this crustacecum sometimes issues from the water, to seek its subsistence on the strand. A great number of them are taken daily during the summer at Charleston.

"All the portuni which inhabit our sea," (the coast of Nice) says M. Risso, "live united in society, and each species chooses a dwelling conformable to its wants and habits. The *bimaculatus* takes up its sojourn in the region of the cortici-ferous polyparia. The *puber* and *plicatus* prefer rocks several hundred feet under water. The *depurator* delights in pebbly plains, always mixing with the columns of little *clupeæ*, such as the anchovy and pilchard. Another, imperfectly described by Rondelet, whose name it bears, conceals itself under the mud. The *guttatus* inhabits the middle of the algæ, which grow at a considerable depth; and the *P. longipes* frequents the holes of the compact limestone which edges the coasts. The portuni feed on mollusca and small crustacea, which they break to pieces, and grind by means of the osselets of their stomach. Their flesh has not the same taste in all the species, and it is only those which live in the rocks that are made use of as food; the others serve as bait for fish.

"Many of these crustacea are tormented by little aselottæ, which insinuate themselves under their corslet, and attach themselves upon their gills. The female portuni have many births in the year, and deposit each time from four hundred to six thousand little globular and transparent eggs, which disclose in more or less time, according to the state of the temperature."

On this last observation of M. Risso, M. Latreille remarks, "M. Risso has had better opportunities than myself of studying the manners of these animals. I confess, however, that I have some difficulty in believing that the females of the same species have many births in the course of one year, from the spring to the end of the autumn. Analogy, and the observations of other naturalists, seem to contradict this assertion."

Pison has represented, in his Natural History of Brazil, a portunus bordering on the *hastatus* of Fabricius, and which, in the language of the country, he names *Cirè apoa*. The word *cirè* seems to be a common denomination for crustacea similar to the preceding, which live habitually at the bottom of the sea, and which proceed to the shore only for the purpose of seeking the ambergris which has been cast there by the waves. They are only to be caught at the period of the strong tides. It appears that they are put into vinegar, and though many of them may be eaten, they are rarely found indigestible. Some other species again form an aliment for the inhabitants of the maritime coasts of China, of the East Indies, &c. These crustacea abound in the seas which border on the tropics; but the Northern Ocean furnishes but few species, and those small, or but of middle size.

The known species of the genus THELPHUSA live in the fresh waters. The one which is proper to the south of Europe, and the Levant, has been for a long time known. It enjoyed a great celebrity among the Greeks, a proof of which is to be found in the antique medals of Agrigentum in Sicily, on one side of which it is usually represented, and often with so much truth, that it is impossible to mistake it. Particular mention is made of this crustaceum in the writings of Pliny, Dioscorides, Nicander, &c.

It is the *carcinos potamios* of the Greeks, and the *grancio* or *granzo* of the Italians. It was believed that its ashes were useful, from their desiccative qualities, to those who had been bitten by a mad dog, either by employing those ashes alone, or mixed with incense and gentian.

According to the report of Ælian, the fresh-water crabs *(Ocypodes?)* as well as the tortoises and crocodiles, foresaw the inundation of the Nile, and, about a month previously to that event, they resorted to the most elevated situations in the neighbourhood.

The *Thelphusa fluviatilis* is common in the environs of Rome, and remains in the mud, so that to get at it the fishermen are obliged to dig a trench all round. It removes to a considerable distance from the water, and can live a week, and sometimes even a month, out of this element. It even appears that crabs can be preserved alive in this manner, by keeping them in cellars, or in fresh and somewhat humid places.

In Rome, the thelphusa fluviatilis is eaten on fast days, and at all times of the year. These crustacea, however, are better in summer, after their moulting, and more especially while they are undergoing this change. They are then served up on the tables of the pope and cardinals. Some persons make them die in milk, to sweeten and soften their flesh. They are carried to market, attached by a cord, but placed at a certain distance from each other, because if they were suffered to touch, they would gnaw each other, and lose a portion of their limbs. Belon observes that the females are distinguished by their tail, which is broader and more rounded, in the form of a shield. This difference, however, here, is not so sensible as in most other species of this family. The same naturalist has found this decapod in the streams of Mount Athos. The Caloyers, or monks of Mount Athos, eat it raw, and maintain that it has more flavour in this state than when cooked.

M. Menard de Groye, correspondent of the Academy of

Sciences, has collected the following observations respecting this animal.

"It was on the 28th day of July, 1812, that I had occasion to see and observe this curious crustaceum, on visiting the celebrated *degorgeoir* of the Lake of Albano, otherwise the Lake of Castello. It is known that the basin of this lake is considered by most travellers, and even naturalists, to be the crater of an ancient volcano. It is five miles in circuit, and as much as four hundred and eighty feet of depth is given to the water which fills the lower part. This water is limpid, perfectly sweet, and is inhabited by divers kinds of fresh-water fish, common frogs, &c. The redundancy of water runs off incessantly in a great stream by this admirable subterraneous canal, which is almost two miles long, and which has been preserved, without deterioration, from the earliest times of Rome. The heat which reigned in the atmosphere when I was in this part of the country, the purity of the water, the solitude, the shade, and freshness of the shore, the bottom of which could be discovered there, to a very great distance from the edge, like a strand, induced me to bathe, and it was thus that I chanced to catch three or four individuals of the species of crab in question.

"I was very much surprised at the first appearance of these crabs, not being by any means prepared for it. They appeared to me so similar in figure, size, gait, &c. to those which are commonly found on maritime shores, in fine, to the *Cancer mœnas*, that I imagined at first that they might be those identical crabs, which had been brought from the sea, which is not, in fact, very distant, as an attempt to naturalize them in this lake, and that this attempt had succeeded. Nevertheless, I remarked that they had a whitish or livid colour, whereas the marine ones to which I was comparing them are brown. Afterwards, perceiving scattered here and there some carapaces, and other old remains of these crabs, and observing that they were spread over a considerable extent of shore, where they appeared altogether at home in their habits, plunging under water if they were out of it, concealing themselves also under stones, &c. and showing a great degree of vivacity, I no longer doubted that they were then in their element, and that on the other hand they would find themselves very ill at ease, if they were carried into the salt-water. It also appeared to me that these fresh-water crabs were more cunning and more alert than those of the sea, which allow themselves to be taken with facility. I could not catch the former but by drawing them towards the shore with the end of my stick, which was not easy, so well did they know how to steal away. They also defended themselves vigorously when they could do no better; and I could perfectly well feel, from the force with which they strained the stick between their claws, that it would not be safe to take them with my hand. A fisherman, whom I found on re-ascending, also informed me that they drew blood. He confirmed me in the opinion that these crabs were perfectly natural in this lake, and that they were known there always. But he added, that they withdrew during winter into the bottom, and did not reappear upon the banks until summer. He added likewise that they were very excellent eating."

The crustacea of the genus GELASIMUS are similar to the ocypodes, in the general form of the body and in habits. One of their most striking characters is the extraordinary disproportion in the length of their claws: one, sometimes the right, sometimes the left, for this varies in individuals of the same species, is enormously large, while the other is very small, and often concealed. One would say that these animals were one-handed. They have the habit of raising the thickest of these claws in the air, as if they wished to make a signal, and call somebody. For this reason, Linnæus has designated one species under the name of *vocans;* but no doubt this claw serves them either as a buckler or offensive weapon.

These crustacea, which, as well as the ocypods, more particularly inhabit warm climates, make their dwelling in humid soils near shores.

M. Bosc has seen one species *(vocans)* in Carolina rushing in crowds upon carcases, covering them, and disputing the pieces with the vultures. The burrows formed by another species *(pugillator)* are so numerous that they touch. They are cylindrical, usually oblique, and very deep. Rarely do many individuals enter into the same one, for in that case the peril would be very imminent. These animals do not fear the water, which sometimes covers them, but they do not seek to enter it, and never remain there a long time of their own accord, except at the time of laying, and until the eggs disclose. M. Bosc has found from the end of February some females in which the under part of the tail was furnished with eggs; but he has never met any young ones in their first age, and he suspects that they pass the first year of their existence in the water or in the earth. The males are distinguished from the females, by being smaller, more coloured, and having the tail triangular. The pincers do not indicate sexual difference. This species remains during winter, or during three or four months, in its burrow, which is almost always closed, so that the animal is obliged to re-open it, when the heat of the sun is strong enough to force it to quit its dwelling.

This crustaceum is not an article of food. It has many enemies, and such are the otters, the birds, the bears, the tortoises, and other reptiles; but its multiplication is so excessive, that no sensible diminution results from their devastations.

The *Gelasima maracoani* is found at Cayenne and in Brazil, running on the sea-shore after a reflux, but remaining concealed at every other time. It forms an article of food.

Ocypode is another genus of crustacea, very natural, though previously to the time of our author, but imperfectly characterized.

These crustacea are proper to the hot climates of the two hemispheres, and remain on the sandy tracts of the edges of the sea, or of rivers near their mouth. They there hollow burrows, into which they retire in case of danger, and where they pass the night. It is presumable that they also shut themselves up there during the period of the moulting Olivier in vain attempted by running to catch the species which he observed on the coasts of Syria, which is probably the same of which Pliny makes mention, and which the Greeks designated, by reason of the celerity of its course, under the name of *Hippeus*—cavalier or knight.

"They ran," says Olivier, "towards the sea, or repaired into their hole, according as one or the other was most within their reach. They almost always trace, in running, an oblique line." M. Bosc also tells us, with respect to a species proper to Carolina *(alba)*, that he had some difficulty in overtaking these crustacea on horse-back, and killing them by musket-shot. The ocypodes, like the analogous crustacea, in all probability feed upon the carcases of animals. Some of the facts reported by travellers, relative to the *earth-crabs*, should be applied to them, but some only, because they have confounded under this latter name many crustacea of different genera, such as *gecarcinus, grapsus,* &c. Pere Labat, in his "Nouvelle relation de l'Afrique Occidentale," speaks of a species of *tourlourou,* which is found at the point of Barbary, and which is there called *crab.* It is reported that this animal cannot be eaten without danger of poison, and that these same crustacea cut in pieces and devour the individuals of their own species, which have been lamed by any accident.

In this part of Africa, it does not appear, from all researches, that any species of *gecarcinus* or tourlourous, properly so called, has as yet been found. But the habitation of the *ocypode hippeus* of Olivier extends from Syria and Egypt as far as Cape de Verde, from which M. Latreille suspects that to this species must be applied what Pere Labat has related concerning the crabs of the point of Barbary.

According to Artus, earth-crabs are also to be seen on the Gold-coast similar to those of the Leeward Islands, and which supply the inhabitants with an excellent food. They dig themselves holes, which serve as a retreat. There are found in the island of Java other earth or land-crabs, but which are not eaten. They quit their burrows during the day, and feed, according to report, on plants. These, perhaps, may be the *ocypode ceratophthalma*, which is very much extended on the maritime coasts of the East Indies. The collection of crustacea of New Holland, formed by Peron and M. Lesueur, although very numerous, presents no species of ocypode and gecarcin.

Linnæus informs us that the *ocypode hippeus*, which he calls *cursor*, and under which name he includes this species and ceratophthalma, is to be found in the Mediterranean sea, as well as in the Indian Ocean; and that after the setting of the sun it quits the water for the sandy shores, and runs with an extreme velocity.

Belon asserts that the lizards succeed in catching it for the purpose of food. Olivier has had no opportunity of verifying this assertion.

The ancient naturalists, and some among the moderns, have represented the crustacea named PINNOTHERES, and also *pinnoter* and *pinnophylax*, as the sentinels and guardians of the mollusca of the genus *pinna*, as their commensals, and in some sort their sutlers. It was believed that they were born at the same time and along with them; that they were even essential to their preservation; that these mollusca, destitute of eyes, and of but little energy of sensation, having held their shell for some time open, so that the little fishes might enter there, were advertised, when there was a sufficient quantity, by a bite made by the pinnotheres; that they then closed their shell, and partook of the booty together.

This opinion, however absurd, appears to have been entertained by the ancient Egyptians, among whom the symbolic representation of the *pinna* and the *cancer* designated a man or father of a family, whose existence depended only on the assistance of his children. But whatever may be the source of these erroneous traditions, the naturalist should not the less desire to know what the animals are which have given rise to them. Camus, in his Commentary on Aristotle's History of Animals, forming the second volume of his French translation of this work, had presented on this question some very judicious reflections. This investigation has equally engaged the attention of M. Cuvier; and in his critical dissertation, the object of which is to ascertain the astaci, or in fact decapod crustacea, mentioned by the ancients, he has discussed, with that sagacity so peculiarly his own, the testimonies relative to the pinnotheres. Not only does he consider the history which has been given of them as the pure fruit of imagination, but he seems to think that the ancients had no very correct notion concerning the crustacea which are the object of these fabulous narratives.

It is well known, that divers crustacea, by reason of the more feeble consistence of their testa, or otherwise, are instinctively impelled to choose particular domiciles, usually moveable, and especially shells, both univalve and bivalve. The pinnotheres are of this number; but they differ from the paguri, in inhabiting none but bivalve shells, and always in company with their proper owners. The same shell may present many individuals of the same species of these parasite crustacea. " The pinnæ," says Aristotle, " have in their shell the animal called the *guardian of the pinnæ;* it is a small squilla or cancer, which they cannot lose without perishing speedily." He adds a little lower, " There are born in some testacea, certain white crabs, and very small. The greatest number is found in those species of muscles whose shell is inflated. After this comes the pinnæ; its crab is named *pinnotheres.* Some are also found in cockles and oysters. These little crabs have no sensible growth, and the fishermen assert that they are formed at the same time as the animal with which they inhabit." He again tells us, a little further on, " that there are born in the cavities of sponges some small crabs, similar to the guardian of the pinnæ; that they are there like the spider in its retreat; and that by opening and closing these cavities at proper times, they can catch there the little fishes. They keep them open to cause the prey to enter, and close them immediately when it has got in."

It is certain that the muscles, cockles, and oysters of our coasts, previously cited by Aristotle, enclose, at least at a certain period of the year, some small crustacea, generally known, similar in all respects to those which, according to him, and many other ancient authors, inhabit the shell mollusca, and with which the genus pinnotheres has been established; that these shell-fish do not habitually present other animals of the same class; that there are often found in certain pinnæ, either other pinnotheres larger than the preceding, or small *caridiæ.* We also know that some paguri and porcellanæ lodge in sponges; and it is probably to these crustacea that Aristotle alludes, in the passage last cited. But it is not proved that the father of zoology was right in advancing that the guardian of the pinnæ was either a little crab, or a little squilla, or *caridion.* According to M. Cuvier, this disjunctive expression would indicate, or rather clearly announce, that Aristotle had not himself observed the fact which he relates, or that he would testify some uncertainty on the point.

Pliny has confounded, under the common name of *pinnotheres*, the paguri, and the species of our genus pinnotheres, properly so called.

We must not apply, as some authors have done, to these last crustacea, a passage of Appian, in his Halieutics, in which he relates that the crab, when the oyster comes to open its shell, puts a stone between the two valves, so that it cannot close, and he thus penetrates into it with facility, and devours its inhabitant. In another passage, which directly concerns the pinnotheres, Appian does not explain himself respecting the habits of these animals, in a manner different from that of his predecessors.

Hasselquist, in his voyage to the Levant, says, respecting his pinna muricata, that the cuttle-fish is the most irreconcileable enemy of the molluscum, which inhabits this shell. But, happily for it, there are always inside, one, or several crab-fish, which remain at the entrance of the shell when the animal opens it, and give warning to it to close on the approach of the cuttle-fish. "Thus," continues he, "the pinna permits, in compensation, the crab to reside within its shell." We may well believe that this crustaceum has no need of such permission to establish itself there, and that the quick movements which it makes to withdraw itself from the danger by which it is also threatened, are sufficient to frighten the animal of the pinna, and induce it to keep its shell close. This crab-fish of Hasselquist is probably a species of salicoque, (*caridion*) or pinnophylax, considered in the same light as the ancients considered it. Linnæus, after the authority of his disciple, but from very vague information, at first ranged this crustaceum in his division of *cancri macrouri*, but subsequently, whether he wished to omit it as a species too uncertain, or whether he believed that Hasselquist was deceived on the subject of the division to which it appertains, he made a brachyurous crab of it, referring to it the characters which Forskael had imposed upon it, and which he had communicated in one of his letters.

We frequently find, more particularly in winter, pinnotheres in the muscles which are brought to market; but the females are always observed to be in a state analogous to that in which a craw-fish is that has just changed its skin. The occasional deleterious quality of muscles has been attributed, though probably without reason, to the presence of these parasites.

The GECARCINI are terrestria crustacea, to which the French colonists of the Antilles commonly give the name of *tourlourous* and *land-crabs*.

Pere Labat, in his Voyage to the French American Islands, &c., has collected several observations on these crustacea He distinguishes four species of them; the *tourlourous*, the *violet-crabs*, the *white-crabs*, and the *ceriques*. What he says of these last may answer for grapsus. They are found in rivers and on rocks, at the edge of the sea; they are much flatter than the others; their shell is thicker and harder; their biters, or pincers, though much smaller, do not pinch less; they have also much less flesh and fat than the others, which causes them to be in less esteem, or at all events, they constitute the last resource of the negroes. Chauvalon, in his Voyage to Martinique, says, that the cericæ of the sea, and which are not taken in the fresh waters, are the *cirè-apoa* of the Brazilians, or the *xirika* of Guiana. It is evident, according to the figure of the *cirè-apoa* given by Marcgrave, that this crustacecum is a portunus. It is possible that the denomination of cerique may be common to crustacea of different genera. Those which he represents are certainly grapsi. According to Peron, the last species, and which appears to be *G. pictus*, is employed as an excellent alexipharmic; it is pulverized and mixed with wine. We see from passages of Aristotle, of Pliny, of Galen, &c., that the same virtues were for a long time attributed to divers crustacea. The grapsus pictus is found upon the sea-shore concealed under the roots of the manchineel, (*Hippomane*, Lin.) A species of the same genus burrows in the sand.

Rochefort, author of a natural history of the Antilles, and, anterior to Pere Labat, does not speak of the *cériques*. But he equally distinguishes three sorts of land crustacea; the *tourlourous*, the *white crabs*, and the *painted crabs*. M. Latreille was of opinion, that by the last term, he meant the grapsus; but their habits are so different, that our naturalist was led to presume that the first crustacea are no other than the *violet-crabs* of Pere Labat.

The species most commonly named *tourlourou* is the smallest of the three. It is a deep red, bordering on brown or black, in the middle of the back; its claws are unequal, and the left is always smaller than the right. These animals use them in cutting roots, fruits, leaves, &c., on which it is reported that they feed. They pinch very strongly, and do not let go when they are seized; their flesh is delicate, wholesome, except, as it is said, when they have eaten of the fruit of the *mancinella hippomane*, Lin. But Jacquine, however, denies that they attack this fruit, and it is much more probable, that like others of the tribe, they subsist on animal substances. It would be important to ascertain from what cause proceeds the deleterious quality, which, under certain circumstances, they possess; this has been attributed to the submarine veins of copper upon which they live; but this opinion requires the corroboration of facts.

What we shall presently relate respecting the singular instinct of the painted crabs, is specially applied by Pere Labat, to the tourlourous. But he gives it as a general rule, that the crabs, craw-fish, cericæ, the paguri, the lizards, and even serpents, descend every year to the sea for the purpose of bathing, of changing skin, or shell, that the tourlourous, crabs and cériques, have also the object of depositing their eggs there. He equally gives the name of tourlourou to the crustacea which we have already mentioned, and which are exclusively found at the point of Barbary. The habit of devouring their disabled comrades, he sapiently conceives to be the cause of the maleficent qualities which they possess. He is much more exact and judicious in his remarks on the sexual and external differences of the touroulous and crabs: these he establishes on the proportions and form of the tail.

The white crabs are the most bulky of all. Some have been seen, one of whose pincers could contain a man's fist. They remain at the foot of trees, and in lone and marshy situations near the sea shore; they make holes in the earth, and retire thither, as the rabbits do, into their burrows, seldom appearing during the day. On searching in the sand to discover them, it is found that they always have one half of their body in the water. They are hunted principally in the night, the people carrying flambeaux of candle-wood in their hands; and as they do not remove to any distance from their retreats, and even withdraw into the first hole which they can find, they must be seized the moment they are perceived.

Labat has observed of the white crabs, that the necessity of changing air, and the fear of being covered by the waves of the sea, oblige them sometimes to issue forth by daylight from their retreats. The holes which they inhabit are then marked, and a stick is, on their return, passed into them, which retains the animals captive. When the tide has flowed, the stick is removed, and the crab is then found smothered at the edge of the hole. The negroes unite their claws, fixing them one in another, then string these crustacea by means of the ring thus formed, and thus carry them to market.

The third sort of land crustaceum of which Rochefort speaks, is that which he designates under the name of *pictus*, and which appears to be the violet-crab of Labat and other travellers. Those animals, intermediate in size between the preceding two, are remarkable for the beauty and agreeable mixture of their colours. Some are of a violet, variegated with white; others of a fine yellow, varied with purple or greenish lines; others have a tan-coloured ground, striped with red, yellow, and green.

They gnaw, in open day, under the trees to find their food, and are usually met with in numerous troops in the morning and evening after the rains. The hollow of some rotten tree, the cavities which are under its roots, or the cleft of a rock, usually constitute the asylum where they take refuge, and withdraw themselves from the view of their enemies. Maugé informed M. Latreille that they climb sometimes on trees, to take the young birds in their nests.

Rochefort states that they repair every year, towards the month of May or June, in the rainy season, to the sea shore to lay their eggs there, and perpetuate their race, travelling from the mountains, in which they make their habitual dwelling, in such numbers that the roads and woods are entirely covered by them. They possess the instinct of directing their course towards the parts, which, by their descents or natural declivities, facilitate their journey, and permit their more convenient approach to the sea shore, which constitutes the limit of their migrations. This migration has been assimilated to an army marching in order of battle, with unbroken ranks and undeviating line. They scale the houses and rocks, and surmount every obstacle which they may encounter in their path; the gardens situated on their passage often suffer greatly from their depredations. Sometimes they even penetrate into houses, especially in the night, and disturb the inmates by the noise they make. The males being well fed, and the females surcharged with eggs, their flesh forms some little compensation to the inhabitants for their troublesome visits, and the mischief which they occasion. It is reported that they halt twice a day, as well for the purpose of feasting, as for a short repose; but they travel principally by night.

When arrived at the sea shore, they bathe themselves there, as it is said, three or four times, and then retire into the neighbouring plains and woods, where they repose for some time. The females then return a second time to the water, and having washed themselves a little, they open their tails, let fall the eggs which are there attached, and take a fresh bath; after this operation, they seek to regain, in the same order, the places from which they had proceeded, and by the same route. But the most vigorous individuals are alone destined to revisit their former mountain-dwelling; most of them are, at their return, so feeble and thin, that they are forced to stop frequently to recover themselves.

The eggs thus deposited in the sea are thrown back upon the fine sand of the beach, and after having been for some time warmed by the rays of the sun, the young are seen to issue from them. These speedily proceed to establish themselves in the neighbouring woods, until they have acquired strength enough to repair to the mountains, and to form other families.

When they return into their habitations, these crustacea have new trials to undergo. It is then the time of their moulting. They all conceal themselves in the earth for some weeks, so that there are none of them to be seen; even the entrance of their burrows is then closed. It is said that they are then, as it were, enveloped in the leaves of trees. The flesh of those which have but just thrown off their old covering is in great estimation. The inhabitants of the islands then name them *crabes boursières (pouched crabs)*, for, as in the moultings of the other crustacea, their teguments then form only a red, tense pellicle, similar to wet parchment. It is for this reason, that they are more delicate, or at all events more profitable.

Rochefort supposes that this change of skin takes place only on their return from the sea. It precedes it, according to Labat, and is made immediately after the laying. We shall neither follow this author nor Rochefort, into the details which they present us upon this subject, having already alluded to them in speaking of the generalities of the class.

The eggs are very small, and of a peculiarly fine flavour. In cooking they assume a red colour, like those of craw-fish and lobsters. They are collected under the tail of the female, into two tufts or pellets, separated from each other by a membrane, and accompanied with a thick substance, of the colour of the eggs, but which becomes white when it has undergone the action of fire. The interior of the body of the males presents, besides the substance which may be compared to fat, a greenish matter, the *taumalin*. Both these substances serve to make the sauce with which these animals are eaten.

These crustacea, with the exception of the white crab, which is in no great estimation, and is, moreover, an object of suspicion, because it lives in places where the mancinella and the mimosa are found, constitute a true manna for the country. They form a large portion of the food of the Carribs. The negroes also eat a great deal of them, nor are they rejected by the whites themselves.

Impressed with the notion that the crabs owe their hurtful qualities to the fruit of the mancinella, Sloane imagined that he explained the fatal accidents which have occurred to some persons after eating these animals, from the neglect of the precaution of cleansing their interior, and removing the ill-digested particles of this fruit. But besides that it by no means appears that these crustacea feed upon this fruit in the island of Grenada, where they are often taken under the mancinella, it has never been discovered that they disagreed with any body. The black colour of the *taumalin* should indicate, according to the report of some writers, that these crustacea are poisoned. But very positive experiments would be necessary to convince us that the fruits of this tree and other dangerous vegetables, are the only agents which can communicate this black colour to the *taumalin*. The violet crabs have been in a great measure destroyed in the island of Martinique; but the Carribs take them there from the neighbouring islands.

Linnæus informs us, on the authority of Brown, that the *cancer ruricola*, a variety in all probability of the same species, sacrifices, to save itself, the pincer or claw with which it has seized the fingers of the person who has attempted to seize it; and that the strong action of this pincer lasts for a minute after it has been detached from the body.

Other land crustacea are to be met with on the Gold coast, in the Island of Java, &c., but they are probably ocypods; for it appears that in general the gecarcini are proper to South America.

The GRAPSI, which are named in the French Antilles "*crabes de paletuviers*," are easily distinguished by their port and appearance. These crustacea are extended through all parts of the world, but more particularly through those which are situated near the tropics. They appear to be the same which in Martinique are called *Ceriques*, as we have before mentioned.

"I have seen," says M. Bosc, "many of the *grapsus pictus* in America, and I have observed that they always keep themselves concealed during the day, under stones and other bodies which are found in the sea. I have further remarked, that although they do not swim, they have the faculty of supporting themselves for a short time on the water, by reason of the breadth of their body and feet, and this by means of repeated leaps. They make these movements always sideways, sometimes to the right, sometimes to the left, according to circumstances. They live, like the other crustacea, on the flesh of other animals which they find dead, or which they can seize alive, catching them with their claws or pincers."

"The *grapsus cinereus*, which I have likewise observed, lives in rivers to which the tide of the sea ascends, or to speak more correctly, on their edges; for it is more frequently out of than in the water. When any one appears in the places where they have assembled, and their assemblies are in general very numerous, they escape into the water, making a great noise with their feet, which they strike one against the other.

"The females of these two species of grapsi lay their eggs in spring, the period in which they commence to re-appear, for during the winter the first remains at the bottom of the sea, and the second, doubtless, buried in the mud."

The *grapsus cinereus* often comes on the shore, or on rocks, to sport or bask in the sun, according to the report of the same writer.

This genus is tolerably numerous in species, spread over the maritime coasts of the warm climates of both continents. New Holland furnishes a good many.

The *grapsus varius* lies in the holes of rocks on the shores of the Atlantic and the Mediterranean. It issues forth from them to receive the rays of the sun, but regains its retreat on the slightest appearance of danger, and fastens itself with so much force with its feet, that it is very difficult to make it let go.

Among the orbicular crabs, the MAIA is very common on the coasts of the Mediterranean, where it generally receives the name of *sea-spider*. It is one of the largest of the crustacea, and is the *maia* of the ancient Greeks, which is figured on some of their medals. They attributed to it a great degree of wisdom, and believed that it was sensible to the charms of music.

In General Hardwick's collection is a crab which Mr. Gray has appropriated to a new genus named GOMEZA. It is apparently intermediate between Corystes among the orbicular crabs, and Leach's genus Atelecydus, having the long antennæ of the former, with the short claws, and somewhat of the shape of the latter. The thorax is ovate and convex; over the eyes are two spines with denticulated margins; the outer antennæ, longer than the body, are ciliated above and below; the eyes are enclosed in a pit, larger than the stems. The specimen, *G. bicornis*, is pale yellow (when dry); the thorax covered with white granules. This species inhabits the Indian ocean.

The DROMIÆ seize with their hind feet alcyones, the valves of shells, and other bodies, under which they shelter themselves, and which they carry along with them. Some have pronounced the *cancer dromia* to be venomous.

Our *Dromiæ Indica* has the thorax subglobose; the back smooth, evenly convex, covered with short hairs; the front bifid, with an obscure central inferior tubercle; each side with five roundish tubercles, distant; the upper edge of the front, and of the carpus, tubercular. Our figure is from a specimen in possession of General Hardwick, which inhabits the Indian ocean.

According to Rumphius, the RANINÆ come to land, and climb even to the tops of houses; but from the form of their feet, this appears to be impossible, or at all events of very small probability.

We have now to speak of the *macrourous decapods*, and shall first make a few general remarks on the genus ASTACUS.

This genus, in consequence of the fresh-water species which is found throughout all Europe, and universally used as food, is one of the most generally known, and one of the most studied, in the whole class of the crustacea. There are few

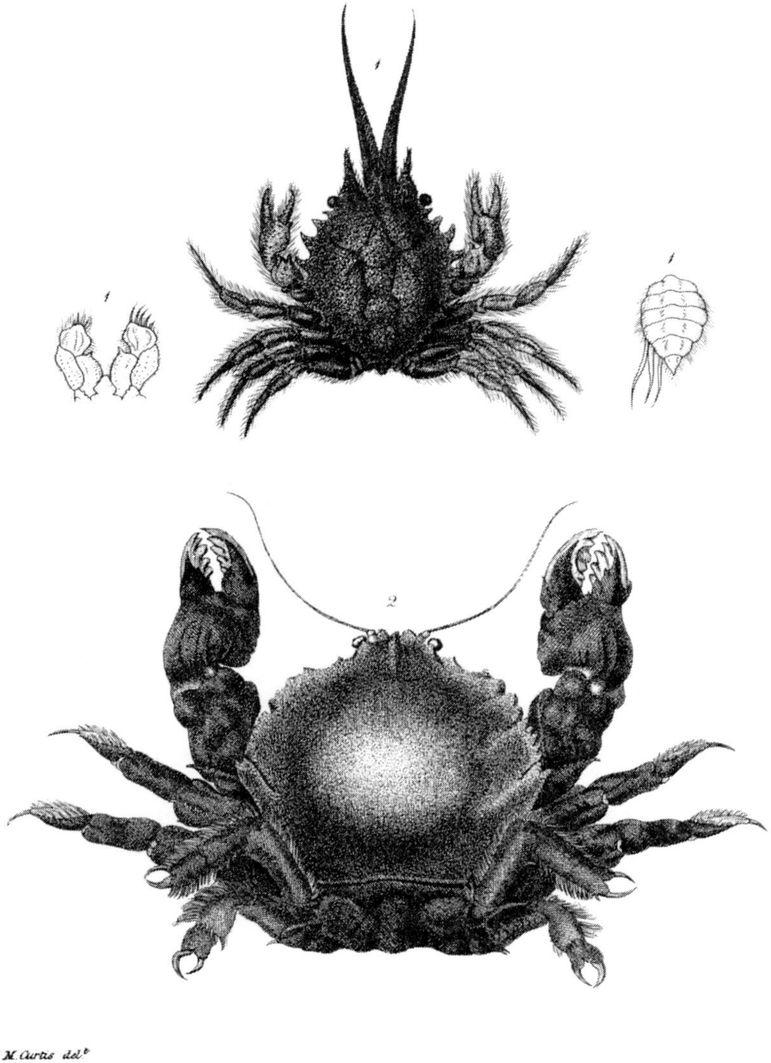

C. M. Curtis del.t

1 *Gomeza bicornis.*

2 *Dromia Indica.*

London; Published by Whittaker & C.o Ave Maria Lane 1833.

works on the history of fishes, or that of insects, in which a particular chapter has not been devoted to it, from Aristotle, who was the first to treat of it, down to the present day.

There are threads attached to the tail of both sexes, varying in number and figure, and mobile at the base. To these threads the female fixes her eggs, and it is probable that those of the male may serve some purpose in the act of generation; but as the coupling of the astaci has not been yet observed, nothing positive is ascertained respecting this subject.

It is with the claws that these astaci catch their prey; they also employ them as a weapon of defence, as may easily be proved by presenting any thing to them, when they are in the water. The animal grasps with so much force, that to make it let go it is necessary to break the foot or burn the tail.

A remarkable peculiarity in the feet of the astaci of both sexes is, that they are the seat of the organs of generation. Buster, on the authority of another observer, reports thus concerning the sexual intercourse of these animals:—When the male attacks the female, she throws herself backwards, and then they embrace closely by means of the feet and tail. After this, at the end of two months, the female is found charged with eggs.

The astaci lay a great number, which, as I have before observed, they have the art of attaching to the mobile threads underneath their tail, and which they constantly carry there, until the little ones are disclosed. There is some reason to believe that these eggs grow and augment in volume, while they are thus attached to those threads. They are enclosed in a species of sac, which is a continuation of their membranaceous pedicle: each thread, thus charged, represents a bunch of grapes, the more exactly too, as the colour of these eggs is of a reddish brown.

When the young astaci are disclosed, they are transparent and extremely soft, but in all other respects similar to the old ones. As their delicacy would expose them the first days after their birth to dangers without number, which they have even some difficulty in escaping at a later period, the wisdom of Nature has again provided for them, for some time, a retreat under the tail of the mother. It does not appear that any one has ever happened to eat of the astaci when thus provided with their little ones. When the mother is tranquil in the water, these little shell-fish may be seen coming out from between her legs, and venturing to creep around her, and then, in the moment of danger, withdrawing altogether into their asylum. It appears that the mother warns them of what they ought to fear, for it is never without a motive that they fly in this manner. They nevertheless abandon the mother by little and little, in proportion as they increase in size; and they are seen but little with her towards the end of the first fifteen days after their birth.

The colour of the astaci is of a greenish brown in those of the rivers (craw-fish), and of a reddish brown, spotted with blue, or some other colour, in those of the sea. But whatever be their colour during life, it always becomes a deep red by boiling or the action of acids. The common lobster, however, sometimes occurs red in the living state. A specimen, which was caught on the coast of Norway, was lately presented by the editor to the British Museum: it was alive, perfectly red, and had all the appearance of having been boiled.

One of the most astonishing facts which the history of the astaci, and probably of almost all the other crustacea, presents to our observation, is the regeneration of their claws and feet when broken or torn off by any accident. There are even some species whose limbs are so slightly attached, that it is sufficient to touch them, to put them near the fire, or even simply to inspire an apprehension of danger into the animals, to determine the latter to abandon them in part or altogether. This fact is so generally known, that no person has ever thought of casting a doubt upon it. The ancients, at least

Aristotle and Pliny, make mention of it; but it has only been in later times that any explanation of it has been attempted.

The time necessary to the reproduction of the new legs is not fixed: they grow faster in proportion as the season is more hot, and the animal is better nourished. Various circumstances, again, render this reproduction more or less prompt: one of the most essential is the place in which the rupture has been made.

If the limb of a crab or lobster has been broken during the summer season, and that in a day or two after the changes that have taken place be examined, we shall find a reddish sort of membrane covering the flesh. In four or five days this membrane assumes a convex surface, like the segment of a sphere; afterwards it becomes conical, and is elongated more and more, in proportion as the foot which pushes it from underneath is developed. Finally, the membrane is torn, and the leg appears: it is soft at first, but in a few days becomes covered with a shell as hard as that of the old one. It now wants nothing but thickness and length, which it acquires in time, for with each change of skin it augments in a more rapid proportion than the feet which are at their full growth.

Reaumur has attempted to explain the causes of this reproduction of parts in the astaci. He inquires, if at the base of each leg there may not be a provision of new legs, as in children there is a tooth under the milk-tooth, which is one day destined to fall? if a lobster can repair the loss of its limbs to an indefinite extent, or if after a certain number of reproductions it be capable of no more? with some other questions of the same kind. It is perfectly obvious that all these are mere conjectures, on which experiment throws no light, and concerning which we shall in all probability ever remain in obscurity.

The antennæ, antennulæ, and jaws, regerminate in the same

manner as the feet. But this is not the case with the tail, any more than in the crabs.

The crustacea which live for many years, and appear to increase in bulk during their whole life, are invested, as has been already said, with a solid crust, incapable of distension without being broken, and consequently calculated to put an insurmountable obstacle to their growth, if nature had not provided a means, by the moulting or changing of this crust, which, if less surprising than the reproduction of individual parts, is not less worthy of the meditations of the observers of nature.

When at the end of spring the birth of a multitude of animals has furnished prey for the crustacea, easy to procure, when they find themselves too much confined in their ancient envelope, there is found between their testa and their flesh an empty interval, which increases, so much so, that if at this period their back be pressed with the finger, it will be found to bend perceptibly, and a little after they are to be found with a soft skin, and the remains of the old one are to be seen in the neighbourhood.

These facts have been known from all time; but it is again to Reaumur that we are indebted for having them confirmed by direct experiment.

When we inspect the spoil of an astacus, nothing is wanting to the completion of its exterior. Even the cartilage which serves to the movement of the mobile finger is to be found there. Each hair was a sheath which covered another hair. The lower articulations of the limbs, which are smaller than the upper, are divided into two in their length, by a suture which separates in the operation, but which is not observed while the animal is living.

The chemical analysis of the testa of the astaci proves that its composition is gelatine united to calcareous earth: the only difference between it and that of the mollusca is, that in the latter there is much calcareous earth, and but little gelatine; and in the former there is much gelatine, and but little calcareous earth.

The astacus being thus left covered with a soft membrane, does not, however, long remain in this state: the skin within four and twenty hours often assumes all the consistence of the old one. The general period, however, for this to take place is from two to three days.

The astaci, ready to moult, have always two stones, which are placed at the sides of the stomach, but which are no longer to be seen in those which have moulted. It appears that these stones are destined to furnish the matter, or a portion of the matter of the testa; for, if on the day after the moulting, when the testa is as yet but half hardened, an astacus be opened, it will be remarked that those stones are diminished one half. If opened on the third day, but a mere atom will be visible, and afterwards no trace of them whatsoever. This method, employed by nature to consolidate promptly the envelope of an animal exposed, while naked, to so many dangers, is extremely worthy of observation.

Reaumur has measured astaci before and after the moulting, and has attained the proof that they augment about one-fifth in bulk. He does not say that this augmentation is the same at all ages; but it seems probable that it gradually diminishes, hence it may be concluded that these animals grow slowly, and the reports of fishermen confirm the supposition.

The fresh-water astaci principally delight in the running and gravelly waters of the mountains. They are also found in lakes and ponds; but there their flesh, unless these collections of waters be augmented by neighbouring sources, is not so good. They conceal themselves, during the day, in holes which they excavate, or under stones, the roots of trees, &c.

It is extremely difficult to stock a stream with astaci, and still more a reservoir, in which there have been none before. Few aquatic animals are more delicate respecting the nature of the water in which they are to live; after being transported in this way, they have been observed to come out of the water and die upon land. It is especially when they are taken from running water, and put into stagnant, that this phenomenon is observed, although this water be not mortal to them, since they are often naturally to be found therein. The only waters which are really mortal to them, are those which are in an actual state of putrefaction.

The astaci, like all the other crustacea, live only upon animal substances. It is most probably in consequence of incorrect observation, that they have been said to eat vegetables; every thing in the animal way is to their taste, whether living or in a state of corruption. In case of famine, and especially when they change skin, they will eat one another. Small fish, small mollusca, the larvæ of insects, and every thing that is drowned in the waters, form the basis of their subsistence in the summer season. They remain during the entire winter without eating, or almost without eating any thing. They have, as enemies, almost all the animals which frequent the waters, or which constantly inhabit there, such as otters, water-rats, aquatic birds, voracious fishes, and even the larvæ of insects. Nevertheless, as they multiply greatly, and the number of their enemies diminishes as they advance in age, that is to say, when they acquire strength, it is sufficient to avoid fishing for some years, in a stream nearly exhausted of them, and to keep watch upon the otters and herons, to have as many there as in the first instance.

The sea astaci are fond of rocky coasts, and rocks, in the fissures of which they can conceal themselves. They are found in almost all seas, and are by no means rare upon the coasts of Europe. Some individuals attain to a gigantic size; not a few have been seen three feet in length.

The fishing for the river astaci is carried on in various ways. The most common consists in merely taking them by day with the hand, in the holes, and under the stones, where they conceal themselves; or during the night with flambeaux, when they are seeking their food. The most agreeable method, and that by which the finest are procured, is the one in which baits are employed. For this purpose a net is attached to an iron circle, or any other heavy matter, and in the middle of this net a piece of meat is fixed; the circle is attached to a long stick by means of three pack-threads; it is put into the water at dusk, the time in which the astaci quit their holes; it is not long before they are seen running at the odour, or the sight of the meat, on which they fling themselves with the greatest avidity. Then the stick is raised, the net withdrawn, and the largest are selected. This fishing often produces very abundant results. Sometimes this plan is modified by placing the meat at the centre of a faggot of thorns; the astaci, desirous of attacking it, get entangled in the branches, and when the faggot is raised, there are sometimes several dozens inside. It is principally in summer that this manner of fishing is advantageous.

The astaci may be preserved many days, when the weather is not too hot, in baskets, in which fresh plants are placed, or in a bucket, or trough, in which there is water to some depth; if only sufficient should be left to cover them, they would perish in a few moments, because the great consumption of air which they make, would not permit them to live in water which was not in considerable mass, or continually removed.

The sea astaci are seldom eaten, but when boiled in sea water, and then seasoned with oil, vinegar, and pepper. But those of the fresh water are transformed, on the tables of the rich, into a great number of viands.

In cooking them, we are told, they are obliged to be put

before the fire in cold water, while alive, for were they to be thrown into the liquor already boiling, the moment they felt the strong impression of the heat, they would break their claws, the preservation of which is a necessary condition in the art of cooking them. When put to the fire in the cold water, they seem to perish without feeling any pain intense enough to cause violent movement or agitation.

In the large rivers of Russia, such as the Don and the Volga, there are astaci of prodigious size, which are never fished but for the sake of getting the stones above mentioned. When a certain quantity have been taken, they are heaped together to cause them to rot, and when their decomposition is almost complete, they are washed with water. The stones, as being the most heavy, fall to the bottom. These stones, which, for many ages, have enjoyed a very high reputation, and which are still considerably sought after in countries where prejudice and superstition predominate, are no longer esteemed in Europe, but as a little bit of chalk, and if it may still be found in the shops of some apothecaries, it is merely from a remnant of ancient custom.

The various species of sea astaci have vulgar names, differing from the scientific, in consequence of an error of Linnæus. Thus the lobster, (*homard*, in French) is not the *cancer homarus* of this naturalist, but the *cancer marinus*. The *cancer homarus* is a part of the genus *Palinurus*, of which we shall presently speak.

We shall next briefly notice that very curious genus the PAGURUS.

The Greeks named generically *carcinion*, the parasite crustacea, which lodge themselves in the empty shells of the mollusca, and the Latins designated these same animals under a synonymous name *cancelli*. Aldrovandus, Gesner, Rondelet, Swammerdam, and other modern naturalists, preserve this last denomination; but Fabricius has bestowed that of Pagurus upon this genus, a name by which the ancients designated a sort of crab, or one of the brachyurous crustacea. The inhabitants of the maritime coasts of France, who are also acquainted with the habit which these animals have of enclosing themselves in univalve shells which they find empty, call them *hermits*, or *soldiers*, because they compare this shell, which serves them as a dwelling, to the cell of a hermit, or the sentry-box of a soldier. Linnæus placed them in the genus cancer, but they have been ascertained strictly to belong to the macrourous decapods.

Aristotle had already mentioned the fact, that the shell serving as an habitation to the carcinion, or pagurus, was not of its own formation; that it had possessed itself of it after the death of the molluscous animal which had formed it; and that its body was not adherent to it, as is that of the last mentioned animal. Belon, Rondelet, and many other naturalists, had confirmed these facts; Swammerdam has, nevertheless, pretended, contrary to so many and such well founded authorities, that the pagurus was born with its shell, and that it even possessed the faculty of enlarging it, in proportion to its own growth. It is positively known, that on its issuing from the egg, its body is naked or without a shell; that its form does not then essentially differ from that which it presents in the adult state; and, finally, that it is without the mantle and the secretory organ, which nature has accorded to the mollusca for the formation of their shells.

It has also been falsely advanced, that the pagurus puts to death the natural proprietor of the shell in which it is desirous of establishing itself. It only takes possession of one that is empty; and that the posterior extremity of its body may fasten there, it always selects one, the summit of which finishes in a spiral. It is but once a year, at the period of moulting, that, its body, having increased in bulk so as to be too much confined in its domicile, it is obliged to choose another more spacious. For this purpose it enters successively, and backwards, into almost all the empty shells with which it meets. It endeavours to discover that in which the hinder part of its body will be most at ease; and unless it be favoured by chance, it is frequently unable to lodge itself until after many trials and examinations.

In their youth, these crustacea sink down almost entirely in their shells, and scarcely can the extremity of their feet be perceived; but when more advanced in age, and increased in bulk, their claws and the two or four following feet always show themselves, in a great measure, outside. When their pincers are of a very unequal size, the largest often closes the entrance of the shell, in the manner of a lid. The same species of pagurus lodges in univalve shells of different species, and even of different genera. "But," says Olivier, "what appears to us not to have been sufficiently observed, though well worthy of observation, is, whether the same individual, on quitting its shell now become too small for it, proceeds constantly to lodge in a shell similar to the first; whether it confines itself to certain species of the same genus; or takes indifferently all which present themselves, no matter to what species they may appertain. Might it not be possible that the individual which at first inhabits a buccinum, and in which its body is in some sort modelled, could not afterwards lodge conveniently but in another buccinum, and that it would find itself incommoded or constrained if it wished to fix itself in a murex or a tonna?" It does not, however, appear, according to the opinion of this skilful naturalist, that the form of the body of the pagurus is intimately adapted to that of the cavity of its dwelling; for, were it so, the individuals of the same species of pagurus, inhabiting shells of divers species, would also present notable differences, which has not been remarked, and which even cannot take place, since the trunk, although of a consistence less solid than that of the other crustacea, is nevertheless capable of a certain degree of resistance, and that a change in its external form would induce others in the principal organs of life. The embarrassment of these crustacea in choosing their retreats would be still greater, and they would be too much exposed to perish if they were obliged to lodge in shells analogous to those which they had abandoned. All the conditions which nature seems to require are, that the shells shall be univalve, of a capacity proportioned to that of the bulk of the body of the pagurus; that they should be turbinated at their extremity; and that their mouth or aperture should be accommodated to the form, to the thickness, and to the action of the claws, and of the anterior feet of the parasite animal. It moves and walks at the bottom of the sea, or on the shore, by means of its organs of locomotion, and by the pincers of its claws it seizes the little marine animals on which it feeds.

When menaced with any danger, it retires, as far as is possible, into the interior of its dwelling, and does not show itself until long after the peril has ceased. When seized, it is said to utter a little cry; but it resists all efforts which can be made to draw it out of its shell, and this cannot be done until after its death. The moments devoted by the paguri to the catching of their prey, those of their amours, and the periods of their change of domicile, are to them times of crisis and danger: they have then to dread a crowd of enemies which devour them, and particularly the fish, which are very eager in their pursuit. These crustacea, according to the testimony of Belon, furnish even an excellent bait to take the fish which frequent the rocks, or which approach the shore.

But all the paguri do not live in the sea. Le Pere Nicholson, in his Essay on the Natural History of St. Domingo, describes a species which inhabits the dry places of the seashores, and of the hills; which, when plunged into the water, and even into the fresh water, uses every effort to get out, perishes there in a short time, and which inhabits univalve terrestrial shells. These last being more rare than the marine shells, the animal does not enjoy the same advantages as the sea-paguri. It is not always in its power to choose, and its habitation is less convenient. This fact corroborates the reflections made above on the passage quoted from Olivier.

Mangé, who visited some of the Antilles, and who collected there a great number of animals, informed M. Latreille, that he found land paguri escape him, at the moment in which he was about to seize them, by rolling with their shells from the top of the rocks, or of elevated places, to the bottom. This species is probably identical with that of Pere Nicholson.

According to M. Bosc, there is in the islands of America a very large pagurus, which lives habitually on land, and which only goes to the sea to lay its eggs, and afterwards seeks a new shell, with which it returns into the mountains and the woods, where it habitually resides. When it is taken, it utters a little cry, and endeavours to pinch the hands. The inhabitants eat it, and derive from its body a yellowish oil, which they esteem a sovereign remedy in rheumatic cases. The shell of the same animal yields them, through the medium of fire, about half a table spoonful of clear water, which these people consider as an excellent cure for the pustules produced upon the skin by the juice of the mancenilla.

It is believed that the paguri issue forth pretty generally from their shells when they proceed in search of their prey. But may they not secure it without employing such a method, and are not their claws, as well as their other front feet, sufficient for this purpose? It appears more certain that they quit their houses at the period of their amours, otherwise it would be impossible to explain, according to the position of the sexual organs, how these animals could have intercourse. The authors, such as Aristotle, Belon, Ulloa, &c. who have asserted that they issue from their shells to seek their subsistence, must have seen them under this particular circumstance. According to the report of the last-mentioned writer, the pagurus, which has for a short time quitted its shell, runs fast, when any danger threatens, back to the place where it has left it, re-enters it quickly, going backwards, endeavours to close its entrance against the enemy, and defends itself with its claws. According to him, its bite produces for two days the same effect as the sting of the scorpion; but the pincers of the pagurus, being similar to those of the other decapod crustacea, cannot act in a different manner, and like them can produce nothing but a pressure, more or less strong, on the body which they have seized.

Some authors have spoken of the combats in which the paguri engage for the possession of a shell: it does not always fall to the lot of the conqueror; for, during the struggle, another individual has sometimes the address to possess himself of the object in dispute.

Other crustacea, which are placed in the same genus, but little known, and some of which perhaps do not belong to it, have no need of shells, and make their retreat in the holes of rocks, in sponges, in the tubes of the serpulæ; others remain, as is reported, in the sand.

Like the other decapod crustacea, the females of the paguri carry their eggs under the tail, and attached to small barbed nets, or to the false feet; but it appears that these oviferous appendages occupy but one of the sides of the tail. According to M. Risso, these animals lay eggs two or three times a year, and always approach the sea shore where a collection of little empty shells is accumulated, so that the young ones may choose, as soon as they are born, a suitable retirement. After their first growth, they possess themselves of columbellæ, of tupiæ, of fresh water shells, which have been carried into the sea; afterwards, of buccina, of cerithiæ, and of rocks. Whether they walk upon the rocks outside the water, or draw themselves into this fluid itself, their antennæ and their palpi are in continual motion. The same observer informs us, that they live in society, and that when dead bodies approach them, they are heaped one upon the other to dispute for the pieces.

Their flesh is of no use for eating, but fishermen occasionally employ it as bait. Some large species of America and the East Indies are nevertheless sought after for the flavour of their flesh. Seba tells us that the *pagurus latro* is good for eating, and that its entrails especially, being properly dressed, constitute an agreeable food. Linnæus says, on the contrary, that it is only good for eating when those parts are removed. According to Rochefort, they are sometimes eaten by the inhabitants of the Antilles.

It is only in the summer that these animals can be observed. They are, during winter, remote from our coasts, or they keep themselves concealed there. They are extended into all parts of the globe, but more particularly in the equatorial regions. It is there that the largest individuals are found.

The PALINURI have very strong relations with astacus, and indeed are sometimes termed *gigantic* lobsters. The palinurus of the European seas is sought after as a delicate meat, especially from the month of May to that of August. The females at this time, having not yet laid their eggs, are preferred to the males. Their eggs, which are named *coral,* form in the interior of the body two elongated masses, of the thickness of the tube of a quill, of a very fine red, diverging towards their two apertures, and situated one on each side, near the basis of their intermediate feet. These eggs are very small on issuing from the body of the mother; but they grow insensibly during the twenty days in which they remain attached to the leaflets of the under part of the tail. After this time, they are all detached, together with their envelopes. They are often found fixed against the rocks, or wandering, and carried along by the waves. Fifteen days are still required before the young palinurus issues from its egg. The female, according to Aristotle, folds back the broad part of her tail to compress the eggs, at the moment when they issue from the body, and elongates the inferior leaflets, so that they may receive and retain them. This is her first laying. The females, after the second, or that in which they get rid of all the rest of their eggs, are thin, and but little esteemed, and the males are then in greater request. Coupling takes place at the commencement of spring, and then more males are caught than females, while the latter are, on the contrary, more abundant on the coasts at the end of spring and the commencement of summer. Aristotle also describes the moulting, which he had well observed, and says that it takes place in spring, and sometimes in autumn. These crustacea abandon our shores towards the end of this last season, or on the approach of winter, gain the high sea, and proceed to conceal themselves in the clefts or caverns of the rocks. It is there also that they change skin. They seldom frequent any but rocky or stony places, live there on fish and divers marine animals, and attain, after some years, to the length of a foot, measured from the head to the extremity of the tail. In some places, but little favourable to fishing, these crustacea, being less exposed, and more tranquil, may live a very long time, and acquire a very large size; some have been observed nearly three feet in length.

M. Risso tells us, that the males seek their females in April and in August; that in coupling the two sexes are face to face, and press so strongly, that they are separated with difficulty, even when they are out of the water; and that the eggs descend along the belly, and issue forth through the anus.

This naturalist informs us, that on the coasts of Nice they

fish for the palinuri with what we call bow-nets or *weels*. They put in cages of osier the feet of sepiæ burnt, with small fishes, crabs, &c. They let them down during the night into rocky places, from fifty to two hundred fathoms deep, and they take in the morning the palinuri which are found there. Their weight is sometimes very considerable. The fishermen are persuaded that they have more flesh at the full moon than at any other time. The extreme fecundity of these crustacea compensates for the great consumption which is made of them for the table.

In the maritime towns, they are brought to market still alive; but they are always cooked when they are to be sent to any great distance, because they die in a short time after they have been taken out of the water, and their flesh quickly putrifies, especially in summer.

We have figured two undescribed species of decapod crustacea belonging to the Porcellanæ of the text. The first

Porcellana hirsuta, is red-brown, rugulose, velvety; the legs and abdomen fringed with long hairs; the carpus above flat, rugulose, front edge with five acute triangular teeth, the hinder edge with a series of conical, incurved, short spines; the front edge of the claw is crenulate at the base; the forehead is triangular, bent down, with a small spine over the front edge of each eye.

The other *Porcellana*, which is named *polita*, by *Gray*, is purplish-brown, much polished, and punctulate; the carpus above is flat, the front edge has three long serrated teeth; the hinder edge has a spiny ridge near the end; the forehead is triangular, produced, with the margin rather concave.

To these we have added a figure of the *megalope sculpta* of Leach, and the *megalope maculata* of the same naturalist, both of which are yellowish white, and were taken in the Gulph of Guinea.

Porcellana hirsuta.

C. M. Curtis del.

London; Published by Whittaker & Co. Ave Maria Lane. 1833.

1

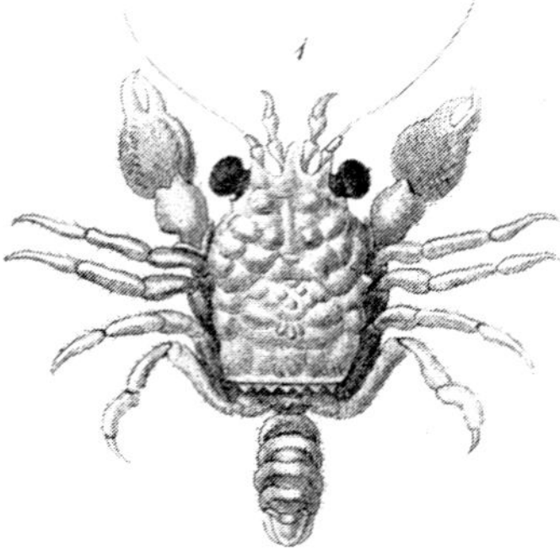

2

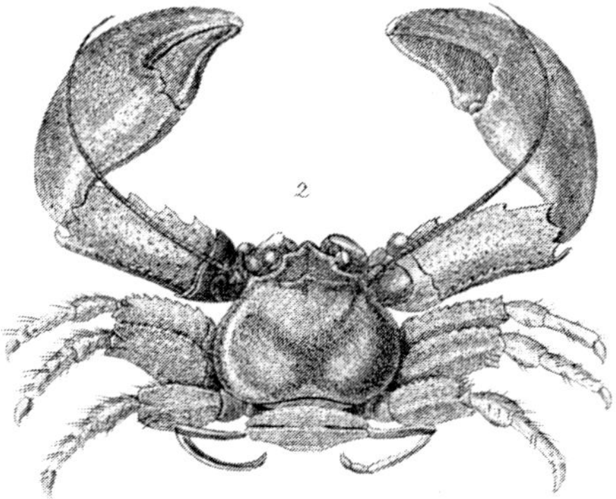

3

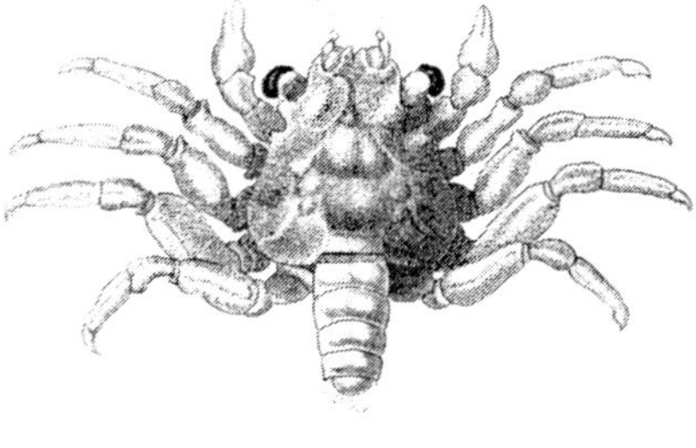

C. M. Curtis del.

1. *Megalopus sculpta.*
2. *Porcellaria polita.*
3. *Megalopus maculata.*

London, Published by Whittaker & C° Ave Maria Lane 1833.

We shall now take leave of the decapod order of crustacea, by a brief notice of the *prawns* and *shrimps*, PALÆMON and CRANGON.

The latter are very distinct from the former by several characters. They are, as is well known, in great estimation for the table, and they are also used as bait in fishing. Their ordinary movements are forward, and by jumps; but when they fear any danger, they escape by running backwards. They live on little animals, which they seize with their claws, and on such as are dashed by the waves against the rocks; but they are themselves the prey of a great number of marine fishes, aquatic birds, echini, asteriæ, &c. Their flesh is less esteemed than that of the prawns or palæmon, with which they have often been confounded.

The *palæmons*, or prawns, appear to belong to the division of decapod crustacea, which the Greeks named *Karis*, and the Latins have rendered by the word *Squilla*. They must, however, be carefully distinguished from the squilla of Fabricius, which belongs to the order of Stomapods. They are marine crustacea, which in the summer frequent the mouths of rivers; they are also found in salt and brackish marshes; they are fished for by means of a net, in the form of a sac, attached squarely to the end of a pole, or with large nets with close meshes, which are thrown to a distance into the sea, and which bring back innumerable quantities of them to the shore. As these animals approach very closely to the beach, it is sufficient, if the first means be employed, to enter into the water as far as the waist, to plunge the net there, and to drag it before one in returning to land.

Olivier tells us, that in the Levant they salt the large species, which they preserve in large baskets, constructed principally of the leaves of the palm-tree, and that they are sent in this state to Constantinople, Smyrna, and into all the towns of Turkey, where the Greeks and Armenians consume a great quantity of them during their Lent, and other days of abstinence; but these species, for the most part, belong to the genus *Penæus*.

The flesh of these crustacea is tender, sweet, agreeable, and regarded as a nutritive aliment, and easy of digestion. The use of it is recommended to persons attacked with marasmus, or threatened with pulmonary consumption. A great quantity of them is caught in the mouths of the Seine, the Loire, and the Garonne. "The mode in which they are cooked there," says M. Bosc, "is by putting them on the fire, with salt and vinegar." Every part of them may be eaten, in consequence of the thinness of the testa. The flesh of these animals corrupts very speedily after their death, which takes place almost on their issuing from the water, and the odour which they then exhale, is, like that of other crustacea in the same state, of the most insupportable kind. They must, therefore, be cooked immediately, to preserve them for some days. The females, when they are charged with eggs, which takes place in spring, are more esteemed and more delicate. These crustacea are also employed in line-fishing, and in some places, as in the United States, according to M. Bosc, this is the only use which is made of them.

Several fish are extremely fond of them, and devour a prodigious quantity. Accordingly they repair in great numbers to the coasts, and the mouths of rivers, a little time after the arrival of the prawns, and afterwards disappear with them, on the return of fine weather.

Nature compensates for the destruction of these crustacea, by a prodigious fecundity. The females lay thousands of eggs, and the species is preserved. These little animals, besides, swim with so much celerity, that many of them escape from the pursuit of their enemies. They usually proceed forward, and swim by means of the false fin-like feet which they have under the tail; but in danger they accelerate their movements, vary their direction, going sideways and backwards, especially by means of the leaflets of the extremity of their tail, which, forming a fan, appear to be more particularly destined to strike the water in front, and to carry the body backwards. The two scales with which their external antennæ are accompanied, are also useful to them under these circumstances. The sort of rostrum, or advanced and denticulated bill, which their front presents, is probably a defensive weapon; but we cannot believe, with Rondelet, that it is capable of stopping fish of small bulk, and still less of killing them.

With respect to the order of STOMAPODA, we have very little to say. They are all marine, and inhabit, in preference, the countries situated between the tropics, and do not ascend beyond the temperate zones. M. Latreille, though he has seen a great number of individuals, has never met with one that carried eggs. Their habits are totally unknown; only, that it is without doubt, that those which are provided with claws, make use of them to seize their prey, after the manner of those orthoptera called *mantis*. In consequence of this conformity, these Stomapods have been called, *sea mantes*. According to the evidence of M. Risso, they remain at great depths, on the sandy and muddy bottoms, and couple in spring; but some other stomapods less favoured as to natatory appendages, having, moreover, the body very much flatted, and much more extended in surface, live habitually at the surface of the waters, and move there very slowly.

In the order AMPHIPODA, which is composed of the genus GAMMARUS, the crustacea called PHRONIMA present a remarkable peculiarity. They are small animals, which have for their domicile the interior of the body of divers soft radiata, such as the medusæ; "Similar," says M. Risso, "to the *argonauts* and *carinariæ*, these crustacea may be seen when the waters are calm, voyaging along in those living wherries, without giving themselves the trouble to swim; nevertheless, when they are desirous of diving, they re-enter their dwelling, and suffer themselves to fall by the mere effect of their weight. These animals, which live on animalculæ, do not show themselves at the surface of the water, until the end of spring, and remain in the somewhat muddy depths, during the rest of the year. Their mode of propagation is yet unknown; but it is certain that they do not carry their eggs on one of their sides, like the paguri, though, like these, they have the habit of lodging in the spoils of living bodies.

The TALITRI, according to M. Risso, remain in troops, and conceal themselves in the plants which the sea heaps upon the shore. Their number is always very considerable in the places which they frequent, and the rapid leap which they make at the moment when they move, causes them easily to be remarked.

M. Bosc informs us, that the talitri, unlike gammarus proper, are oftener out of the water than in, at least during the summer months. They love to remain upon the line of the ordinary tides, that is, in places which are covered with water only at the flow. Every time, that on divers maritime coasts, both in Europe and America, he removed the stones, or the debris thrown out by the sea, under which these little crustacea remain during the day, sheltered from the sun, and in a humidity necessary to their existence, he observed that they made their escape with much activity, so that out of many hundreds which presented themselves to view, he could scarcely catch above two or three individuals. To execute these active movements, they bend back, under their body, the appendages of their tail, and then let them go at will, exactly like the *podura* among the insects. They give, if we may employ such an expression, continual fillips to the ground on which they are placed.

The talitri live on animals smaller than themselves, or on

dead bodies cast on shore by the waves; they are themselves also the prey of several fishes and aquatic birds; they also form an excellent bait for taking small fishes by the line. Like the other crustacea, they change skin in summer, an operation which they execute very promptly. "The males," says M. Bosc, carry their females, which are smaller than themselves, between their feet, and this burthen does not hinder them from leaping." According to M. Risso, the females lay several times in the year, a fact, which, however, appears to need further confirmation. They carry their eggs under the scales of the breast, and when the young are disclosed, they remain there until they are strong enough to seek their food themselves.

In the subgenus Corophium, a very curious species is the *longicorne*. This is called *pernys* on the coasts of La Rochelle: it lives in holes which it forms in the mud. The animal does not begin to appear until the commencement of May. It carries on a continual war against the nereids, the amphinomæ, the arenicolæ, and other marine annelides, which make their dwelling in the same places. There is nothing more curious than to see, at the rising of the tide, myriads of these crustacea agitating themselves in all directions, striking the mud with their long arms, and thinning it for the purpose of discovering their prey. When they find one of these annelides, often ten or twenty times larger than themselves, they unite for the purpose of attacking and devouring it. They do not give over their carnage until they have smoothed and thoroughly searched all the mud and slime. The muscle fishers even pretend that they cut the threads which retain these shell-fish, so as to make them fall into the mud, and then that they devour them. They appear to multiply during the whole of the fine season, since females are found at different times charged with eggs. The grallæ and many fishes devour them in their turn.

On the habits of the order Læmodipoda, we have nothing of any interest to insert.

On those of the Isopod order, there is but little to add to the text.

We shall first notice Oniscus, the subgenus proper of which name is the type of the order, and is composed of the little animals commonly called *wood-lice,* in this country, and *cloportes* in France.

The name *cloportides,* or *oniscides,* is given by M. Latreille to a group which comprehends the *oniscus* of Linnæus, respiring the air in an immediate manner, or which have gills, analogous, as to their properties, to the lungs of vertebrated animals. These crustacea, with the exception of *ligia,* are all terrestrial, and if plunged into the water, will perish there, after a greater or less time.

The onisci are, in general, very small crustacea, which seldom appear during the day. They usually remain in humid places, under stones, in the clefts of walls, in cellars, and often bury themselves in the earth. They appear to dread the light and heat of the sun. They walk slowly; but when they are pursued, they endeavour to save themselves by flight, and then they run tolerably fast.

They feed on different substances, attack and gnaw fruits of all kinds which have fallen to the earth, and also eat the leaves of plants. Degeer has observed small onisci devour a large one of their own species, which had been shut up with them, which proves that they are carnivorous.

The females lay eggs which disclose the young, as it were, in their bodies; they carry them in a sort of oval sac, slender and flexible, placed underneath their body, and extending from the head towards the fifth pair of feet. When the young are entirely formed, to give them a free issue, the mother opens the sac, or ovary, in which are formed one longitudinal cleft, and three transverse ones; then the little ones issue forth in a crowd, pressing one against the other, and after they have gone out, the mother closes the ovary. According to some writers, these crustacea are oviparous; but Geoffroy appears disposed to believe them oviparous and viviparous; that is, not that the living young are formed in the body of the mother, but the eggs; and instead of excluding these externally, she makes them pass into a sort of membranaceous pouch, which she has under the body, that there she hatches them, as it were, until these young ones being completely formed, issue forth from this pouch.

The *bopyrus cymothoe*, and some others of this order, are parasitical. The former inhabit the head of the *palœmon squilla;* those of the latter genus live altogether at the expence of fishes, whose blood they suck. The fish appear not to mind them; but they are not numerous in any individual.

SECOND GENERAL DIVISION.

THE ENTOMOSTRACA, *Müll.*

Under this denomination, formed from the Greek, and signifying *insects with shells*, Otho Frederick Müller comprehends the genus *monoculus* of Linnæus, to which we must add some of his *lernææ*. His researches on these animals, the study of which is so much the more difficult, as they are for the most part microscopic, and those of Schœffer and Jurine the elder, have excited the admiration and merit the acknowledgment of all naturalists. Other, but more partial labours, such as those of Ramdohr, Straus, Herman the younger, Jurine the younger, Adolphe Brogniart, Victor Audouin, and Milne Edwards, have extended our knowledge of these animals, especially in anatomical points; but in this respect, M. Straus,—though anticipated, as well as the elder Jurine, as to many important facts of organization by Ramdohr, whose memoir on the monoculi, published in 1805, they do not seem to know,—has surpassed them all. Fabricius has confined himself to the adoption of the genus *limulus* of Müller, which he has placed in his class of Kleistagnatha, or our family of brachyura, order decapoda. All the other entomostraca are united, as in the Linnæan system, into a single genus, that of *monoculus*, which he places in his class of polygonata, or our isopoda.

These animals are all aquatic, and for the most part inhabit the fresh waters. Their feet, the number of which varies, and in some exceeds an hundred, are in general adapted only for swimming, and sometimes ramified or divided, sometimes furnished with pinnulæ, or composed of lamellate articulations. Their brain is formed but of one or two lobules. The heart has always the form of a long vessel. The gills, composed of hairs or setæ, either isolated or joined together like beards, combs, or aigrettes, constitute a part of these feet, or of a certain number of them, and sometimes of the mandibles or of the upper jaws. (See *Cypris.*) Hence is the origin of the word *branchiopoda,* which we have applied to these animals, of which at first we had formed but a single order. Almost all of them have a testa, or shell, composed of one or two pieces, very thin, and most frequently nearly membranaceous and diaphanous, or at least they have a large anterior thoracic segment, often confounded with the head, and appearing to be a substitute for the testa. The teguments are generally rather corneous than calcareous; which approximates these animals to the insects and arachnida. In those which are provided with the usual jaws, the lower or exterior ones are always uncovered, all the jaw-feet performing the office of feet properly so called, and none of them being attached to the mouth. The second jaws, those of the phyllopoda, at most, excepted, even resemble these latter organs. Jurine has sometimes designated them under the name of hands.

Such are the characters which distinguish the *grinding* or masticating entomostraca from the malacostraca; the other entomostraca, those which compose our order of pæcilopoda, cannot be confounded with the malacostraca, because they are destitute of organs adapted for mastication, or because the parts which appear to serve as jaws are not assembled anteriorly, and preceded by a labrum, as in the foregoing crustacea and the masticating insects, but simply formed by the haunches of the locomotile organs furnished for this purpose with small spines. The pæcilopoda represent in this class those which in that of the insects are distinguished under the name of *suctoria.* They are all parasite, and seem to conduct by gradual shades to the lernææ; but the presence of eyes, the faculty of changing skin, or even of undergoing a sort of metamorphosis, the power of being able to transport themselves from one place to another by means of the feet, appear to us to establish a positive line of demarcation between these last animals and the preceding. We have consulted with respect to these transformations, different well informed naturalists, who have had frequent occasions of observing the lernææ, and none have ever witnessed any change of skin among them. It is true, indeed, that the young of daphniæ, and of some other neighbouring subgenera, those probably also of cypris, and of cytherea, on issuing from the egg, scarcely differ from their parents, except as to size; but those of cyclops, of phyllopus, of argula, undergo notable changes in their early age, either in the form of the body, or the number of the feet. In some, such as the argulæ, these organs even undergo transformations, which modify their uses.

The antennæ of the entomostraca, the form and number of which vary much, answer, in many of these animals, the purpose of swimming. The eyes are very rarely supported on a pedicle, and when they are, this pedicle is but a lateral elongation of the head, and never articulated at its base. They are often very closely approximated together, and often compose but a single eye. The organs of generation are situated at the origin of the tail; and it is erroneously that the antennæ of some males have been considered as the seat of them. This tail is never terminated by a fan-like fin, and does not present those false feet, which we have observed in the malacostraca. The final feet, indeed—if we except the phyllopoda—are thoracic, or jaw-feet. The eggs are accumulated under the back, or they are external, and under a common envelope, in the form of one or two small clusters, situated at the base of the tail. It appears that they may be preserved a long time in a state of desiccation, without losing their vital properties. At most, it is not until after the third moulting, that these animals become adult and capable of reproduction. In the case of some of them, it has been verified, that a single act of sexual intercourse can fecundate several successive generations.

We shall here close our supplement on the Crustacea in general, and on the first principal section of that class, the Malacostraca in particular, and shall now proceed to the translation of the Entomastraca.

THE FIRST ORDER OF ENTOMOSTRACA, OR SIXTH OF THE CLASS OF CRUSTACEA.

THE BRANCHIOPODA.

The characters of this order are, a mouth composed of a labrum, of two mandibles, of a ligula, and of one or two pairs of jaws; gills—or the first of these organs, when there are several—always anterior.

These crustacea are invariably erratic, generally covered by a testa in the form of a shield, or of a bivalve shell, and provided with four or two antennæ. Their feet, with some few exceptions, are solely adapted for swimming. The number of their feet varies; in some it is only six, in others, from twenty to forty-two, or even more than a hundred. Many of these animals have but a single eye.

The majority of these crustacea, being, as we have already observed, almost microscopical, it is obvious that the application of one of the characters which we have hitherto employed, namely the presence or absence of mandibulary palpi, would, under such circumstances, present difficulties almost insurmountable. We shall, however, place at the head of the entomostraca, all the branchiopoda whose mandibles are provided with palpi. They will compose the first two divisions of the *lophyropa;* but the form and the number of the feet, number of the eyes, and the testa, will furnish us with characters more easily to be employed, and more within the reach of general observation.

The order of the branchiopoda in the methods of Degeer,

Fabricius, and of Linnæus, was composed, with the omission of a single species (*M. Polyphemus*), only of the genus

MONOCULUS, *Linn.* *,

Which we shall divide into two principal sections.

The first, that of LOPHYROPA, is distinguished by the number of the feet, which never exceeds ten. Their articulations, besides, are more or less cylindrical or conical, and never entirely lamelliform or foliaceous. Their gills are not numerous, and the majority of these animals have but a single eye. Several, moreover, have mandibles furnished with a palpus. M. Straus, indeed, appears to attribute this character exclusively to *cypris* and *cytherea*, which compose his order of ostrapoda; but the observations of the elder Jurine, and of M. Ramdohr, prove that it is also proper to the *cyclopes*. The antennæ are almost always four in number, and serve for the purpose of locomotion.

In the second section, that of PHYLLOPA, the number of the feet is twenty at least, and in some, much more considerable. Their articulations, or at least the last ones, are flatted, in the form of ciliated leaflets. Their mandibles never exhibit palpi. They all have two eyes, situated in some at the extremity of two movable pedicles. Their antennæ, the number of which, in several, is but two, are generally small, and not adapted for swimming.

We shall divide the LOPHYROPA into three principal groups, very natural, and the first two of which approximate to the crustacea of the first three orders, in consequence of their mandibles having each one palpus, and likewise from some other characters.

1st. Those (CARCINOIDA, *Lat.*) whose testa, more or less ovoïd, or ovaliform, is not bent in two like a bivalve shell,

* In the system of Geoffroy, the genus *Binoculus* is included.

and leaves the lower part of the body uncovered. They never have antennæ in the form of ramified arms. Their feet are ten in number, and more or less cylindrical or setaceous. The females, whose gestation has been observed, carry their eggs in two species of external sacs, situated at the base of the tail. Some have two eyes.

2nd. Those (OSTRACODA, *Latr.*) *ostrapoda*, Straus, whose testa is formed of two pieces, or valves, resembling those of the shell of a muscle, united by a hinge, and enclosing the body in a state of inaction. They have but six feet, none of which terminate in the manner of a digitated fin, accompanied with a branchial plate. Their antennæ are simple, filiform, or setaceous. They never have more than one eye. Their mandibles and upper jaws are provided with a branchial plate. The eggs are placed under the back.

3rd. The last, (CLADOCERA, *Lat.*, *Daphnides*, Straus,) have also but a single eye, and the testa folded doubly, but without a hinge (Jurine), terminated posteriorly in a point, and leaving the head, which is covered with a sort of buckler in the manner of a beak, exposed. They have two antennæ, usually very large, in the form of arms, divided into two or three branches, at the end of the pedicle, furnished with filaments, always projecting, and serving the purpose of oars. Their feet, ten in number, are terminated by a digitated, or pectinated fin, and accompanied, with the exception of the first two, by a branchial plate. The character is particularly applicable to daphnia, the most numerous subgenus of this division, and by analogy, to the polyphemi, and lyncæi.

Their eggs are also situated under the back. Their body is always terminated posteriorly, in the manner of a tail, with two setæ or filaments at the end. The anterior extremity of the body is sometimes elongated into a beak, and sometimes forms an approximation to a head, almost entirely occupied by a large eye.

The first division of the branchiopodous lophyropa, that of carcinoïda, may be subdivided into two, according to the number of the eyes; some have two.

Here the testa entirely covers the thorax; the eyes are large and very distinct; the intermediate antennæ are terminated by two filaments.

ZOEA, *Bosc,*

Have the eyes very large, globular, entirely uncovered; and certain prominences in the shape of horns on the thorax.

Zoe pelagica, Bosc. (Hist. Nat. des Crust. II. xv. 3. 4.) has the body semitransparent, four antennæ inserted below the eyes, the exterior ones elbowed and bifid; a sort of long beak on the front of the thorax, between the eyes, and a long and pointed prominence on the hinder part of the back. The feet are very short, and scarcely visible, with the exception of the last two, which are elongated, or terminated like a fin. The tail is of the length of the thorax, curved, formed of five articulations, the last of which is large, crescented, and spinous. This crustaceous animal was found by M. Bosc in the Atlantic ocean.

The *monoculus taurus* of Slabber, (Microsc. V.) and the *cancer germanus* of Linnæus, appear to have some relations with it. (*See* the Natural History of Crustacea and Insects, by Latreille, and the work of M. Desmarets on the former animals.) This genus has not yet been described in a complete, or at least in a satisfactory manner, and we have not been able to procure a single individual of it.

NEBALIA, *Leach,*

Have the eyes triangular, flatted, and partly covered by a triangular and vaulted shell.

The feet are forked, and the appendages of the end of the

tail are in the form of setæ. *Nebalia Herbstii*, Leach., Zool., Miscell. xlv. Desm. Consid. xl. 5; Rambd. monoc. i. 8, 9?

The *nebalie ventrue* of M. Risso, (Journ. de Phys. Octobre, 1822) probably constitute in the section of the Schizopoda, a subgenus proper. In the *cyclops exiliens* of Viviani, the thorax is divided into several segments, which circumstance excludes it from the nebaliæ. It also forms a new subgenus, intermediate between the preceding and the following.

There the thorax, or the testa, seen from above, is divided into five segments, the first of which, much the largest, supports the antennæ, the eyes, and the jaw-feet. The second and third have each a pair of feet; the fourth supports the two following pairs, and the fifth the last. Their eyes are small, and not projecting, all the antennæ are terminated by a simple filament.

CONDYLURA, *Lat.*,

The lower antennæ are much longer; the anterior sides of the first segment are elongated into a point, and form two shells, or scales, approximating together like a bill; the feet are terminated in a silky point. Some of the intermediate ones have, as in the Schizopoda, an external appendage near their base. The tail is narrow, composed of seven rings, the last of which, elongated and conical, advances between the two lateral appendages, which are slender, in the form of stylets, with two articulations, the last of which is setaceous. *Condylure de Dorbigny*, Lat., on the maritime coasts of La Rochelle.

Nota. The genus *Nicothoë*, of MM. Audouin, and Milne Edwards, on the supposition that it has mandibles and jaws, would belong to this section; but as the crustaceous animal on which it has been established, is parasite, and as I think that I have perceived in it the vestiges of a sucker, I have placed it in the order of pœcilopoda. I remark, nevertheless, that the feet, with the exception of the anterior ones, much resemble those of cyclops, and that the females also carry their eggs in two sacs, situated at the base of the tail, in the same manner as the latter.

The second of these naturalists has just published, in the 13th volume of the Annals of the Natural Sciences, some new researches on the Nebaliæ, and the characters of those other new genera of crustacea. Our labours on the animals of this class having been terminated at the moment when the memoir of M. Milne Edwards was communicated to the academy, and not having then the time to return to the subject, we have transferred our account of those genera, as well as of those established in the family of the Araneïdes by M. Savigny, and of some others recently introduced by Count Dejean, into that of the carnivorous coleoptera, in the supplement to this work. We shall also give them the characters of some other generic sections, established by MM. Guerin, Lepetier de St. Fargeau, and Serville. I could not have introduced them into my present researches, without hurrying an examination, which should be so much the more careful, as it is more easy to multiply generic groups.

The other lophyropa of our first division, in which the thorax, as well as in condylura, is divided into several segments, the first of which is much the largest, present but a single eye, situated in the middle of the forehead, between the upper antennæ. Such are

CYCLOPS, *Mull.*,

So accurately observed by the elder Jurine and M. Ramdohr. Their body is more or less ovaliform, soft, or gelatinous, and is divided into two portions, one anterior, and composed of head and thorax, and the other posterior, or the tail. The segment immediately preceding the sexual organs, and which, in the females, supports two appendages in the form of little feet (*fulcra*, Jurine), may be considered as the first of the tail, which is not always very exactly distinguished from the thorax. It is formed of six segments or articulations; the second supports, underneath, in the males, two articulated appendages, sometimes simple, sometimes having, at the internal side, a small division, or branch, of various forms, and constituting wholly, or in part, the organs of generation. The vulva is situated, in the other sex, on the same articulation; the last is terminated by two points, or stylets, forming a fork, and more or less furnished with setæ, or penniform filaments. The other, or anterior portion of the body, is divided into four segments, the first of which, much the largest, composes the head and a portion of the thorax, which are thus covered by a common scale. It supports the eye, four antennæ, two mandibles (*internal mandibles*, Jurine), furnished with a simple palpus, or divided into two articulated branches, two jaws (external mandibles, or labrum with barbles, Jurine), and four feet, each divided into two cylindrical stems, furnished with hairs, or barbed filaments; the anterior pair, representing the second jaws, differs a little from the following ones: it is compared to sorts of hands by Jurine. Each of the three following segments serves as an attachment to a pair of feet, composed like the two last of the preceding. Two of the antennæ, superior to the others, are longer, setaceous, simple, and composed of a great number of small articulations; they facilitate by their action the movements of the body, and almost perform the office of feet; the inferior (*antennulæ*, Jurine) are filiform, present most generally but four articulations, and are sometimes simple, sometimes forked. They make by their rapid movements a whirlpool in the water. In the males, the upper antennæ, or one only, present strangulations and a swelling, followed by a hinge-joint. By means of these organs, or of one of them, they seize either the last feet or the end of the tail of their females, in their amorous preludes, and retain them in spite of themselves in situations appropriate to the manner in which they fix themselves. The females carry off the males when they do not at first wish to yield to their desires. Intercourse takes place as in the preceding crustacea by prompt and reiterated acts. Jurine has witnessed three in the space of a quarter of an hour. It was believed until his time that the generative organs of the males were situated at the upper antennæ; and this erroneous opinion appeared to receive some confirmation from analogous facts observed in the araneides. On each side of the tail of the females is an oval sac, filled with eggs (external ovary, Jurine), adhering by a very slender pedicle to the second segment, near its junction with the third, and where also the orifice of the deferential canal of the eggs is visible. The pellicle forming these sacs is but a continuation of that of the internal ovary. The number of the eggs which they contain augments with age. At first, brown or obscure, they afterwards assume a reddish tint, and become almost transparent when the little ones are ready to come forth, but without growing larger. If they are isolated or detached, at least at a certain period, the germ will perish. A single fecundation (but that is indispensable) may suffice for successive generations. The same female can have ten broods of eggs in the space of three months. Counting but eight, and supposing each of them to consist of forty young, the sum total of births would amount to nearly four thousand five hundred millions. The duration of the stay of the fœtus in the ovary is from two to ten days, which depends on the temperature of the seasons, and divers other circumstances. The oviferous sacs sometimes present elongated glandiform bodies, more or less numerous, which appear to be assemblages of infusory animalcules.

At their birth, the young ones have but four feet, and their body is rounded, and without a tail. Müller had formed with these young individuals his genus *Amymone*. Some time after (in fifteen days, from February to March) they acquire another pair of feet. This is the genus *Nauplius* of the same. After the first moulting, they have the form and all the parts which characterize the adult state, but with smaller proportions. Their antennæ and their feet are proportionally shorter. At the end of two other moultings, they are fit for generation. Most of these entomostraca swim upon the back, leap with vivacity, and can go backwards as well as forwards. In default of animal matters, they attack vegetable substances; but the fluid in which they live habitually does not pass into the stomach. The alimentary canal extends from one extremity of the body to the other. The heart, in the cyclops *castor* is immediately situated under the second and third segment of the body, and ovaliform. Each of its extremities gives birth to a vessel, one of which goes to the head and the other to the tail. Immediately under it is another analogous organ, but pyriform, producing also, at each end, a vessel, perhaps representing the branchio-cardiac canals, of which we have spoken in treating of the circulation of the decapod crustacea. It would result from several experiments of Jurine upon cyclopes, alternately asphyxiated, and restored to life, that in this sort of resurrection the extremity of the intestinal canal and the fulcra give the first sign of life; and that the irritability of the heart is less energetic. That of the antennæ, and more especially of those of the males, of the palpi, and of the feet, is inferior. When a portion of the antenna is cut away, no organic change is effected. The reparation takes place under the skin, since this organ re-appears perfectly entire at the next moulting. The cyclops *staphylinus* forms a particular division, by reason of its shorter antennæ, the upper of which have much fewer articulations than in the other cyclops, while the lower, on the contrary, present more; also by reason of its body, which gradually grows slender towards its posterior extremity, so that it appears to have no tail, at least abruptly formed, and its under part is armed, in the female, with a sort of horn, curved backwards. The cyclops *castor*, and some others, whose lower antennæ and mandibulary palpi are divided beyond their base into two branches, may also compose another group. That which Dr. Leach designates under the generic name of CALANUS, might, indeed, form a subgenus proper, if it were true that the animal of which it is the type had no inferior antennæ. But has he ascertained this point himself, or does he speak of it only after Müller? This I know not.

Cyclops quadricornis, Monoculus quadricornis, Linn., Müll., Entom. xviii. 1—14; Jurine. Monoc. i. ii. iii., has all the antennæ simple, or without divisions. The lower have four articulations, and their length scarcely equals the third of the upper; the body properly so called is tolerably inflated, and almost ovoïd; the tail is narrow, and composed of six segments. The colour varies much; some are reddish, others whitish or greenish. The total length is two lines. This species is very common.—Desmarets' Consid. p. 364. See, for the other species, the same work, p. 361—364; Müll. Entom. *G. cyclops,* Jurine Hist. des Monoc. p. 1—84, first family of the Monoculi with univalve shell, Ramd. Monoc. i. ii. iii.

The second general division of the branchiopodous lophyropa, those whose testa is formed of two valves, united by a hinge (our OSTRACODA, or the order of *Ostrapoda* of M. Straus), is composed of two subgenera, the first of which, that of Cytherea, appears to us, since the valuable researches of this naturalist on the second (cypris), to require, for the purpose of establishing its characters in a less equivocal manner, a profounder study than has been bestowed upon it by Müller, our only authority on the subject. According to him,

CYTHEREA, *Müll.* CYTHERINA, *Lam.*,

Would have eight simple feet, finishing in a point, though it is probable that there are but six; two antennæ, equally simple, setaceous, composed of five or six articulations, with some scattered hairs.

They are found in the salt waters and the brackish waters of the sea-shores, among the sea-weed and confervæ.

If these entomostraca are exclusively marine, it is not surprising that Jurine and other observers, whose researches, in consequence of the places of their residence, could extend only to the entomostraca of the fresh water, should not have spoken of the Cythereæ.

CYPRIS, *Müll.*,

Have but six feet, though M. Ramdohr says four, and M. Jurine eight. The first considered the last two as appendages of the male sex; and the second took the palpi of the mandibles, and the branchial plate of each upper jaw, for so many feet. Nor did the latter reckon in this number, those which the former presumed to be sexual organs. He regards them as filaments of five articulations, issuing laterally from the pouch of the matrix, and the use of which he is ignorant.

The two antennæ are terminated in the manner of a fasciculus of setæ, like a brush.

The testa, or shell, forms an ovaliform body, arched and gibbous on the back, or on the side of the hinge, almost straight, and a little emarginated and reniform, on the opposite side. In front of the hinge, in the medial line, the eye forms a thick blackish and round point. The antennæ, inserted immediately underneath, are shorter than the body,

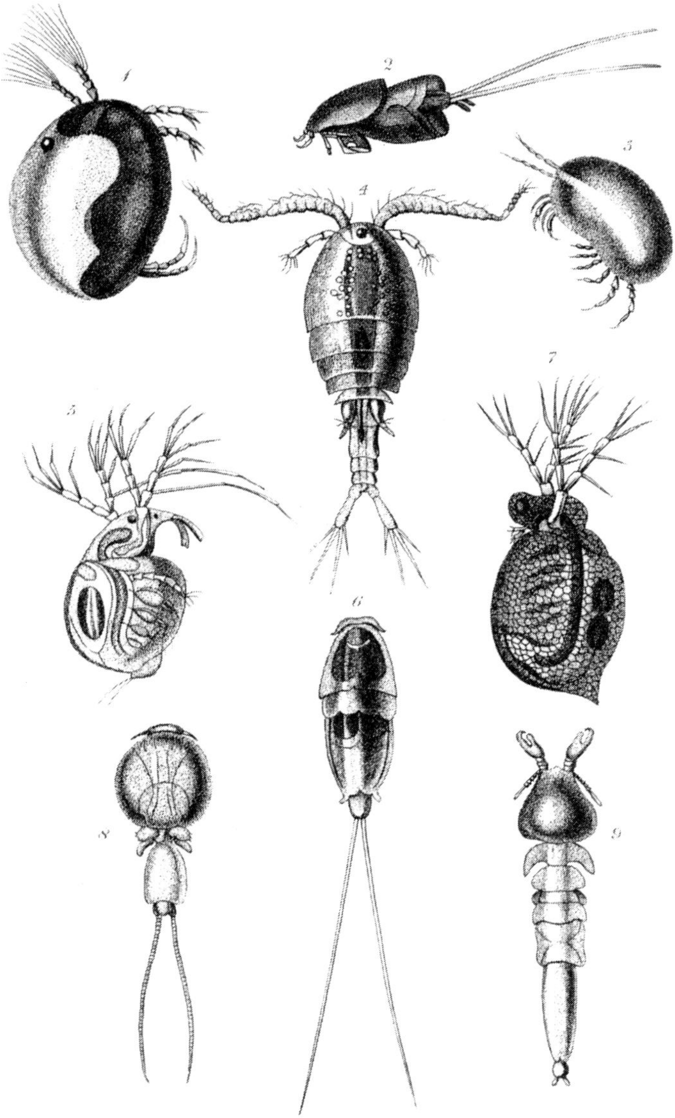

1 *Cypris religiosa*. 5 *Lynceus roseus*.

2 *Anthosoma Smithii*. 6 *Pandarus bicolor*.

3 *Cytherea fulva*. 7 *Daphnia clathrata*.

4 *Cyclops communis*. 8 *Caligus Mulleri*.

9 *Dichelestium sturionis*.

London, Published by Whittaker & Co. Ave Maria Lane, 1833.

setaceous, composed of seven or eight articulations, the last of which are shortest, and terminated by a fasciculus of ten or fifteen setæ, serving as fins. The mouth is composed of a carinated labrum, of two large denticulated mandibles, each supporting a palpus divided into three articulations, and to the first of which adheres a small branchial plate, presenting five digitations (interior labium, Ramd.), and of two pairs of jaws; the upper two, much larger, have at the internal edge, four mobile and setaceous appendages, and at the external side a large branchial plate, pectinated at its anterior edge; the second are composed of two articulations, with a short palpus (forked in *cypris strigata*, Ramd.), almost conical, inarticulated, silky at the end, as well as the extremity of their jaws. A sort of compressed sternum performs the office of lower labium (external labium, Ramd.) The feet are divided into five articulations, the third of which represents the thigh, and the last the tarsus; the anterior two are inserted below the antennæ, much stronger than the others, directed forwards, with stiff setæ, or long hooks, gathered into a fasciculus, at the extremity of the last two articulations; the following four feet are without them. The second, situated at the middle of the under part of the body, are at first inclined backwards, arched, and terminated by a long and strong hook, which goes forward; the last two never appear externally, are raised, and applied on the posterior sides of the body, to support the ovaries, and are terminated by two very small hooks. In the figure of Ramdohr, these feet have but three articulations, and the last is a little dilated and emarginated at the end, with a hook in the middle of this emargination. The body presents no distinct articulation, and is terminated posteriorly in a kind of tail, soft, folded underneath, with two conical or setaceous filaments, furnished with three setæ or hooks at the end, directed backwards, and issuing from the testa. The ovaries form two large simple and conical vessels, closed at their origin, situated at the posterior parts of the body, under the testa, and open, one by the side of the other, at the anterior part of the abdomen, where the canal formed by the tail establishes a communication between them. The eggs are spherical. The times of laying and moulting with these crustacea are not less numerous than with the cyclops and other entomostraca, and their manner of living is the same. Ledermuller says he has witnessed their sexual intercourse. Nevertheless, none of the modern naturalists, who have observed them most closely, could ever positively discover their sexual organs, nor witness their union. M. Straus has seen below the origin of the mandibles, the insertion of a thick conical vessel, filled with a gelatinous substance, appearing to communicate with the œsophagus by a narrow canal, which he suspects to be a testicle or a salivary gland. The individuals subjected to this observation having ovaries, the cypris would be, on the first of these suppositions, hermaphrodite. But that is so much the more doubtful, as he himself remarks, that the males may very probably exist only at a certain time of the year, and that the vessel of which he speaks, communicating with the œsophagus, appears to have more relations with the digestive functions than with those of generation.

According to Jurine, the antennæ are true fins, the filaments of which these animals develope, and re-unite at will, according to the degree of rapidity which they are desirous of communicating to their progress. Sometimes they only allow a single one to appear, at others they unfold them altogether. We also think that these filaments, and those of the two anterior feet, may just as well concur in respiration as those laminæ of the mandibles and of the two upper jaws, which M. Straus calls branchial. The last, or those of the jaws, appear to me to be a true palpus, but much dilated, and the two others an appendage of the mandibulary palpi.

According to the Genevese naturalist just cited, these animals, when they swim, move the two anterior feet with as much rapidity as the antennæ; but they move them slowly when they walk on the surface of marshy plants. These feet, conjointly with the two terminated by a long hook, or the penultimate feet, then support the body. He supposes that those which, according to him, form the second pair, are intended to produce an aqueous current, and direct it towards the mouth. This would assimilate their functions to those of the lower antennæ, which he names antennulæ. The two filaments composing the tail are united, and seem to form but one, when they come forth from the testa. They answer the purpose, as is presumed, of cleansing its interior. The female deposits her eggs in a mass, fixing them, by means of a gluten, on plants or mud. Hooked, at this time, with the aid of the second feet, and so as not to fear the shocks of the water, she employs almost twelve hours in this operation, which in the larger species furnishes as many as four and twenty eggs. M. Jurine has collected some of these packets of eggs, when they came forth, and after having isolated them, he has seen them disclose the young, and he has obtained another generation without the intervention of the males. A female which had laid its eggs on the twelfth of April, up to the eighteenth of May inclusive, changed skin six times. The twenty-seventh of the same month she laid a second set, and two days after, on the twenty-ninth, a third. He concludes from this, that the number of moultings from infancy is in relation with the gradual development of the individual; that this development cannot manifest itself but by the general separation of an envelope become too small to lodge the animal; and that there is a determinate limit of size to which the latter must attain.

The lophyropa of our third division, (our *Cladocera)* or the Daphnides of M. Straus, compose, in the history of the monoculi of Jurine, his second family. The form of two of their antennæ, which resemble two ramified arms, serving as oars, and the faculty which they have of jumping, have caused one of their most common species to be called the *arborescent aquatic flea.*

The first of these naturalists, who has given us an excellent monograph of Daphnia, a subgenus of this division, has established two new ones, one under the denomination of LATONA, having as character, the antennæ in the form of oars, divided into three branches, and of a single articulation; (*Daphnia Setifera*, Müll.) and the other that of SIDA, approximating to the known subgenera of the same division, with relation to the same antennæ, divided only into two branches, but one of which has two articulations, and the other three. (*Daphnia cristallina*, Müll.) According to him, the daphnia should be distinguished from the preceding, and from the lynceæ, by one of the two branches of the oars being composed of three articulations, and the other of four. Nevertheless, Jurine (Hist. des Mon. p. 92.) says, that each branch is composed of three articulations; but it appears that he took no account of the first of the posterior branch, which is indeed very short. The last in all those lophyropa, is terminated by three filaments, and each of the preceding gives out another. These filaments are simple or barbed. There exist also two other antennæ, but very short, especially in the females, situated at the anterior and inferior extremity of the head, and which have but a single articulation, with one or two setæ at the end.

POLYPHEMUS, *Müll.*,

Have, as well as daphnia and lynceus, their antennæ in the form of oars, divided into two branches; but each of them is composed of five articulations. Moreover, their head, very distinct, and rounded, supported on a sort of neck, is almost entirely occupied by a large eye; their feet are completely uncovered.

As yet we know but a single one, the *Polyphemus of the ponds, Monoculus pediculus*, Linn. Deg. Insect. VII. xxviii. 6—13; *Polyphemus oculus*, Müll., &c.

According to Jurine, the feet in no respect resemble those of the monoculi of this division. They are composed of a thigh, a leg, a tarsus with two articulations, and from the extremity of which come forth (that of the last pair excepted), some small filaments. From the anterior extremity of the head some small antennæ project, of a single articulation, terminated by two filaments. The shell is so transparent, that all the vis cera may be distinguished through it. The matrix, when it is full of eggs, occupies the major part of the interior of this shell. Their largest number does not exceed ten. When we trace the gradual development of the fœtus, we are struck by the prompt appearance of the eye, comparatively to that of the other parts of the body: it is at first greenish, and passes insensibly to deep black. The abdomen, after being convoluted on itself, from behind forwards, folds suddenly backward to form a long, narrow, and pointed tail, from which come two long articulated filaments. The animal always swims upon its back, and most frequently horizontally, communicating to its arms or oars, and to its feet, lively and reiterated movements. It executes, with considerable nimbleness and agility, all sorts of evolutions. It is subject, in its youth, and after its first moultings, to a malady called *ephippium*, but this ephippium or saddle has always a determinate figure, and never contains the two oval balls, which it presents in daphnia. In a state of captivity, this animal will not live long, and the young cannot be reared, at least Jurine could not preserve them after the first moultings, nor observe the progress of their generations. He did not recognize males in any of the individuals which he kept; in fact, he was able to procure but a very small number of individuals, this species being rare in the environs of Geneva. But it appears that it is very common in the marshes and ponds of the north, where it forms very considerable troops.

DAPHNIA, *Müll.*,

Have their oars always uncovered as far as their base or the origin of their pedicle, as long, or almost as long, as the body, divided into two branches, the posterior of which has four articulations, the first very short, while the anterior has but three. Their eye is small, or in the form of a point, and, if we except some species, we do not see in front of it, as in lynceus, a small black spot, in the form of a point, which Müller took for another eye. This is also the opinion of Ramdohr, and, as he has discovered it in *daphnia sima*, it might be possible that this character was common to this sub-genus and the lyncei, although but little visible in the different species. Schœffer had already observed this spot.

Although the organization of these crustacea, from the extreme smallness of the animal, would seem to escape the inspection of the observer, yet there is scarcely any that is better known. Without speaking of those who have especially occupied themselves with microscopic researches, four most profound naturalists, Schœffer, Ramdohr, Straus, and Jurine the elder,—but the third more particularly,—have studied these animals with the most scrupulous attention. If some details of organization have escaped the last, the researches of MM. Ramdohr and Straus supply that deficiency. Jurine, besides, completes their observations as to habits, which, for a long time, he most accurately traced and observed.

The mouth is situated underneath, at the base of the bill. We consider, with M. Ramdohr, as a hood of an elongated form, the lower portion of the head, which M. Straus calls labrum, and we apply this last denomination to the part which he names the posterior lobule of the labrum. Immediately underneath are two mandibles, (interior jaws, Ramdohr), very strong, without palpi, directed vertically, and applied on two horizontal jaws, (external jaws; Ramdohr), terminated by three robust corneous spines, in the form of curved hooks. Then come ten feet, all having the second articulation vesicular; the first eight are terminated by an expansion in the form of a fin, furnished on its edges with setæ or barbed filaments, disposed in the manner of a crown or comb; the two anterior appear more especially adapted for prehension. Accordingly, M. Ramdohr takes them for double palpi (internal and external). These are the same pieces that Jurine elsewhere calls hands. In the figure which he has given of them, the terminal setæ appear to be barbed. We do not see why these appendages might not serve for respiration, a property which M. Straus grants only to the following, because these latter have, in addition, a lamina on the internal side, which, with the exception of the last two, is bordered with a range of setæ, in the manner of a comb, and equally barbed, to judge from the figures given by Jurine and M. Ramdohr. The last two feet have a structure a little different, and M. Ramdohr distinguishes them under the name of claws. The abdomen, or the body properly so called, is divided into eight segments, perfectly free between its valves, narrow, elongated, curved underneath at its extremity, and terminated by two small hooks turned backwards. The sixth segment presents, on its upper part, a range of four nipples, forming denticulations, and the fourth, a sort of tail. The ovaries are placed along the sides, between this segment and the first, and open separately near the back, into a cavity (matrix, Jurine), situated between the shell and the body, where the eggs remain for some time after being laid.

Müller has given the name of *ephippium* or saddle to a large, obscure, and rectangular spot, which at certain periods of the year, and particularly in summer, shows itself, after the moulting of the females, at the upper part of the valves of the shell, and which Jurine attributes to a malady. According to M. Straus, this ephippium presents two ovaliform bubbles, transparent, placed one before the other, and forming with those of the opposite side two small oval capsules, opening like a bivalve capsule. It is divided, as well as the valves of which it forms a part. into two lateral moieties, united by a suture along their superior edge. Its interior presents another similar one, but smaller, with the edges free, except the superior one, which attaches to the valves, and the two moieties of which, playing like a hinge one upon the other, present the same ampullæ, as the exterior valves. Each capsule encloses an egg, with a horny and greenish shell, similar in other respects to the common eggs, but remaining longer without being developed, and passing the winter under this form. At the period of moulting, the ephippium, along with its eggs, is abandoned, with the exuvia of which it forms a part. It serves as a shelter for these eggs during the cold. The heat of spring causes them to disclose, and little ones come forth, absolutely similar to those produced from the common eggs. Schœffer has asserted that they may remain a long time in a state of desiccation, without the germ being injured; but none of those which M. Straus preserved in that state disclosed. They are absolutely free, or without adhering to each other, in the cavities which are proper to them. According to Jurine, they may disclose in summer at the end of two or three days. Under the climate of Paris, where M. Straus has observed them at all periods of the year, they require at least one hundred hours. The fœtus, twenty hours after the egg is laid, presents nothing but a rounded and unformed mass, on which may be remarked, on close examination, the obtuse rudiments of arms, in the form of very short and imperfect stumps, cemented against the body. Neither head nor eye are visible. The body, green or reddish, is punctated with white, like the eggs, and as yet makes no movement. It is not until the ninetieth hour, and when the eye has appeared, that the arms and the valves are elongated, and the fœtus commences to move. At the hundredth hour, it is already very active. In fine, at the hundred and tenth, it differs from the young just born, only by having the setæ of the oars glued against their stem, and that the tail of the valves is bent underneath, and received between the inferior edges of these pieces. Towards the end of the fifth day, the tail which terminates the valves in youth, and the setæ of the arms, let go, like a spring, and the feet then only begin to stir. The young ones being then in a state to appear, the mother immediately lowers her abdomen, and they shoot forth. Some eggs, newly laid, and placed in a glass vessel, where they were closely observed by M. Straus, were developed in the same manner. Jurine has also given us, on the progressive changes of the fœtus of daphnia, some analogous observations, but which were made in winter; and as the little ones were not disclosed until the tenth day, he had the advantage of being able more accurately to observe and define those developments. The first day, the egg presents a central bubble, surrounded with other smaller ones, with coloured molecules in the intervals. These molecules and bubbles appear intended to form the organs by agglomerating, approximating to the centre, and finally disappearing. The sixth day, the form of the fœtus begins to be defined: the seventh, the head and feet are distinguishable. On the eighth, the eye appears, as well as the intestine. The following day, the net-work of the same eye commences to be visible. The bubbles have almost entirely disappeared, with the exception of the central one, which occupies the alimentary canal under the heart. On the tenth, the development of the fœtus is terminated, the little one issues from the matrix, and remains for a moment motionless.

The males, at least in the species observed by M. Straus, are very distinct from the females. The head is proportionally shorter; the back is less projecting. The valves are less broad, and less gibbous superiorly, and they gape in front, so as to present, in this place, a broad and almost circular aperture. The antennæ are much larger, presenting the appearance of two horns directed downwards, and have been considered by Müller as the sexual organs in the male. These sexual organs, M. Straus was unable to discover, but he has remarked that the onglet, terminating the last articulation of the two anterior feet, (the second, supposing the oars to be the first) is much larger than in the female, that it has the form of a very large hook, strongly curved outwards, and that the setæ of the third articulation is also much longer. These hooks answer the purpose of seizing the female. The nipples of the sixth segment of the abdomen, are much less visible, and have the form of tubercles in early age. With the exception of the lower antennæ, much longer in the males, the two sexes nearly resemble, and the two valves of the shell are terminated in both, by a stylet, denticulated underneath, arched towards the bottom, and of a length almost equal to that of the valves. At each moulting, this stylet grows shorter, so as to form, in the adults, only a simple obtuse point.

The males are very ardent in the pursuit of their females, and often of the same individual.

A single act fecundates the females for several successive generations, as far as six at least, as has been proved by M. Jurine; M. Straus remarking that the orifices of the ovaries are placed very deeply under the valves, and that, therefore, no part of the body of the male could reach them, suspects that there is no copulative organ in the latter, and that he only ejaculates the fecundating fluid under the valves of the female, whence it is introduced into the ovaries; but analogy seems hostile to such a conjecture. Jurine has witnessed their intercourse, which lasts, at most, from eight to ten minutes. The male, placed at first on the back of the female, seizes her with the long filaments of his anterior feet; then getting towards the inferior edge of her shell, and approximating his own to its aperture, he there introduces these filaments, as well as the hooks or harpoons of those feet; he then approaches his tail to that of his companion, who at first rejects his addresses, runs with great swiftness, carrying him along with her, but ends by yielding. Some small bodies, in the form of grains, of a green, rose, or brown colour, according to the seasons, composing the ovaries, ascend gradually into the matrix, and there become eggs. Jurine observes, that the males of the *D. Pulex* are of small number, in comparison with the females, that in spring and summer hardly any are to be found, but that they are less rare in autumn.

About eight days after their birth, the young daphniæ change their skin for the first time, and subsequently continue the same operation every five or six days, according to the greater or less elevation of temperature; not only the body and the valves, but also the gills and the setæ of the oars, are stripped of their epidermis. It is not until the third moulting, that these crustacea begin to reproduce; they at first lay but a single egg, then two or three, and increase progressively, even up to fifty-eight in one species, *(D. Magna)*. In a single day after the laying, the female changes skin, and in the teguments which she has abandoned, are found the shells of the last laid eggs. In a moment afterwards she lays again. The young of one and the same birth, are almost always of the same sex, and it is rare to find in a birth of females, two or three males, and *vice versâ*. But in five or six births during the summer months, one at the most, consisting of males, takes place. Individuals are often to be met with, whose teguments are of a milky white, opake and thick, without the animals appearing to be affected by this modification. On the renewal of their testa, but slight traces of this alteration are perceptible, and those are manifested by rugosities.

These crustacea cease to reproduce, and to moult, on the approach of winter, and end by perishing before the commencement of the frosts. The eggs contained in the ephippia, which had been laid during the summer, disclose on the first return of warm weather, the following spring. The pools are then speedily repopulated with an infinite number of daphniæ. Many naturalists have attributed the sanguine colour which these waters sometimes assume, to the presence of myriads of the *Daphnia Pulex*. But M. Straus declares that he never observed this fact, and that at all times this species is but slightly coloured. In the morning and evening, and even during the day, when the sky is cloudy, the daphniæ remain habitually at the surface of the water. But during the very warm weather, and when the sun shines strongly on the pools or stagnant waters which they inhabit, they sink into the water; and remain at a depth of six or eight feet, or more. Frequently, not a single one is to be seen at the surface. They swim by little springs, of greater or less extent, according as their oars are more or less long, and as the buckler, which covers the front of their body, projects outwards, more or less, the largeness of this projection being capable of impeding their movements. According to M. Straus, their nourishment consists exclusively of small portions of vegetable substances which these animals find at the bottom of the water, and very often of confervæ. They constantly rejected the animal substances which he presented to them. He has frequently seen them swallow their own excrements, drawn in by the current of water produced by the action of their feet, and which conveys their ordinary aliment towards their mouth. The hooks which terminate the extremity of their tail, serve to cleanse their gills.

The *Daphnia pulex*, the most common of all, (*Monoculus pulex*, Linn.); *pulex aquaticus arborescens*, Swamm. Bib. Nat. xxxi; *Perroquet d'eau*, Geoff. Hist. Ins. ii. p. 455; Schœff. Die. Grün. arm. polyp. 1755, i. 1. 8.; Straus, Mem. du Mus. d'Hist. V. xxix. 1—20; Jurine, Mon. viii.—xi. has, according to M. Straus, the beak large and convex; the setæ of the oars plumose; the first nipple of the sixth segment, in the form of a little tongue; the valves denticulated at the inferior edge, and terminated by a short tail, obtuse in the females. This last character distinguishes it from another species, with which it has been confounded, the *D. longispina*, Str. Deg. Insect. VII. xxvii. 1—4.

See for the other species, the afore cited memoir of M. Straus: Muller, Entom. and Jurine Hist. des Monocles, second family. pag. 185. 58. and p. 181—200. *See* also, for *D. sima*, and *longispina*, Ramd. Monoc. v.—vii.

The last subgenus of the lophyropa is that of

LYNCEUS, *Müll.* CHILODORUS, *Leach.*

Which is but little distinguished from the preceding, except by its oars, evidently shorter than the shell, and the lower portion of which makes little or no projection. According to M. Straus, the articulations of their branches should be more numerous than in the preceding subgenera; all have in front of their eye a small spot, which has the appearance of another eye; the beak is proportionally more elongated than that of daphnia, curved and pointed.

See Müll. Entom. *G. lynceus.* Jurine, Monoc. pag. 151—158; and Desmar. Consid. 375—378.

The second section of branchiopoda, that of PHYLLOPA, is distinguished, as we have said, from the first, by the number of the feet, which is at least twenty, and by the lamellate or foliaceous form of their articulations; the eyes are always two in number, and sometimes pedunculated; several of them have, besides, a simple eye.

These animals represent, in the class of the crustacea, the myriapoda of that of the insects.

These crustacea are distinguished into two principal groups.

The one, (CERATOPHTHALMA, *Lat.*,) have ten pair of feet, at the least, and twenty-two, at most, without a vesicular body at their basis, and the anterior of which are never longer than the others, nor ramified; their body is either enclosed in a testa, in form of a bivalve shell, or naked, with each of the thoracic divisions supporting a pair of feet uncovered; the eyes are sometimes sessile, small, and very much approximated; sometimes, and most frequently, they are situated at the extremity of two mobile pedicles. The eggs are either interior or exterior, and enclosed in a capsule at the base of the tail.

Here the eyes are sessile, immoveable, and the body is enclosed in an oval testa, having the form of a bivalve shell. The ovaries are always internal. Such are

LIMNADIA, *Adolph. Brogn.*,

Which are so closely connected with the preceding, that the only known species has been placed among the daphnia by the younger Hermann. The testa is bivalve, oval, and encloses the body, which is linear, and inflected in front. At the head, and almost confounded with it, are, first, two eyes placed transversely, and very closely approximated; secondly, four antennæ, two of which are much the largest, each composed of a peduncle of eight articulations, and of two branches or filaments, setaceous, divided into eight articulations, and a little silky; the other two intermediate ones, are small, simple, and broad at their extremity; thirdly, the mouth, situated underneath, consisting of two mandibles, swelled, arched, and truncated at their lower extremity, and of two foliaceous jaws. These parts form, when united, a sort of inferior beak. The body, properly so called, is divided into twenty-three segments, each supporting, with the exception of the last, a pair of branchial feet. All these feet are similar, very much compressed, bifid, with the external division simple, ciliate at the external edge, and the other quadriarticulate, and strongly ciliated at the internal edge. The first twelve pairs are of the same length, and larger than the others. The length of the latter diminishes progressively. The eleventh pair, and the two following, have at their base a slender filament, ascending into the cavity, which is between the back and the testa, and serving as a support for the eggs. The last segment, or tail, is terminated by two filaments. The ovaries are interior, and situated on the sides of the alimentary canal, from the base of the first pair of feet as far as the eighteenth, and their issue appears to be situated at the root of some of them. The eggs, after the laying, occupy the dorsal cavity, of which we have spoken, and are attached there, by means of small filaments, themselves adhering to those of the supports. They are at first round and transparent; they then assume a yellowish tint, which afterwards grows obscure at the centre, and their figure becomes irregular and angular.

All the individuals observed by M. Adolphe Brongniart, were provided with them. The males, supposing any to exist, do not appear at the same time as the females, that is, in the month of June, and are unknown.

Limnadia Hermani, Adol. Brong. Mem. du Mus. d'Hist. Natur. VI. xiii.; *Daphnia gigas*. Herm. Mem. Apterol, v., has been found in great numbers, in the small pools of the forest of Fontainbleau.

In other cases, each eye is situated at the extremity of a pedicle, formed by the lateral and horn-like elongation of each side of the head. The body is naked, without testa, and annulated in its entire length. The females carry their eggs in a capsule, situated towards the base of the tail, in those in which the body thus terminates, or at the posterior extremity of the body, or of the thorax, in those which have no tail.

These have a tail.

ARTEMIA, *Leach*,

Whose eyes are supported on very short pedicles; the head is confounded with an oval thorax, supporting ten pair of feet, and terminated by a long and pointed tail. The antennæ are short and subulate.

Artemia Salina, Cancer. Salinus., Lin., Montag., Linn., Trans. XI. xiv. 8—10.; *Gammarus Salinus*, Fab., Desmar., Consid., pag. 393, is a very small crustaceous animal, commonly found in the saline marshes of Lymington, in England, when the evaporation of the waters is very much advanced, but concerning which, we have, as yet, but very imperfect information.

BRANCHIPUS, *Lat.* CHIROCEPHALUS, *Benedict Prevost, Jurine*,

Have the eyes supported on very projecting pedicles, the body narrow, elongated, and compressed; the head distinct from the trunk, variously appendaged, according to the sexes, with two projections in the form of horns between the eyes; eleven pairs of feet, and the tail terminated by two leaflets, more or less elongated, and edged with ciliæ.

Although Schœffer, and Benedict Prevost, have given very detailed monographs on two species of this genus, their labours are, nevertheless, still very imperfect, as to the complete and comparative knowledge of the buccal organization, and of some other parts of the head. Considered in the two sexes, these animals present us with the following generalities:

the body is almost filiform, composed of a head separated from the trunk by a sort of neck; of a trunk or thorax, hollow underneath, in its length, divided, at least above, the neck not comprized, into eleven segments, each supporting a pair of branchial feet, very much compressed, generally composed of three lamellate articulations, with the edges furnished with a fringe of hairs, or barbed filaments; and of an elongated tail, going into a point, consisting of nine segments, terminated by two leaflets, more or less elongated, and bordered with ciliæ. The under part of its second segment presents the male sexual organs, and in the female an elongated sac, containing the eggs which she is ready to lay. The head presents, first, two reticulated eyes, apart, situated at the extremity of two flexible peduncles, formed by lateral extensions of the head. Second, two antennæ, at the least, frontal, scarcely longer than the head, slender, filiform, and composed of very small articulations. Third, two projections underneath these antennæ, sometimes in the form of horns, and having a single articulation, sometimes digitiform, (the first finger of the hands, according to Prevost) and of two articulations. Fourth, a lower mouth, composed of two sorts of denticulated mandibles, without palpi, and of some other pieces. We presume that these projections in the form of horns, are only an appendage, or division of the frontal antennæ, but increased in size and altered in shape in the males. The two other antennæ may be wanting or obliterated in the females, and form in the other sex of one of the species (*Chirocephale diaphane*, Prevost), those singular tentacula, appendaged, and denticulated, in the form of a flabby proboscis, capable of being spirally rolled, which Benedict Prevost designates under the name of fingers of the hands. It is probable that the mouth has, as in apus, two pairs of jaws, a ligula, and a labrum, but the forms and respective situations of which have not yet been well recognized. It appears to me indubitable, that this piece, in the form of a beak, of which Schœffer speaks, and which M. Prevost calls sucker, is not the labrum; that the four bodies or nipples, placed on the sides, and mentioned by the first, are not the mandibles, and the two upper jaws; and that the pieces considered by the second as barbles, are also not maxillary. The first two feet which, according to Schœffer, are composed but of two articulations, the last going into a point, might represent the first two jaw-feet of the decapod crustacea, and the two large antenniform feet of apus. The principal male sexual organs, or at least those which are regarded as such, consist of two conoïd bodies, biarticulate, and coming forth only on pressure, situated on the under part of the second ring, at which some vessels end, that proceed from the first. M. Prevost presumes that the two vulvæ of the female are at the extremity of the tail, but do not give issue to the eggs. This issue (two apertures, according to Schœffer,) is at the second ring, and communicates interiorly with the sac enclosing the eggs, and serving as an external matrix. But we are not acquainted with any crustaceous animal in which the female sexual organs are placed at the posterior extremity of the body, and therefore this opinion appears to us to possess but slight foundation.

The observations of Schœffer on the hairs of the feet of these crustacea demonstrate, that they are so many aerial canals, and even the surface of the feet of which they are composed, seems to absorb a portion of the air, which is attached to it in the form of small bubbles.

The *Chirocephale diaphane* of Benedict Prevost, and which appears to us to have great relations with our *branchipe des marais*, if indeed it differ from it at all, has, on issuing from the egg, the body divided into two masses, pretty nearly equal, and almost globular. The first presents a simple eye, two short antennæ, two very large oars, ciliated at the end, and two feet, rather short and slender, consisting of five articulations. At the end of the first moulting, the two composite eyes appear, the body is elongated posteriorly, and terminates in a conical articulated tail, with two filaments at the end. The following moultings gradually develope the feet, and the oars vanish. The sucker, which in the early age extends over the belly and covers it, also diminishes in proportion.

These branchiopoda are found, and usually in great abundance, in small pools of fresh and muddy waters, and often in those which are formed after great rains, but particularly, as it would seem, in spring and autumn. The first frosts cause them to perish. They swim with the greatest facility on the back, and their feet, incapable of any service in walking, then present an undulatory movement, very agreeable to witness. This movement produces a current of water between them, and which, going along the canal of the chest, carries to the mouth the small corpuscula on which the animal is sustained; but when it is desirous of advancing, it strikes the water quickly with its tail, on the right and left, which causes it to proceed, as it were, by bounds and leaps. When taken out of this element, it moves its tail for some time, and curves itself circularly. If deprived of a sufficient quantity of moisture, it makes no further movement.

According to the report of Benedict Prevost, the male of the species, which is the subject of his memoir, when desirous of coupling, swims around his female, seizes her by the neck with the corniform appendages of his head, and holds himself fixed there until she receives the posterior extremity of his tail, so as to draw near the two valves of the copulative organs. This coupling thus resembles that of the libellulæ. The eggs are yellowish, spherical at first, and then angular, with a thick and hard shell, which contributes to their preservation. It even appears that desiccation, unless carried to too great an extent, will not injure the germ, and that the young are disclosed, if a sufficient quantity of rain should fall. M. Desmarest has often observed branchiopoda in the small standing pools of rain water on the summits of the *grès* of Fontainbleau. The females of the chirocephali lay several distinct sets of eggs, after a single coupling, each at several times. These operations last some hours together, and sometimes as long as an entire day. Each laying produces from one hundred to four hundred eggs: they are sent out with much quickness, by casts of ten or twelve, and with sufficient force to sink a little in the mud.

Benedict Prevost has observed that his *Chirocephale diaphane* was subject to some maladies, the description of which he gives. This species, as we have said, appears to differ but little from our *Branchipe des marais*. The two horns situated below the upper antennæ are composed, in both sexes, of two articulations, the last of which is large, and arched in the male, very short and conical in the other sex. In the *Branchipus stagnalis* the horns present but one articulation, and those of the male resemble, in their form, direction, and teeth, the mandibles of the males of *Lucanus cervus*.

The following have no tail. Their body is terminated almost immediately at the end of the thorax and of the last feet. Such are,

EULIMENE, *Latr.*

Their body is almost linear, and presents four short antennæ, almost filiform, two of which are smaller, and almost similar to palpi, placed at the anterior extremity of the head; a transverse head, with two eyes supported on peduncles tolerably large and cylindrical; eleven pair of branchial feet, of which the first three articulations and the last are smaller, going into a point; and immediately after them a terminal piece, almost semi-globular, replacing the tail, and from which issues an elongated filament, which is perhaps an oviduct. I have observed, towards the middle of the fifth pair of feet, and of the four following, a globular body, analogous perhaps to the vesicles which these organs present in the following subgenus, that of Apus.

The only known species, *Eulimène blanchâtre,* Latr. Règne Animal, par M. Cuvier, iii. p. 68; Nouv. Dict. d'Hist. Nat. x. p. 333; Desmar. Consid. p. 353, 354; is very small, whitish, with the eyes and the posterior extremity of the tail blackish. It is found in the river Nice.

The others, and last phyllopa, ASPIDIPHORA, *Latr.*, have sixty pairs of feet, all furnished externally, near their base, with a thick ovaliform vesicle, and the two anterior of which, much larger and branching, resemble antennæ; a large testa, covering the major portion of the upper part of the body, almost entirely free, clypeiform, emarginated posteriorly, supporting anteriorly, on a circumscribed space, three eyes, simple, sessile, and the anterior two larger and lunulated; and two bivalve capsules, enclosing the eggs, and annexed to the eleventh pair of feet. These characters distinguish

APUS, *Scop.*,

Which form a part of the genus BINOCULUS of *Geoffroy*, and of that of LIMULUS of *Müller*.

Their body, including the testa, is ovaliform, broader, and rounded before, and narrowed posteriorly, in the manner of a tail. But, when we do not include the testa, but consider the body naked, it is at first almost cylindrical, convex above, concave and divided longitudinally by a furrow underneath, and afterwards terminates like an elongated cone. It is composed of thirty rings, diminishing much in size towards the posterior extremity, and which, with the exception of the last seven or eight, support the feet; the first ten are membranaceous, soft, without spines, present on each side a small eminence, in the shape of a button, and have each but one pair of feet. The others are more solid or corneous, with a range of small spines at the posterior edge; the last is larger than the preceding, almost square, depressed, angular, and terminated by two filaments or articulated setæ. In some species composing the genus LEPIDURUS of Dr. Leach, we observe in their interval a corneous lamina, flatted, and elliptical. If the number of feet is about a hundred and twenty, the last rings, commencing with the eleventh or twelfth, must support more than one pair, which, in this point of view, approximates these crustacea to the myriapoda. The testa, perfectly free from its anterior attachment, covers a large portion of the body, and thus protects the first segments, which, as we have observed, are of a softer consistence than the following: it consists of a large corneous shell, very slender, almost diaphanous, representing the upper teguments of the head and of the thorax united, and forming a large oval buckler, convex, notched in the manner of an angle, and denticulated at its posterior extremity. It is divided at its upper face by a transverse line forming two united arches, into two areas, the anterior of which, almost semi-lunar, answers to the head, and the other to the thorax. The first presents, in the middle, three simple eyes, or without any sensible facettes, very closely approximated, the anterior two of which are larger, almost in the form of a kidney, and the posterior of which is much smaller and oval. A duplicature of the anterior portion of the testa forms underneath a sort of frontal buckler, flatted, in the form of a half-moon, and serving as a basis to the labrum. The posterior area, that which corresponds to the thorax, is carinated at the middle of its length. This testa is fixed only by its anterior extremity, so that from this point the entire back of the animal may be uncovered. The sides of this shell, viewed underneath, and by the light, present each a large spot, formed of a great number of lines, describing concentric ovals, and which appear to be tubes, filled with a red liquor. Immediately under the buckler, or frontal disk, are situated the antennæ and the mouth. The antennæ are two in number, inserted on each side of the mandibles, very short, filiform, and consisting of two articulations almost equal. The mouth is composed of a squared and advanced labrum; of two strong, corneous mandibles, bellied inferiorly, compressed and denticulated at their extremity, without palpi; of a large ligula, deeply emarginated; and of two pairs of jaws, in the form of leaflets, applied one upon the other, the upper of which are spinous and ciliated at the internal edge, and the lower almost membranaceous, are similar to little false feet; they terminate by a slender elongated articulation, and are lengthened externally, at their base, into a sort of auricle, supporting an appendage with a single articulation, and ciliated, which may be considered as a sort of palpus. The ligula presents, according to M. Savigny, a ciliated canal, which conducts directly to the œsophagus. The feet, the number of which is about one hundred and twenty, diminish insensibly in size from the second pair; they are all very much compressed, foliaceous, and are composed of three articulations, not comprehending the two long filaments of the end of the two anterior ones, nor the two leaflets terminating the following,—pieces which may be considered as forming, when united, a fourth articulation, pincer or forceps like, or with two fingers, lengthened and connected into sorts of antenniform filaments. On the posterior side of the first articulation is inserted a large triangular, branchial membrane, and the following, or the second, supports also, on the same side, a sac, which is ovaliform, vesicular, and red. The opposite edge of these feet presents four triangular and ciliated leaflets, the upper of which is very much approximated to the fingers of the forceps, and appears to form a third upon the second feet and the following, as far as the tenth pair. In proportion as the size of these organs diminishes, the leaflets approximate one to the other, the forceps is less defined and less sharp, and the first finger is widened at the expence of its length, and rounded; the two anterior, much larger, in the form of oars, resemble ramified antennæ, and have been considered as such by some writers. They present four setaceous filaments, composed of a great number of articulations, and of which the two of the end (one more especially), are much longer than the two others, which are situated at the internal or anterior side. It is evident that the two of the extremity are the analogues of the two fingers of the forceps, and that the others also represent two of the lateral leaflets. Of this we may be convinced, by comparing these feet with their analogues, and the two or three following in the young individuals. After the sixth or seventh moulting, the latter much resemble the two anterior, and the antennæ are even proportionally longer than in the adult state, and terminated by setæ or hairs. The eleventh pair is very remarkable: the first articulation presents, behind the vesicle, two circular valves, applied one upon the other, formed by two leaflets, and enclosing the eggs, which resemble small grains of a very lively red. All the individuals which have been studied up to the present day, having been found provided with similar feet, it has been suspected that they fecundate themselves, and that there are no males.

These crustacea inhabit ditches, pools, and dormant waters of all kinds, and almost always in innumerable societies. When thus assembled, they are swept up by very violent winds, and have then been seen to fall like showers of rain. They appear more commonly in spring, and at the commencement of summer. Their food consists principally of tadpoles They swim very well on the back; and when they sink into the mud, they keep their tail elevated. When born, they exhibit but one eye, and but four feet, in the form of arms or oars, having aigrettes of hairs, and the second of which are the largest. Their body has no tail, and the testa forms only a plate, covering the anterior portion of the body. Their other organs are developed by degrees, in the course of successive moultings. M. Valenciennes, employed in the Museum of Natural History, has remarked that these animals were often devoured by the bird known vulgarly under the name of *wag-tail.*

The known species being by no means numerous, it is not necessary to form, as Dr. Leach has done, with those which have a lamina between the filaments of the tail, a genus proper (LEPIDURUS). Such is the *Monoculus apus*, Linn.; Schœff. Monoc. vi.; *Limule serricaude*, Herm. fils.; Desm. Consid. lii. 2. The keel of the buckler is terminated posteriorly in a small spine, which is not seen in the following, *Apus canciformis* (*Binocle a queue en filet*, Geoff. Insect. xxi. 4; *Limulus palustris*, Müll.; Schœff. Monoc. i.—v.; *l'Apus vert.* Bosc.; Desm. *ibid*, li. 1.); the latter has no lamina between the filaments of the tail. It is the type of the genus *apus* proper of Dr. Leach. He has figured another species, *Apus Montagui*, Edin. Encyc. Suppl. i.—xx.

THE SECOND ORDER OF ENTOMOSTRACA, OR THE SEVENTH AND LAST OF THE CLASS CRUSTACEA.

PŒCILOPODA

Is distinguished from the preceding by the diversity of form in the feet, the anterior of which, of an indeterminate number, are ambulatory, or adapted for prehension; and the others lamelliform or pinnated, are branchial and natatory. But it is more particularly in the absence of the usual mandibles and jaws, that they differ from all the other crustacea; sometimes these parts are replaced by the haunches, furnished with spines of the first six pairs of feet; sometimes the organs of manducation consist, either in an external siphon, in the form of an inarticulate beak, or in some other instruments adapted for suction, but concealed, or scarcely distinguishable.

The body is almost always covered, either wholly, or in a great measure, by a testa, in the form of a buckler, consisting of a single piece in the majority, of two in some, and always presenting two eyes, when these organs are distinct. Two of their antennæ (*Cheliceres*, Lat.,) are, in several, shaped like a forceps, and perform the functions of one. The greater number of them have twelve feet, almost all the others ten or twenty-two. They live for the most part on aquatic animals, and most commonly on fishes.

We shall divide this order into two families,

The first, that of

XYPHOSURA,

Is distinguished from the following by several characters. There is no siphon; the haunches of the first six pairs of feet are provided with small spines, and perform the office of jaws; the number of feet is twenty-two; the first ten, with the exception of the anterior two of the males, are terminated in the manner of a forceps with two fingers, and inserted as well as the following two, under a large semi-lunar buckler. These latter support the sexual organs, and have the form of large leaflets, as well as the following ten, which are branchial, and annexed to the under part of a second testa, terminated by a very hard, ensiform, and mobile stylet. These animals are, moreover, erratic. They compose the genus

LIMULUS, *Fab.*

The species of which have received in trade the name of the *Molucca-crab.* The body, suborbicular, a little elongated, and narrowed posteriorly, is divided into two parts, covered by a solid testa of two pieces, one for each division, very hollow underneath, and presenting above, two longitudinal furrows, one on each side, and a keel at the middle of the back. The first piece of the testa, or that which covers the front of the body, is much larger than the other, forms a large semi-lunar buckler, with an edge, supporting on its upper part two oval eyes with very numerous facettes, in the form of small grains, situated one on each side, on the external side of a longitudinal keel; and at the anterior extremity of that of the middle, which is common to the two pieces of the testa, two small simple eyes approximated together. These keels are armed with some teeth, or sharp tubercles. The duplicature of this testa forms, underneath, at its anterior extremity, a plane border, or reflected edge, very much arcuated, and terminated inferiorly by a double arch, advanced like a tooth at the centre of re-union. Immediately below this projection, in the concavity of the buckler, is a small inflated labrum, carinated in the middle, terminating in a point, and above which are inserted two small antennæ, in the form of small didactylous claws, and elbowed in the middle of their length, at the junction of the first articulation, and of the following, or of the forceps properly so called. Immediately underneath, are inserted and approximated by pairs, upon two lines, twelve feet, of which the first ten, the anterior two or four of the males alone excepted, are terminated in a didactylous forceps, while their radical articulation is advanced interiorly, in the manner of a lobe, furnished with small spines, and performs the office of a jaw. The size of these feet increases progressively. If we except those of the fifth pair, they are composed of six articulations, comprehending the mobile finger of the forceps. These have an additional articulation, and differ besides from the preceding, in having externally at their base, an arched appendage, inclining backwards, of two articulations, the last of which is compressed and obtuse; also their fifth articulation is terminated at the internal side, by five small mobile leaflets, corneous, narrow, elongated and pointed, and, moreover, the two fingers of the forceps are mobile, or articulated at their base. The two pieces situated in the interval of these feet, which M. Savigny considers as a ligula, appear to me to be but two maxillary lobes of these organs, but detached or free. The pharynx occupies the interval comprised between all these feet. The males are distinguished from the females by the form of the forceps, which terminate the two or four anterior ones. They are swelled, and destitute of the mobile finger. The last two feet of this buckler are united, and in the form of a large membranaceous leaflet, almost semicircular, supporting the sexual organs at its posterior face, and presenting in the middle of an emargination of the posterior edge, two small triangular divisions, elongated and pointed, which appear to represent the internal fingers of the forceps. Some sutures indicate the other articulations. The second piece of the testa, articulated with the preceding, at the middle of its posterior emargination, and filling the vacancy which it forms, is almost in the form of a triangle, truncated and emarginated angularly at its posterior extremity. Its lateral edges are alternately emarginated and denticulated, and the emarginations, beginning from the second, present each in their middle an elongated and mobile spine. There are six on each side. In the inferior concavity are enclosed, and disposed by pairs, ten fin-feet, almost similar, in form, to the last two feet, but united simply at their base, applied one upon the other, and supporting at their posterior face, the gills, which appear composed of very numerous and crowded fibres, disposed on a single plane, one against the other. The anus is situated at the inferior root of the stylet, terminating the body. According to an observatión which has been communicated to us by M. Straus, the interior of the first buckler presents, besides the brain, but a single ganglion, the sub-œsophageal. The two nervous cords are afterwards lengthened into the interior of the second buckler, and form there at the origin of the branchial feet, but faintly marked ganglions, which throw out branches over those organs. According to M. Cuvier, the heart, as in the stomapoda, is a thick vessel, furnished internally with fleshy columns, running along the back, and giving off branches on both sides. A rugose œsophagus, ascending forwards, conducts into a very fleshy gizzard, furnished internally with a cartilaginous coat, all bristled with tubercles, and followed by a wide and straight intestine. The liver pours the bile into the intestine by two canals on each side. A great portion of the testa is filled by the ovaries in the female, and by the testicles in the male.

These crustacea sometimes attain to the length of two feet; they inhabit the seas of warm climates, and are found most frequently on their shores. It appears to me that they are peculiar to the East Indies, and to the coasts of America. Here we designate the species found there (*Limulus cyclops*), *pan*, or *sauce-pan-fish*, because it has something of a similar form to this article, and by removing the feet, the testa may serve to draw water. According to the testimony of M. Leconte, a most accomplished naturalist, and who has so greatly contributed by his researches and discoveries to the progress of entomology, it is given to pigs to eat. The savages employ the stylet of the tail to make arrows; the point is considered formidable; their eggs are eaten in China. When these animals walk, their feet are not visible. They are found in a fossil state in certain strata of a middle formation.

Knorr. Monum. du Déluge, i. pl. xiv. Desm. Crust. Foss. xi. 6, 7. It would seem, according to these figures, that the lateral spines of the second piece of the testa, form only smaller teeth, instead of spines articulated by their base; but these articulations have, perhaps, disappeared.

Some have the four anterior feet terminated, at least in one of the sexes, by a single finger.

We know but a single species of this division, which I have seen figured on the Chinese vellums, it is the *Limulus heterodactylus*, serving as type to the genus *tachypleus* of Dr. Leach. This limulus is probably the *Kabutogani*, or *Unkia* of the Japanese, and representing on their primitive zodiac, the constellation of Cancer.

In the others, the two anterior claws at most, are monodactylous. All the ambulatory feet are didactylous, at least in the females.

This division is composed of several species, but which, seeing the little attention which has been given to the detailed form of their parts, to the differences of sex and age, and of the localities which are proper to them, have not yet been characterized in a rigorous and comparative manner. It is thus, for instance, that the limulus, which is found commonly in America, seen in its early age, is whitish, or of a flaxen colour, with six strong teeth, all along the crest of the upper middle of the testa, and two others, equally strong and pointed, on each lateral crest of the buckler, or of the first piece of the testa; while in the more aged individuals which are sometimes more than a foot and a half in length, the colour is of a very deep brown, or almost blackish, and the teeth, particularly those of the middle, nearly obliterated. Here again, the lateral edges of the second piece of the testa, have fine denticulations, which do not exist or are scarcely perceptible in the former. We shall refer to the young individuals, the *Limulus cyclops* of Fabricius, and the *L. Sowerbii* of Leach,

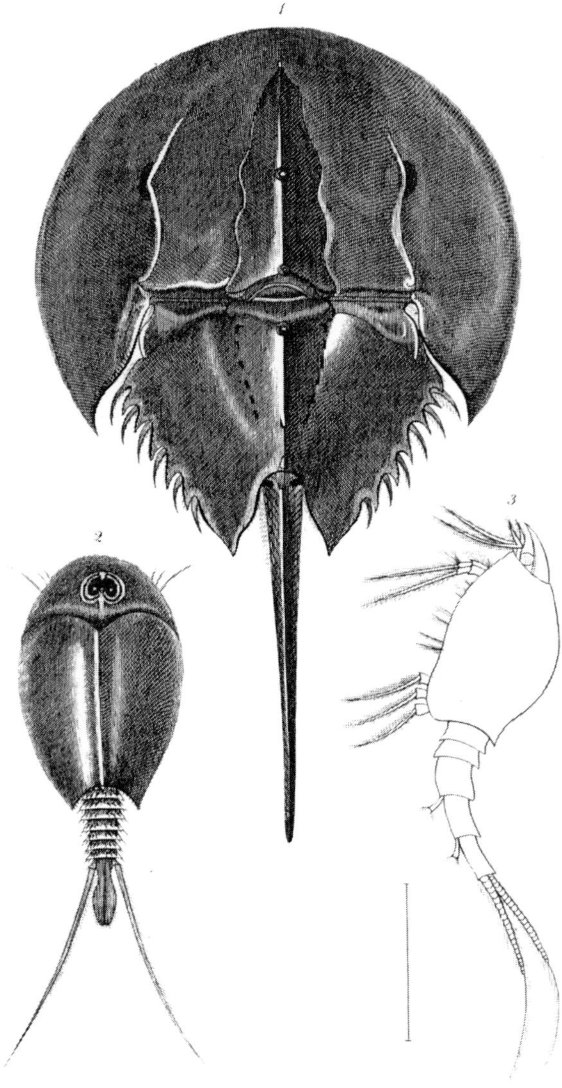

1 *Limulus Polyphemus.*

2 *Apus cancriformis.*

3 *Nebalia Herbstii.*

London, Published by Whittaker & C.o Ave Maria Lane, 1833

(Zool. Misc. lxxiv.); his *Limulus tridentatus*, and the *Limule blanc* of M. Bosc; and to the second individuals, or the largest, my *Limule des Moluques (Monoculus polyphemus*, Lin.; Crus. Exot., liv. 6. cap. 14. p. 128; Rumph. mus. xii. a, b.) which I had at first distinguished specifically, in the belief that these large individuals exclusively inhabited those islands. In both, that is, at all ages, the tail is a little shorter than the body, triangular, finely denticulated at the upper crest, without any defined furrow underneath. We shall designate this species under the name of *limulus polyphemus*. These last characters will distinguish it from some others described by me and Dr. Leach. (*See* the second edition of the Nouveau Dict. d'Hist. Nat. Desmarets. Consid. p. 344—358.)

The second family, that of

SIPHONOSTOMA,

Presents no kind of jaws whatsoever. A sucker or siphon, sometimes external, and in the form of a sharp inarticulated beak, sometimes concealed or scarcely distinct, replaces the mouth; the number of the feet is never beyond fourteen; the testa is very slender, and consisting of a single piece. These entomostraca are all parasitical.

The composition of the beak is not yet well known. It is evident, from the figure given by the younger Jurine, of the *argulus foliaceus*, that it encloses a sucker. But is this the case with that of the others, and what is the number of its pieces? On these points we are ignorant. I presume, however, that this siphon is composed of the labrum, the mandibles, and the ligula, which forms the sheath of the sucker. In the preceding entomostracon, the four anterior feet, the form of which is very different from that of the following, would correspond to the four jaws of the decapods.

We shall divide this family into two tribes.

The first, that of CALIGIDES, *Lat.*, is characterized by the presence of a testa, in the form of an oval or semi-lunar buckler; by the number of visible feet, which is always twelve (or fourteen, if, with Dr. Leach, those may be considered as such, which I think to be the two inferior antennæ); by the form and size of those of the last ten pair, which are sometimes multifid, pinnate, or terminated in a fin, and very well adapted at all periods, and in the adult state, for swimming; sometimes in the form of leaflets, or broad and membranaceous. The sides of the thorax never present expansions in the form of wings, inclining backwards, and they enclose the body posteriorly.

In some the body, presenting several segments above, is elongated, and narrowed posteriorly, to terminate in the manner of a tail with two filaments, or two other projecting appendages at the end. This extremity is not covered by a division of the superior teguments, in the form of a large rounded shell deeply notched at the posterior edge; the testa occupies at least one half of the length of the body; this subdivision will comprehend two genera of Müller.

The first, that of

ARGULUS, *Müller*,

Was at first designated by us under the name of *Ozole*, and described in an incomplete manner. The younger Jurine has since observed the species which serves as its type, with the most scrupulous attention, has traced it in all its ages, and given us a monograph which leaves nothing to desire upon the subject. He has restored to this genus the name which Müller had originally imposed upon it.

The arguli have an oval buckler, emarginated posteriorly, covering the body, with the exception of the posterior extremity of the abdomen, supporting on a middle triangular space, distinguished under the name of hood, *(chaperon)* two eyes, four very small antennæ, almost cylindrical, placed in front, the superior of which, shorter and of three articulations, have at their base a strong hook, toothless and recurved; while the inferior are of four articulations, with a small tooth to the first. The siphon is directed forwards. The feet are twelve in number; the first two terminate by a basement, annulated transversely, widened circularly at the end, striated and denticulated on its edges, presenting at the interior a sort of rosette, formed by the muscles, and appearing to act in the manner of a cupping-glass, or sucker. Those of the following pair are adapted for prehension, with the thighs thick and spinous, and the tarsi composed of three articulations, the last of which is provided with two hooks. The other feet are terminated by a fin, formed of two fingers or elongated pinnulæ, furnished on their edges with barbed filaments; the first two of these, that is, the third pair, comprehending the two preceding pairs, have each an additional finger, but recurved; the last two are annexed to that portion of the body which projects posteriorly out of the testa, *i. e.* the tail. The females have but a single oviduct, and covered by two small feet, situated behind these two palettes; the organ considered as the penis of the male, is placed at the internal extremity of the first articulation of the same feet, near the origin of the two digits. On the same articulation of the two preceding feet, and facing these copulative organs, is a vesicle presumed to be seminal. The abdomen, considering as such that part of the body which extends behind from the ambulatory feet, the beak, and a tubercle enclosing the heart, is entirely free, from its origin, without distinct articulations, and is terminated immediately after the last two feet, by a sort of tail, lamelliform, rounded, deeply emarginated, or bilobate, and without hairs at the end. It is a sort of fin. The transparence of the body allows the heart to be distinguished; it is situated behind the base of the siphon, lodged in a solid tubercle, semi-transparent, and formed of a single ventricle. The blood, composed of small transparent globules, is directed forward in the form of a column, which is soon divided into four branches, two of which go directly towards the eyes, and two others towards the antennæ; these last branches, then reflected backwards, and united to the first, form on each side a single column, which descends towards the cupper, goes round its base, and disappears. A little underneath the two following feet, we distinguish on each side another sanguineous column, which is curved outwards, subsequently extends near the edges of the testa, and, having arrived near the two penultimate feet, is bent forward, and ceases to be visible. Another column, in which the blood, as well as in the preceding, proceeds from front to rear, traverses the middle of the tail longitudinally; it unites posteriorly to two other currents, which are observable on the edges of this tail, but proceeding in a contrary direction, or seeming to bring back the blood to the heart. The younger Jurine has avoided the employment of the term *vessel*, because the blood driven into the anterior part, appears to spread and to be disseminated there, so as to countenance the notion that the globules of the blood are dispersed in the parenchyma of those parts, rather than contained in particular vessels. But after what we have said respecting the circulation of the decapods, it may be seen that in the present case the blood is at first distributed in the same manner; and the currents or columns, of which we have just spoken, appear to indicate the existence of peculiar vessels. Accordingly, we find this skilful observer afterwards allow that the circulation does not take place throughout, in a manner equally diffuse, as in the anterior part of the testa, where to us it appears to be effectuated as in the decapods. The brain, placed behind the eyes, appears to him to be divided into three equal lobes, one anterior and two lateral. The anterior part of the stomach gives birth to two large appendages, divided, each, into two branches, which ramify into the wings of the testa. The alimentary substances, of a bistreous colour, which they contain, render these ramifications perceptible. The cœcum is provided, towards its origin, with two vermiform appendages.

The males are very ardent in love, which causes them to mistake one sex for another, or address themselves to females pregnant or dead. In coupling, they are placed upon the back of the females, to which they hook themselves by means of their cupper-feet, and remain in that situation for several hours. The length of gestation is from thirteen to nineteen days. The eggs are smooth, oval, and of the whiteness of milk: they are fixed with a sort of gluten on stones, or other hard bodies, either in a right line, or in two ranks, to the number of from one to four hundred. When pressed against each other, their form becomes almost hexagonal.

Twenty-five days after the laying, and after having at first assumed a yellowish and opake tint, we commence to distinguish the eyes and some portion of the embryo. Subsequently, at the end of about ten days, or towards the thirty-fifth day after the laying, the shell is cleft longitudinally, and the young, or tadpole, comes into the world. It is then scarcely three-eighths of a line in length. Its form generally resembles what it is to be in the adult state; but its locomotive organs present some essential differences. Müller has described it in this state, under the name of *Argulus charon*. Four oars, or long arms, two of which are placed before the eyes, and the other two behind, terminated, each, by a pencil of flexible and pinnated setæ, which the animal moves simultaneously, and by means of which it swims with facility and by jumps, issue from the anterior extremity of the testa; they do not represent antennæ, as these last organs are also to be seen. The cupper-feet are replaced by two strong feet, elbowed near their extremity, and terminated by a powerful hook, with which this animal can fasten on fishes. Other feet, proper to the adult state, viz. those of the second and third pair, or the two ambulatory and first two of the swimming feet, are the only ones which are developed and free ; the following are, as it were, swathed up, and fixed on the abdomen. The heart, the proboscis, and the ramifications of the appendages of the stomach, are all distinct. When the first moulting, which is performed by means of a rupture of the lower surface, has taken place, the oars disappear, and all the swimming feet are to be seen. Three days after comes the second moulting, which produces no important change. But at the third, which takes place two days after, we commence to perceive, towards the middle of the two anterior feet, the rudiments of the cuppers. On the fourth moulting, which likewise takes place at the end of two days, these same feet are finally transformed into regular cupping-feet, preserving, however, the terminal hook. At the end of six days comes a fresh change of skin, and the appearance of the genital organs of either sex ; but another moulting, which is delayed for six days longer, is necessary to enable these animals to unite and multiply. Thus the duration of their state of infancy, or of their metamorphoses, is five and twenty days. As yet, however, they have arrived to but one-half of their proper size. For this, some other moultings, which take place every six or seven days, are necessary. Jurine has ascertained that the females cannot become mothers without the intervention of the males. Those which he had isolated, perished of a disease, which was indicated by the appearance of several brown globules, disposed in a semi-circle towards the posterior part of the hood, and which are formed, as it would seem, in the parenchyma, since they are not destroyed by the moultings.

The only known species of this genus, *Argule foliacée*, (Jurine fils., Ann. du Mus. d'Hist., Nat., vii. xxvi., *Monoculus foliaceus,* Linn., *Argulus delphinus, et argulus charon.*, Müll., Entom., *Argulus delphinus,* Herm. fils., Mém. Apter. v. 3.

vi. 2; *Monoculus gyrini*, Cuv. Tab. Elem. de l'Hist. Nat des An. p. 454; *Ozolus gasterostei*, Lat.; Hist. Nat. des Crust. et des Insect. IV. xxix. 1—7; Desm. Consid. L. i.; *Pou du gastéroste*, Baker, Micros. II. xxiv.) fixes itself on the under part of the body of the tadpoles of frogs, of stickle-backs, &c., and sucks their blood. Its body is flatted, of a clear yellowish green, and about two lines and a half in length. The younger Hermann, who has very well described this animal in its perfect state, and who quotes a manuscript of Leonard Baldaneur, a fisherman of Strasburg, bearing the date of 1666, in which the same animal is figured, says, that it is seldom to be met with in the environs of that city, except on trouts, and that it often causes their death, especially of those which are kept in fish-ponds. It is also found on perch, pikes, and carps. It has never been found upon the gills. Like the gyrini, this animal whirls round like a top. He says that its body is divided into five rings, not very distinct, upon the back.

CALIGUS, *Müll.*,

Have no cupper-feet; those of the anterior pairs are unguiculated, the others are divided into a number, more or less considerable, of pinnulæ, or in the form of membranaceous leaflets. The testa leaves discovered a good part of the body, which is terminated posteriorly, in the majority, by two long filaments, and in the others by appendages, in the form of a fin or stylet. This interval also frequently presents some other appendages, but small, or much less projecting.

The name of *fish-lice*, under which they are collectively designated, indicates that their habits are the same as those of the arguli, and the other siphonostomata. Many naturalists have considered the tubular filaments of the posterior extremity of their body to be ovaries. I have sometimes discovered eggs under the posterior and branchial feet; but never in these tubes. Moreover, we never see any exterior oviducts, similarly prolonged, excepting in such females as lay their eggs in holes or deep cavities. Now the females of the caligi are not in this predicament. Müller, and other zoologists, have remarked, that these crustacea can erect and agitate these appendages. We think with Jurine the younger, and such is also the opinion of his father, that they answer the purpose of respiration, in the same manner as the filaments of the end of the abdomen in apus.

We find in the third volume of the General Annals of the Physical Sciences, printed at Brussels, an extract from the observations of Dr. Surriray, on the fœtus of a species of caligus, which he believes to be the *elongatus*, and which is very common on the operculum of the *esox belone*. This naturalist informs us, that having rubbed the two threads of the tail of this crustaceous animal, he pushed out many transparent and membranaceous eggs, each enclosing a living fœtus, very different from the mother, and the description of which he gives us. From these observations we might deduce, that these filaments are sorts of external oviducts. But may there not be some mistake in the case? for I have studied, with some attention, these same organs in several individuals, preserved, it is true, in spirits, without having discovered a body of any description in them.

Some, whose feet are all free, and annexed, with the exception of the last two, to the anterior part of the body, (*Cephalothorax*, Lat.) covered by the shield; in which some at least of the posterior feet are furnished with numerous and pinnated filaments; and in which the siphon is not apparent, have the abdomen naked above, and terminated by two long filaments, or by two stylets. They compose the subgenus

CALIGUS (proper). CALIGUS RISCULUS, *Leach.*

Caligus piscinus, Lat.; *Caligus curtus*, Müll. Entomost. xxi. 1, 2; *Monoculus piscinus*, Lin.; *Caligus Mulleri*, Leach;

Desmar. Consid. l. 4. The *Oniscus lutosus* of Slabber (Encyclop. Method. Att. d'Hist. Natur. cccxxx. 7, 8), appears, by reason of the fin-like appendages of the tail, to form a subgenus proper. The *binocle à queue en plumet* of Geoffroy might be placed in it.

In all the others, the upper part of the abdomen is imbricated, or this part of the body, is, as it were, enclosed in a sort of case, formed by the hinder feet, which resemble membranes, and are folded upwards.

Among these last there are some whose antennæ are never advanced in the manner of small claws, all whose feet are free, and the last of which do not envelope the body in the manner of a membranaceous case. They form the following subgenera,

PTERYGOPODA, *Latr.* NOGAUS? *Leach,*

Which have the posterior extremity of the body terminated by two fin-like appendages; feet pinnated or digitated on the under part of the post-abdomen, or of the second division of the body not covered by the shield, and a distinct beak.

But a single living species, found upon the shark. *See* the genus *Nogaus,* Desm. Consid. p. 340.

PANDARUS, *Leach,*

Which have two filaments at the posterior extremity of the body; the feet of the first and of the fifth pair unguiculated, and the others digitate, but whose siphon is not apparent.

Pandarus bicolor, Leach, Desmarest, l. 5; *Pandarus Boscii,* Leach, Encyc. Brit. Suppl. I. xx. *See,* for the other species, Desm. *ibid.* p. 339.

DINEMOURA, *Lat.,*

Having likewise two long filaments to the anus, but whose siphon is apparent. Their two anterior feet are unguiculated, the following are terminated by two long digits; the others are in the form of membranaceous leaflets. *Caligus productus*, Müll. Entom. xxxi. 3, 4; *Monoculus Salmoneus*, Fab.

The last subgenus of this subdivision, that of

ANTHOSOMA, *Leach*,

Approaches the preceding, as far as the existence of a siphon, and that of the two filaments at the end of the tail; but is removed from it, as well as from those which precede it, by reason of two of its antennæ, carried forwards, in the form of small monodactylous claws, and of the last six feet, which are membranaceous, joined inferiorly, and folded laterally on the post-abdomen, so as to form a case to envelope it; those of the first and third pair are unguiculated; the second are terminated by two short and obtuse digits. *Anthosoma Smithii*, Leach, Desm. Consid. 1. 3; *Caligus imbricatus*, Risso.

In others the body is oval, without projecting appendages in the manner of a tail, composed of filaments, or of appendages in the form of lines, at its posterior extremity. A portion of the upper teguments forms at first, and in front, a buckler, not covering its anterior moiety, being more narrow, rounded, and emarginated anteriorly, widened, and, as it were, bilobate at the other end; then come successively three other pieces, or scales, rounded and emarginated posteriorly; the second of which, the smallest of all, is almost in the form of an inverted heart, and the last and largest is vaulted. The four posterior feet are in the form of laminæ, and united by pairs; those of the first and third are unguiculated; the second have their extremities bifid. The siphon is apparent. The eggs are covered by two oval pieces, contiguous, coriaceous, placed under the abdomen, and exceeding it in length. Such are the characters of the genus

CECROPS, *Leach*,

Of which but a single species is known, which has been found attached to the gills of the tunny and the turbot, the *Cecrops de Latreille* (Leach, Encyc. Brit. Suppl. i. pl. xx. 1. 3, male, —2, 4, female—5, antennæ, magnified. Desm. Consid. l. 2.)

The second tribe, that of LERNÆIFORMES, *Lat.*, is composed of entomostraca still more approximated than the preceding, from their external forms, to the lerneæ. The number of discernible feet is but ten, and these organs are for the most part very short, and not at all, or but little, adapted for swimming. Sometimes the body is almost vermiform, cylindrical, with the anterior segment simply a little broader, and provided with two advanced didactylous forceps; sometimes, in consequence of two lateral expansions in the form of lobes, or wings, thrown behind the thorax, and of the two posterior ovaries, it forms a small quadrilobate mass. This tribe is composed of two genera. The first, that of

DICHELESTIUM, *Hermann (Son)*,

Presents us with a narrow and elongated body, a little dilated in front, composed of seven segments, the anterior of which (*corslet*, Hermann), is broader, rhomboidal, formed of the head, and a portion of the thorax united. It supports, 1st, four short antennæ, the lateral of which are filiform, and composed of seven articulations, and the intermediate advanced in the manner of little arms, of four articulations, with the last in the form of a didactylous forceps; 2nd, a siphon, inferior, membranaceous, and tubular; 3d, three sorts of shapeless palpi (two multifid feet?) on each side, situated on a prominence; 4th, four feet adapted for seizing, of which the first two are formed of a thigh and leg, and terminated by divers unequal and denticulated hooks, and the second consist of an enlarged thigh, terminated by a strong claw. The second and third segments are almost lunulated, and support, each, a pair of feet, formed of a single articulation, terminated by two sorts of digits, denticulated at the end. To the fourth segment is attached another pair of feet, the fifth and last, but in the form of simple, oval vesicles, divergent, and immoveable, and which Hermann presumes to be ovaries, rather than feet. Both this segment and the following are almost square; the sixth is much longer and cylindrical; the seventh and last is three times shorter, almost orbicular, flatted, and terminated by two small vesicles. The eyes are not distinct.

Dichelestium Sturionis, Hermann, fils, Mem. Apterol, page 125, v. 7, 8; Desmar. Consid., L. v., is about seven lines in length, and one in breadth. The second segment, lengthened on each side into an obtuse papilla, and the following four are red, and of a whitish-yellow laterally. The feet do not appear when the animal is viewed from above. It insinuates itself deeply into the skin, and covers the osseous arches of the gills, but without fixing itself, as would appear, on their combs. Hermann has collected as many as a dozen from a single fish. Two or three of this number, males, perhaps, were one-third shorter than the others, and had a curved body; one of these twelve individuals lived for three days. These crustacea turn round frequently, and with vivacity; they hook themselves very strongly by means of their frontal forceps.

NICOTHOE, *Aud.* and *Miln. Edw.*,

Terminate the class of crustacea, and are distinguished from the rest by their heteroclite form. They present, on a simple view, nothing but a body formed of two lobes, united in the manner of a horse-shoe, and enclosing two others. But when observed through the microscope, we discover that the two large lobes, are large lateral expansions of the thorax, in the form of wings, almost oval, and thrown backwards; that the other two lobes are external ovaries, or clusters of eggs, analogous to those of the female cyclopes, inserted one on each side, by means of a short pedicle, at the base of the abdomen; and that the body of the animal is composed of the following parts: 1st, a distinct head, supporting two eyes, separated, two antennæ, lateral, short, setaceous, of eleven articulations, having each a hair on the internal side, with the mouth formed of a circular aperture, performing the office of a cupper, and accompanied on each side with maxilliform appendages (anterior feet): 2d, a thorax of four segments, having underneath five pairs of feet, the two anterior of which are terminated by a very strong hook, bidenticulate at the internal side, and the other eight are composed of a large articulation, terminated by two stems, almost cylindrical, almost equal, furnished with setæ, and of three articulations each: 3d, an abdomen, proceeding to a point, of five rings, the first of which, being larger, gives birth to the oviferous sacs, and the last is terminated by two long hairs. The lateral expansions appear to be only an excessive development of the fourth and last ring of the thorax. We perceive in their interior two sorts of intestines, proceeding from the medial line of the body, and which may be considered as two cœca, or divisions of the intestinal canal, which had formed a hernia. They are endued with very distinct peristaltic movements. We have seen, in speaking of the arguli, that their stomach also presents two cœca, which ramify into the interior of the wings of the testa, and, perhaps, the thoracic expansions of the nicothoes, are also two analogous lobes. If so, this genus might be placed near the preceding.

We owe our knowledge of the only species composing the genus, namely,

Nicothoe astaci, Ann. des Scienc. Nat. Dic. 1826, xlix.

1—9. to MM. Victor Audouin and Milne Edwards; it is half a line in length, and near three in breadth, comprehending the thoracic extensions. It is of a rose-colour, softer on the oviferous sacs, with the expansions yellowish. It adheres closely to the gills of the lobster, and sinks deeply between the filaments of these organs. The species is not numerous, and merely to be found on some individuals. All the nicothoes observed by these two naturalists were provided with ovaries. It is probable that these crustacea can swim before they fix themselves, and that their thoracic lobes have acquired their ordinary development; this development, as in the case of the Ixodes, may be the product of the superabundance of the nutritive juices.

Of the Trilobites.

In the neighbourhood of the limuli, and of the other entomostraca, provided with a great number of feet, are ranged according to the opinion of one of my compeers of the Royal Academy of Sciences, M. Alexander Brogniart, and of divers other naturalists, those singular fossil animals, confounded at first under the general denomination of *Entomolithus paradoxus*, designated at present under that of trilobites, and of which he has given an excellent monograph, enriched with very fine lithographic figures. M. Eudes Longchamps, professor at the university of Caen, M. le Comte de Razoumowski, M. Dalman, and some other men of science, have since published some new observations on those fossils. We must, on the hypothesis in question, admit as a positive, or at least, a very probable fact, the existence of locomotive organs, though, notwithstanding all researches on the subject, no vestige of them seems to have been discovered. M. Outlines, indeed, in his Oryctology, thinks that he has perceived something of this kind, and that they are unguiculated. M. Victor Audouin, embracing with ardour the opinion of M. Brogniart, has combated in a particular memoir, the one which I put forth on this subject, and according to which, I approximated them to chiton. The most essential part of the difficulty was to authenticate the existence of the feet, which he has not done. As for the application of his theory of the thorax of insects to the trilobites, it appears to me so much the more doubtful, as, according to my views, the first rings of the abdomen of the insects, alone represent the thorax of the decapod crustacea.

Supposing, on the contrary, those animals to be deprived of feet, they come more naturally near the chitones, or rather, perhaps, they formed the primitive source of the articulated animals, being connected on the one side with these last mollusca, and on the other with the above mentioned crustacea, and even with glomeris, to which some trilobites, such as the calymenæ, appear to approximate, as well as to Chiton, inasmuch as they could also assume in contracting, the form of a spheroid. Since the publication of the work of M. Brogniart, some naturalists have not agreed with him in opinion, and have wholly or in part, adopted mine; others still hesitate. Be this, however, as it may, these animals appear to have been annihilated by the ancient revolutions of our planet.

If we except an heteromorphous genus, that of *Agnostus*, the trilobites have, as well as the limuli, a large anterior segment, in the form of a shield, almost semi-circular, or lunulated, and followed by about from twelve to twenty-two segments, all, with the exception of the last, transverse, and divided by two longitudinal furrows, into three ranges of parts, or lobes, and from thence the origin of the denomination trilobites. With some scientific men, these are called *entomostracites*. The squillæ, various amphipod and isopod crustacea, have also several of their segments divided into three, by two sunk and longitudinal lines; but those lines are more approached to the edges, and do not form deep furrows.

It would seem that in many trilobites, and particularly in the Asaphi, that the body is composed, (the buckler not comprehended) of a dozen segments, very much detached over the sides, and of another forming the post-abdomen, or a tail, triangular or semi-lunar, presenting only superficial divisions, and which do not cut its edges. In the paradoxides, on the contrary, the lateral lobes are terminated by sharp and very distinct elongations, of which we reckon nearly twenty-two. A species of trilobite mentioned by Count Razoumowsky, in a memoir on the fossils, and which he presumes ought to constitute a new genus, is, in this particular, very remarkable. Its lateral lobes form sorts of strips, very long, and going into a point. The feet of the nymphs of the gnats are in the form of elongated, flatted laminæ, without articulations, terminated by filaments, and folded back upon the sides. They are in a rudimentary state, and may be analogous to the lateral division of this species of trilobite, which is neighbouring to the paradoxides.

The genus AGNOSTUS, *Brogn.*, is the only one whose body is semi-circular or reniform. In all the other genera it is oval or elliptic, and presents the general characters which we have above indicated.

CALYMENE, *Brogn.*, is distinguished from all the other trilobites, by the faculty of being able to contract the body into a ball, and in the same manner as spheroma, the armadillo, the glomeris, that is, by approximating together, underneath, the two extremities of the body. The buckler, as broad, or broader, than long, presents, as in Asaphus and Ogygia, two oculiform eminences. The segments do not edge out beyond the body laterally, they are united as far as the end, and the body is terminated posteriorly in a sort of triangular and elongated tail.

In ASAPHUS, *Brogn.*, the oculiform tubercles appear to present a lid, or are granular. The sort of tail which terminates the body posteriorly, is less elongated than in calymene, and either almost semi-circular, or in the form of a short triangle. In the *asaphus of Brogniart*, described and figured by M. Eudes des Longchamps, the posterior angles of the buckler, instead of being directed backwards, as in the other species, are recurved.

The buckler of OGYGIA, *Brogn.*, is longer than broad, with the posterior angles elongated like a spine. The oculiform eminences present neither lid nor granulation. The body is elliptical.

These eminences having the appearance of eyes, do not exist, or do not at all appear in the genus PARADOXIDES, *Brogn.* The segments, or at least most of them, out-edge the body laterally, and are free at their lateral extremity.

Such are the characters of the five genera established by M. Alexandre Brogniart, and which may be distributed into three principal groups; the *reniformes,* (genus AGNOSTUS) the *contractiles* (genus CALYMENE) and the *extensiles,* (genera ASAPHUS, OGYGIA, and PARADOXIDES).

We shall refer, for the knowledge of the species and their localities, to the admirable labours of this naturalist, who, with respect to the fossil crustacea, properly so called, or clearly recognized to be such, has associated with himself one of his first pupils, and a correspondent of the Academy of Sciences, M. Anselm Gaetan Desmarest, often cited by us, both for this part, and his work on the living crustacea.

Other naturalists have proposed some other generic sections for the trilobites; but as I must confine myself to the most general considerations, I stop at those presented to us by the last work as yet in our possession, on these singular fossils.

THE SECOND CLASS OF ARTICULATED ANIMALS, AND PROVIDED WITH ARTICULATED FEET.

THE ARACHNIDES

ARE, like the crustacea, deprived of wings, and likewise not subject to change form or undergo metamorphoses, but merely moult or change skin. They have also the sexual organs remote from the posterior extremity of the body, and situated, with the exception of those of several males, at the basis of the belly. But they differ from these animals as well as from insects in many points. In the same manner, as in the latter, their body exhibits at its surface some apertures, or transverse clefts, named stigmata, destined for the entrance of air, but few in number (eight at most, more commonly two), and invariably situated at the lower part of the abdomen. Respiration otherwise is performed, either by means of air-gills performing the office of lungs, enclosed in pouches, of which these apertures form the entrance, or by means of radiated tracheæ. The organs of vision consist in small simple eyes grouped in various ways, when they are numerous. The head, usually confounded with the thorax, presents, in place of antennæ, only two articulated pieces, in the form of small didactylous or monodactylous claws or forceps, which have been erroneously compared to the mandibles of insects, and similarly designated. They move in a different direction, or from above downwards, assisting, nevertheless, in manducation, and replaced in the Arachnides, whose mouth is in the form of a siphon or sucker, by two pointed laminæ, which serve as lancets. A sort of labium, or rather tongue, produced by a pectoral elongation; two jaws formed by the radical articulation of the first articulation of two small feet or palpi, or by an appendage or lobe of this same articulation; a piece concealed under the mandibles, called *sternal tongue*, by M. Savigny (Description and figure of the *phalangium copticum)*, and which is composed of a projection in the form of a bill, produced from the union of a very small epistoma, or hood, terminated by a very small triangular labrum, with a lower longitudinal keel, usually very well furnished with hair; all these, with the pieces called mandibles, constitute generally, with some trifling modifications, the mouth of the arachnides. The pharynx is placed in front of a sternal projection, which has been considered as a labrum, but which, from its immediate situation behind the pharynx, and the absence of palpi, is rather a tongue. The feet like those of insects, are commonly terminated by two hooks, and sometimes by one more, and all annexed to the thorax (or rather cephalothorax), which, a small number excepted, is formed but of a single articulation, and very often closely united to the abdomen. This last part of the body is soft or but ill defended in the majority.

Considered in the relation of the nervous system, the arachnides are strikingly remote from the crustacea and insects. For if we except the scorpions, which, in consequence of the knots or articulations forming their tails, have some additional ganglia, the number of these swellings of the two nervous cords, is three at most, and even in these last animals, taken altogether, it is no more than seven.

Most of the arachnides live on insects, which they seize living, or on which they fasten and suck the juices. Others live parasitically on vertebrated animals. There are some, however, found only in flour, cheese, and even on divers vegetables. Those which are attached to other animals often multiply very considerably. In some species, two of their feet are only developed with a change of skin; and in general it is only after the fourth or fifth moulting at most, that the animals of this class become proper for generation.

DIVISION OF THE ARACHNIDES INTO TWO ORDERS.

The first have pulmonary sacs, a heart, with very distinct vessels, and from six to eight simple eyes. They compose the first order, that of THE PULMONARY ARACHNIDES.

The others respire by tracheæ, and present no organs of circulation, or if they have any, this circulation is not complete. The tracheæ are divided from their origin into divers branches, and do not form, as in the insects, two trunks, extending parallel to each other through the whole length of the body, and receiving the air from its different parts, through numerous apertures or stigmata. Here we find very distinctly but two at most, situated near the base of the abdomen (none in the *pycnogonides*). The number of simple eyes is four at most. These arachnides form our second and last order, that of TRACHEAN ARACHNIDES.

THE FIRST ORDER OF ARACHNIDES.

THE PULMONARIÆ. UNOGATA, *Fab.*,

Present, as we have already said, a well-marked system of circulation, and pulmonary sacs, always placed under the belly, indicated externally by apertures or transverse clefts (stigmata), sometimes eight in number, four on each side, sometimes to the number of four or two. The number of simple eyes is from six to eight, while in the next order there are but four at most, and generally but two; and even these are sometimes very little apparent, or altogether wanting. The respiratory organ is formed of small laminæ. The heart is a large vessel which runs along the back, and gives out branches on each side, and in front. The feet are always eight in number. The head is invariably confounded with the thorax, and presents at its anterior and superior extremity two pincers or forceps (*mandibles* with some authors, *chelicères* or *antenne-pinces* of Latreille), terminated by two fingers, one of which is mobile, or by a single one, in the form of a hook or claw, and always mobile. The mouth is composed of a labrum, of two palpi, sometimes imitating arms or talons, of two or four jaws, formed, when there are but two, by the radical articulation of these palpi; and in addition, when there are four, by the same articulation of the first pair of feet, and of a tongue composed of one or two pieces. In taking as a basis the progressive diminution of the number of pulmonary sacs, and of stigmata, the scorpions, in which there are eight (while the other arachnides have but four or two), ought to form the first genus of this class, and then our family of pedipalpi, to which it belongs, should precede that of araneïdes. But these last arachnides are in some sort isolated, by reason of the male sexual organs, of the hook of their frontal forceps, of their pedicled abdomen, and its spinnents, and of their habits. The scorpions, more-over, seem to form a natural transition from the pulmonary arachnides to the family of the *Pseudo-scorpiones*, the first of the following order. We shall commence, therefore, as we have done, with the araneïdes, or spinning arachnides.

The first family of the PULMONARY ARACHNIDES, that of

ARANEIDES,

Is composed of the SPIDERS (*Aranea*, Linn.) They have palpi in the form of little feet, without pincers at the end, terminated, for the most part, in the females, by a small hook, while the last articulation encloses, or carries, in the males, divers appendages, more or less complicated, serving for the purposes of generation. Their frontal forceps (mandibles of some writers), are terminated by a mobile hook, folded back inferiorly, having underneath, near its extremity, always pointed, a little cleft, for the issue of a poison, enclosed in a gland of the preceding articulation. The jaws are never above two in number. The tongue is of a single piece, always external, and situated between the jaws, either more or less square, or triangular, or semi-circular. The thorax, having usually an impression in the form of a V, indicating the space occupied by the head, is of a single articulation, to which is suspended behind, by means of a short pedicle, a mobile abdomen, usually soft. In all, it is furnished below the anus with four or six nipples, fleshy at the end, cylindrical or conical, articulated, very much approximated one to another, and pierced at their extremity with an infinity of small holes, for the passage of silken threads of an extreme tenacity, issuing from internal reservoirs. The feet, similar in form, but of various sizes, are composed of seven articulations, the first two of which form the haunches, the following the thigh, the fourth and fifth the leg, and the other two the tarsus. The last is terminated by two hooks, usually denticulated like a comb, and in many by an additional one, but smaller, and without denticulations. The intestinal canal is straight: it consists of a first stomach, composed of several sacs, then towards the middle of the abdomen, a second stomachal dilatation surrounded with silk. According to the observations of M. Leon. Dufour (Annal. des Sc. Phys. tom. vi.), it occupies the greater portion of the abdominal cavity, and is immediately enveloped by the skin. It is of a pulpy consistence, formed of small grains, whose peculiar excretory ducts unite into many hepatic canals, pouring into the alimentary tube the product of the secretion. At the middle of its superior surface is a deep line, where the heart is lodged, and which divides this organ into two equal lobes. Its form varies like that of the abdomen, according to the species, thus its contour is festooned in the *Epeira sericea.* In this subgenus, as well as in the *Lycosa tarentula,* its surface is covered with a coat of a chalky white, cut into areolæ, which are easily perceived through the smooth skin of some species. They are observed to obey the motion of the systole and diastole of the heart. The individuals of both sexes often shoot through the anus an excrementitious fluid, composed of one part white as milk, and another black as ink.

The nervous system is composed of a double cord, occupying the medial line of the body, and of ganglia, which distribute nerves to the various organs. M. Dufour has been unable to determine the number and disposition of these ganglia; but, according to the figure which Treviranus has given of this system, the number of ganglia should be but two. The observations of the latter also supply the deficiencies of M. Dufour, relatively to the organ of circulation, which, according to him, appears to consist but of a simple dorsal vessel, and also relatively to the testicles and the spermatic vessels, on which he has afforded no information.

The dorsal region of the abdomen in many araneïdes, especially in those which are smooth or but slightly furnished with hair, presents deep or umbilical points, the number and disposition of which vary. M. Dufour has ascertained that those little orbicular depressions were determined by the attachment of the filiform muscles which traverse the liver, and which he has also observed in the scorpions.

The pulmonary cavities, to the number of one or two pair, are indicated externally by so many yellowish or whitish spots, placed near the base of the belly, immediately after the segment, which, by means of a fleshy thread, unites the abdomen with the thorax. Each pulmonary pouch is formed by the superposition of a great number of triangular leaflets, white, and extremely slender, which meet together around the stigmata, which always answer in number to the pulmonary sacs. Where there are four of them, a sort of fold, or vestige of a ring, existing even in those where there are but two, and placed immediately after them, forms a line, which separates the two pairs.

The female araneïdes have two very distinct ovaries, lodged in a sort of capsule formed by the liver. Not fecundated, they appear of a spongy tissue, as it were flaky, and constituted by the agglomeration of rounded corpuscles, scarcely perceptible, which are the germs of the eggs. In proportion to the progress of fecundation, the cluster formed by these eggs becomes less compact; and we find that they are inserted laterally on many canals. Their great analogy with the ovaries of the scorpion causes the same observer to presume that they form meshes, terminating in two distinct oviducts, which open into one and the same vulva. The configuration of the latter varies much: sometimes it is a longitudinal bilabiated cleft, as in the *Micrommata argelas;* sometimes it is sheltered by a prolonged operculum, and terminated like a tail, as in the *Epeïra diadema;* or it sometimes presents itself in the form of a tubercle.

With respect to the simple eyes, the same observer remarks, that they shine in darkness, or dusk, like those of the cat, and that the araneïdes have probably the faculty of seeing both by day and night.

The abdomen of the araneïdes putrifies and changes after death to such a degree, that its colours and even its form become undistinguishable. M. Dufour has succeeded, by means of a very prompt desiccation, of which he explains the process, in remedying this inconvenience.

According to Reaumur, the silk undergoes a first elaboration in two small reservoirs, having the figure of a Rupert's drop, placed obliquely one on each side, at the basis of six other reservoirs, shaped like intestines, situated one at the side of the other, bent six or seven times, setting out a little below the origin of the belly, and leading to the nipples by a very slender thread. It is in these last vessels that the silk acquires more consistence and the other qualities which are peculiar to it. They communicate with the preceding by branches, forming a great number of bends, and then divers interlacings. On proceeding from the nipples, the silken threads are glutinous; a certain degree of desiccation or evaporation of moisture is necessary before they can be employed. But it appears, that when the temperature is propitious an instant is sufficient, since these animals make use of them almost immediately as they escape from their spinnerets. Those white and silky flakes which we see flying about in spring and autumn, on days in which there has been a fog, and which are vulgarly termed "*the virgin's threads,*" are certainly produced, as we have been convinced by following them from their point of departure, by divers young araneïdes, and especially by Epeïræ and Thomisi. These are principally the large threads which should serve as an attachment to the radii of the web, or those which compose its chain, and which, becoming more heavy in proportion to the degree of moisture, sink, approach towards each other, and end by being formed into pillets. They are often seen to unite near the web commenced by the animal, and where it remains. It is, moreover, probable, that many of these araneïdes, not having as yet a sufficient provision of silk, confine themselves to throwing out some simple threads to a distance. It is, in my opinion, to some young lycosæ, that we must attribute those which are seen in great abundance crossing the furrows of ploughed fields, where they reflect the light of the sun. Chemically analyzed, these threads of the virgin present precisely the same characters as the silk of the spiders. They are not therefore formed in the atmosphere, as has been conjectured, for want of proper observations with his own eyes, by a philosopher, whose authority is of great weight, M. le Chevalier de Lamarck. Stockings and gloves have been fabricated with this silk, but such attempts not being susceptible of application on a large scale, and being subject to many difficulties, are more curious than useful. This substance is of much greater importance to the araneïdes. It is with it that the sedentary species, or those which do not hunt their prey, weave those webs of a tissue more or less compact, whose forms and positions vary according to the habits of each, and which are so many snares to take the insects on which they feed. These are scarcely arrested by means of the hooks of their tarsi, when the spider, sometimes placed in the centre of its net-work, or at the bottom of its web, sometimes in a peculiar habitation situated near it, and in one of its angles, runs up, approaches the insect, makes use of all its efforts to strike it with its murderous dart, and distil into the wound a poison which acts most promptly. When the insect opposes too powerful a resistance, or it might be dangerous to struggle with it, the spider retires for a moment, to wait until the other has lost its strength, or become more entangled. But if there is nothing to fear, the spider hastens to fetter its prey, by winding its threads of silk round its body, which sometimes completely envelope it, and form a covering that withdraws it from our sight.

Lister has asserted, that the spiders ejaculate and dart out their threads, in the same manner as the porcupines shoot their quills, with this difference, that the latter weapons, according to the popular opinion, are detached from the body, while in the spiders, these threads, though pushed to a distance, remain attached to the animal. This feat has been considered impossible. Nevertheless, we have seen threads issuing from the nipples of some thomisi, directed in a right line, and forming, as it were, moveable radii, when the animal moved circularly. Another use of the silk, and common to all the female araneïdes, is in the construction of cocoons, destined to enclose their eggs. The contexture and the form of these cocoons are variously modified, according to the habits of the races: they are generally spheroïdical. Some have the form of a cap, or that of a kettle; some are supported on a pedicle or stem, or terminate in a knob. Foreign substances, such as earth, leaves, &c. sometimes cover them, at least partially. A finer tissue, a sort of wadding, or down, often envelopes the eggs internally: they are either free or agglutinated, and more or less numerous. These animals, being very voracious, the males, to avoid all surprise, and not to be the victims of a premature desire, approach their females at the season of love with the most extreme distrust and the greatest circumspection. The apparatus of generation in the males, or at all events that which is presumed to be so, is usually very complicated and various, formed of scaly pieces, more or less hooked and irregular, and of a white fleshy body, on which vessels are sometimes perceived, of a sanguine appearance, and which is regarded as the fecundating organ, properly so called; but in the arachnides, with four pulmonary sacs, and in some others of the division of those which have but two, the last articulation of the palpi of the same individuals presents but a single corneous piece, in the form of a hook or ear-pick, without the least distinct aperture. Although Müller and some others were wrong, with reference to some entomostraca, in placing the male sexual organs on two of their antennæ, it is not less true that the parts considered as analogous in the araneïdes are very different from those which have been observed on the antennæ of these crustacea, and that we cannot conceive what their use can be if we refuse them this office.

According to the experiments of Audebert, who has given us a history of the Simiæ, worthy of the talents of this great painter, it is proved that a single fecundation can suffice for many successive generations. But as in all the insects, and other analogous classes, the eggs are sterile if the two sexes do not unite. The coupling, in our climates, takes place from the end of summer, until towards the end of September. The eggs, which are first laid, often disclose before the end of autumn; the others pass the winter. It has been remarked that the females of some species of lycosæ, or *wolf-spiders*, tear the cocoon of the eggs, when the young are about to come into the world. When newly born, they climb on the back of the mother and remain there for some time. Other female araneïdes carry the cocoon under the belly, or watch over its preservation, by fixing it near them. The two hinder feet are not developed, in some young ones, until a few days after their birth. There are some which, at the same period, are assembled for some time in society, and appear to spin together. Their colours then are often more uniform, and the naturalist who has but little experience, might very erroneously be induced to multiply the species. One of our fellow-labourers in the *Encyclopedie Méthodique*, M. Amédée Lepelletier de St. Fargeau, has observed, that these animals, as well as the crustacea, possess the faculty of regenerating their lost members.

I have ascertained that a single sting of an araneïd of middle size will kill our domestic fly in the space of a few minutes. It is also certain, that the bite of those large araneïdes of South America, which are known under the name of crab-spiders, and which we range in the genus mygale, will kill small vertebrated animals, such as little birds, colibris, pigeons, &c., and may produce a violent attack of fever in man himself. The wound even of some species of our southern climates, has proved mortal. We may, then, without adopting all the fables which Baglivi and others have put forth respecting the tarantula, be cautious, especially in warm climates, of the bite of the araneïdes, and particularly of the large species. Divers species of the genus *sphex* of Linnæus seize the araneides, pierce them with their sting, and transport them into the holes where they have deposited their eggs, that they may serve as food for their young. Most part of these animals perish in the after season; but there are some which live many years, and of this number are the mygale, lycosa, and probably several others. Though Pliny says that the *phalangium* are unknown in Italy, we nevertheless presume that these last araneides, and other large species, spinning no web, and also the galeodes, are the animals designed collectively by that name, and of which many species are distinguished. Such was also the opinion of Mouffet, who has figured as a species of phalangium, a lycosa, or a mygale of the island of Candia.

Lister, who was the first properly to observe the araneides whose habits fell under his notice, viz. those of Great Britain, has laid the basis of a natural distribution, which all subsequent authors have done little more than modify. The more recent knowledge of some species peculiar to hot climates, such as the *mason spider*, described by the Abbé Sauvages, and some others analogous; the employment of the organs of manducation introduced into the method by Fabricius; a more precise study of the general disposition of the eyes, and their respective sizes; and, further, the consideration of the relative length of the feet, have all contributed to extend this classification. M. Walckenaer has entered in this respect into the most minute details, and it would be difficult to discover a species which does not find its place in some one of the sections which he has established. There is, however, a character, the application of which has not been generalized, the presence or absence of the third hook at the end of the tarsi. M. Savigny has presented us in this point of view, a new method, but of which I am only acquainted with a simple synopsis.

M. Leon Dufour, who has published some excellent memoirs on the anatomy of insects, who has made an especial study of those of the kingdom of Valencia, where he has discovered many new species, and to whom botany is not less indebted, has bestowed particular attention on the respiratory organs of the araneïdes, and it is according to him that we shall divide them into those which have four pulmonary sacs (and externally four stigmata, two on each side, and very near each other), and into those which have but two. The first, which embrace the order of araneïdes theraphosæ, of M. Walckenaer, and some other genera of that which he designates collectively under the denomination of spider, compose according to our method but a single one, that of

MYGALE.

Their eyes are always situated at the anterior extremity of the thorax, and usually very closely approximated. Their forceps and feet are strong; the copulatory organs of the males are always projecting, and often very simple; the majority have but four spinnerets, two of which, lateral or exterior, and situated a little above the other two, are longer, with three articulations, not reckoning the elevation formed by their peduncle. They fabricate silken tubes, which serve them for an habitation, and which they conceal, either in burrows that they have excavated, or under stones, the bark of trees, or between leaves.

The theraphosae of M. Walckenaer will form a first division, having for characters—four spinnerets, the two intermediate and inferior very short, and the two exterior extremely projecting; crooks of the forceps bent underneath along their keel or lower edge, and not within or on their internal face; eight eyes in all, most frequently grouped on a small eminence, three on each side, forming, when united, a reversed triangle, and the upper two approximate, the other two disposed transversely, at the middle of the preceding.

The fourth pair of feet, and after them, the first, are the longest; the third is the shortest.

In some the palpi are inserted at the upper extremity of the jaws, so that they appear to be composed of six articulations, of which the first, narrow and elongated, with the internal angle of the superior extremity projecting, performs the office of jaw. The tongue is always small, and almost square; the last articulation of the palpi in the males is short, in the form of a button, and bearing at its extremity the sexual organs. The two anterior legs of the same individuals have a strong spine, or spur, at their lower extremity. Such are the characters of

MYGALE (proper), *Walck.*

Some do not present at the upper extremity of their forceps immediately above the insertion of the claw, or hook, which terminates it, a transverse series of spines, or corneous and mobile points, disposed in the manner of a rake. The hairs which furnish the under portion of their tarsi, form a thick and tolerably broad brush, edging, and usually concealing the hooks. The masculine sexual organs consist in a single scaly piece, terminated in an entire point, that is, without notch or division. Sometimes it has almost the form of an ear-pick; sometimes, and most frequently, it is globular underneath, and then grows narrow, terminating in a point, and forming a sort of arched hook.

This division is composed of the largest species of the family, and some of which, in a state of repose, occupy a circular space of six or seven inches in diameter, and sometimes seize on humming birds and colibris. They establish their domicile in the clefts of trees, under their bark, in the interstices of stones, or rocks, or on the surfaces of the leaves of divers vegetables. The cell of the *Mygale avicularia* has the form of a tube narrowed to a point at its posterior extremity. It is composed of a white web, compact in its tissue, very fine, semi-transparent, and similar in appearance to muslin. M. Goudot has given me one, which, unfolded, was seven or eight inches long, by about two wide, measured in its greatest transverse diameter. The cocoon of the same species was of the form and size of a large nut. Its envelope, composed of a silk similar in kind to that of its habitation, was formed of three layers. It appears that the young are hatched there, and undergo their first moulting. M. Goudot informed me, that from a single one, he took out a hundred young ones.

This mygale (*Aranea avicularia*, Lin,; Kléem, insect xi. and xii., male) is about an inch and a half in length, blackish, very hairy, with the extremity of the palpi, of the feet, and the lower hairs of the mouth reddish. The genital organ of the males is hollow at its base, and finishes in an elongated and very sharp point.

South America and the Antilles furnish other species, which are known by the French colonists under the name of *araignées-crabes* (crab-spiders). Their bite is supposed to be very dangerous. The East Indies have also another very large species, (*M. fasciata,* Seba., Mus., I. lxix. 1.; Walck. Hist. des Aran., iv. 1. fem.) We also receive from the Cape of Good Hope, a species almost as large as the avicularia. Another of the same division, (*M. valentina*) has been found in the arid soils and deserts of Moxenta, in Spain, by M. Dufour, who has described and figured it, in the fifth volume of the Annals of the Physical Sciences, published at Brussels. M. Walckenaer has made us acquainted with another species found in this peninsula (*M. calpeiana*) which has two eminences above the respiratory organs. These two species form a small particular group, having as character, the hooks of the tarsi projecting or naked.

In the following mygales, the superior extremity of the first articulation of the forceps presents a series of spines, articulated, and mobile at their base, according to the observations of M. Dufour, and forming a sort of rake.

The tarsi are less hairy underneath than in the preceding division, and their hooks are always uncovered. The males of one species, the only one which I have seen, have their copulatory organs less simple than those of the preceding species. The scaly and principal piece encloses in an inferior cavity a particular body, semi-globular, and terminating in a point, bifid.

These species excavate, in dry and mountainous places, situated to the south, in the southern countries of Europe, and some others, subterraneous galleries, tubiform, being often two feet in depth, and so sinuous, that, according to M. Dufour,

Mygale Fasciata. Fem.

London, Published by Whittaker & Co. Ave Maria Lane, 1833.

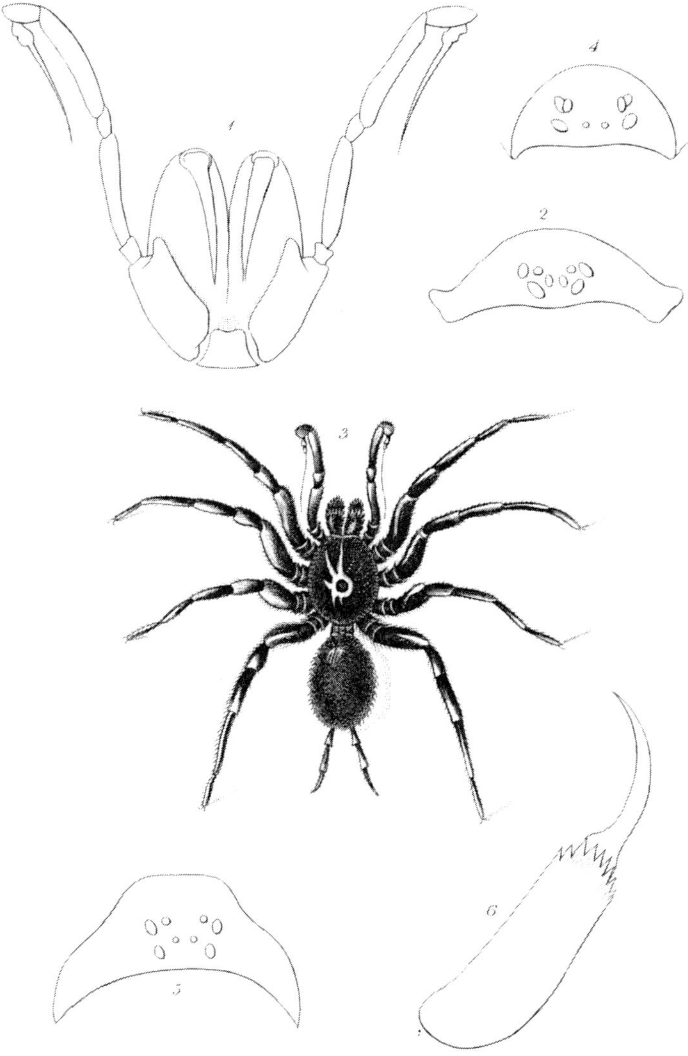

1.2.3. *Mygale calpeiana.*

4. *Mygale notasiana.*

5.6. *Mygale cementaria.*

London. Published by Whittaker & C.º Ave Maria Lane 1833.

the trace of them is often lost. They construct at the entrance, with clay and silk, a moveable cover, fixed by a hinge, and which, in consequence of its form, perfectly adapted to the aperture, of its inclination, of its natural weight, and of the superior situation of the hinge, closes of itself, and in the most exact manner, the entrance of the habitation, and thus forms a trap, which it is difficult to distinguish from the surrounding soil. Its interior face is clothed with a silken layer, to which the animal hooks itself, to draw to it this door, and hinder it from being opened. If it is a little open, we may be sure the spider is in its retreat. When it is discovered, by a fissure made in the conduit, in front of its issue, it remains stupefied, and suffers itself to be taken without resistance. A silken tunnel, or the nest properly so called, invests the interior of the gallery. The naturalist just cited is of opinion that the males do not excavate in this way. Independently of his never having met with them except under stones, they appear to him to be less favoured by nature with organs proper for the execution of such labours. Without pronouncing on this point, we presume, with him, that our *Mygale carminans* is only the male of the following species. Nevertheless, M. Walckenaer has his doubts on this point.

The female *mason-spider* (*M. cœmentaria*, Latr.; *Araignée maçonne*, Sauvag. Hist. de l'Acad. des. Scienc. 1758, p. 26; *Araignée mineuse*, Dorthès Linn. Trans. ii. 17. 8, &c. &c.) is about eight lines in length, of a reddish, bordering on brown, and more or less deep, with the edges of the corslet paler. The forceps are blackish, and have, each of them, above, near the articulation of the hook, five points, the internal one of which is shorter. The abdomen is of a mouse-grey, with deeper coloured spots. The first articulation of all the tarsi is furnished with small spines; the hooks of the last have a spur at their base, and a double range of sharp teeth. The spinnerets project but little. According to M.

Dufour, the presumed male, of which I have made a species, *M. cardense,* differs from the preceding individual by its longer feet, by the hooks of the tarsi, the teeth of which are as numerous again, but without spurs, and by its spinnerets being shorter. But a more apparent character is the strong spine terminating underneath the two anterior legs. This mygale is found in the southern departments of France, situated on the shores of the Mediterranean, in Spain, &c.

The female pioneer-spider (*M. fodiens,* Walck. Faun. Franc. Arach. ii. 1, 2; *M. Sauvagesii,* Dufour. Ann. des Scienc. Phys. V. lxxiii. 3; *Aranea Sauvagesii,* Ross.) is a little larger than that of the preceding species, of a clear reddish brown, and without spots. The external spinnerets are long. The four anterior tarsi alone are furnished with small spines. All have a spur at the end, and their hooks present but a single tooth, situated at their base. The forceps are stronger, and more inclined, than those of the mason-spider. The points of the rake are a little more numerous. The first articulation presents, underneath, two ranges of teeth. The male is unknown. This species is found in Tuscany and Corsica. The Museum of Natural History possesses a small clod of earth, in which four of its nests are to be seen, disposed in a regular quadrilateral figure.

M. Lefevre, so zealous for the progress of entomology, and who has made so many sacrifices for this science, has brought from Sicily a new species of mygale, the body of which is entirely of a blackish brown. The male does not present at the extremity of the anterior legs this strong spine, which appears in general to be peculiar to individuals of the same sex in the other mygales.

There is found at Jamaica another species, *M. nidulans,* represented with its nest by Brown, in his Natural History of Jamaica, pl. xliv. 3.

In others, the palpi are inserted in a lower dilatation of the external side of the jaws, and have but five articulations. The tongue, at first very small (atypus), is afterwards elongated and advanced between the jaws, and this character becomes general. The last articulation of the palpi of the two sexes is elongated, and fined into a point, towards the end. The males have no strong spur at the extremity of their two anterior legs.

ATYPUS, *Latr.* OLETERA, *Walck.*,

Have a very small tongue, almost covered by the internal portion of the base of the jaws, and the eyes close together and grouped upon a tubercle.

Atypus Sulzeri. Latr., Gen. Crust. et Insect. I. v. 2, male; Dufour, Ann. des Scienc. Phys. V. lxxiii. 6.; *Aranea picea*, Sulz.; *Oletère atype*, Walck. Faun. Franç. Arach. ii. 3, has the body entirely blackish, and about eight lines in length. The thorax is almost square, depressed posteriorly, swelled, widened, and broadly truncated in front, which gives it a form very different from what this part of the body presents in the mygales. The forceps are very strong, and their claw, underneath, near the base, has a small eminence, in the form of a tooth. The last articulation of the palpi of the male is pointed at the end. The genital organ below gives origin to a small semi-transparent piece, in the form of a scale, widened, and unequally forked at the end, with a small silky hair at one of its extremities. This species excavates, in inclined soils covered with turf, a cylindrical tube, seven or eight inches in length, at first cylindrical, afterwards inclined, where it spins a tunnel of white silk, of the same form and the same dimensions. The cocoon is fixed with some silk, and by the two ends, to the bottom of this tunnel. It is found in the environs of Paris, and of Bordeaux; and M. de Basoches has observed, near Séez, a variety which is constantly of a clear brown.

M. Milbert, correspondent of the Museum of Natural History, has discovered, in the environs of Philadelphia, another species (*Atypus rufipes*) altogether black, with the feet fawn-coloured.

ERIODON, *Latr.* MISSULENA, *Walck.*,

Differ from the atypi, in their elongated and narrow tongue advancing between the jaws, and their eyes scattered on the front of the thorax.

The only known species (*Eriodon occatorius*, Latr.; *Missulena*, Walck. Tab. des Aran. pl. ii. 11, 12), is an inch long, blackish, and proper to New Holland, from which it has been brought by Péron and Lesueur.

Our second and last division of mygales, or araneides with four pulmonary sacs, presents characters common to eriodon, as that of having the tongue prolonged between the jaws, the palpi composed of five articulations; but the claws of the forceps are bent along their internal edge; their spinnerets are six in number; the first pair of feet, and not the fourth, is the longest of all; the third is always the shortest. Some of these arachnides have but six eyes. The number of pulmonary sacs will not permit us to remove the subgenera of this division from the preceding; and as it conducts us to *drassus*, to *clotho*, to *segestria*, subgenera presenting only two pulmonary sacs, the natural order will not permit us to pass from the mygales to the lycosæ and other hunting or erratic araneides. The mygales are genuine weaving spiders; and, in truth, it was in this division that the avicularia of Linnæus was originally placed.

This second division comprehends the two following subgenera,

DYSDERA, *Latr.*,

Which have but six eyes, arranged in the form of a horse-shoe, with the aperture in front; whose forceps are very strong

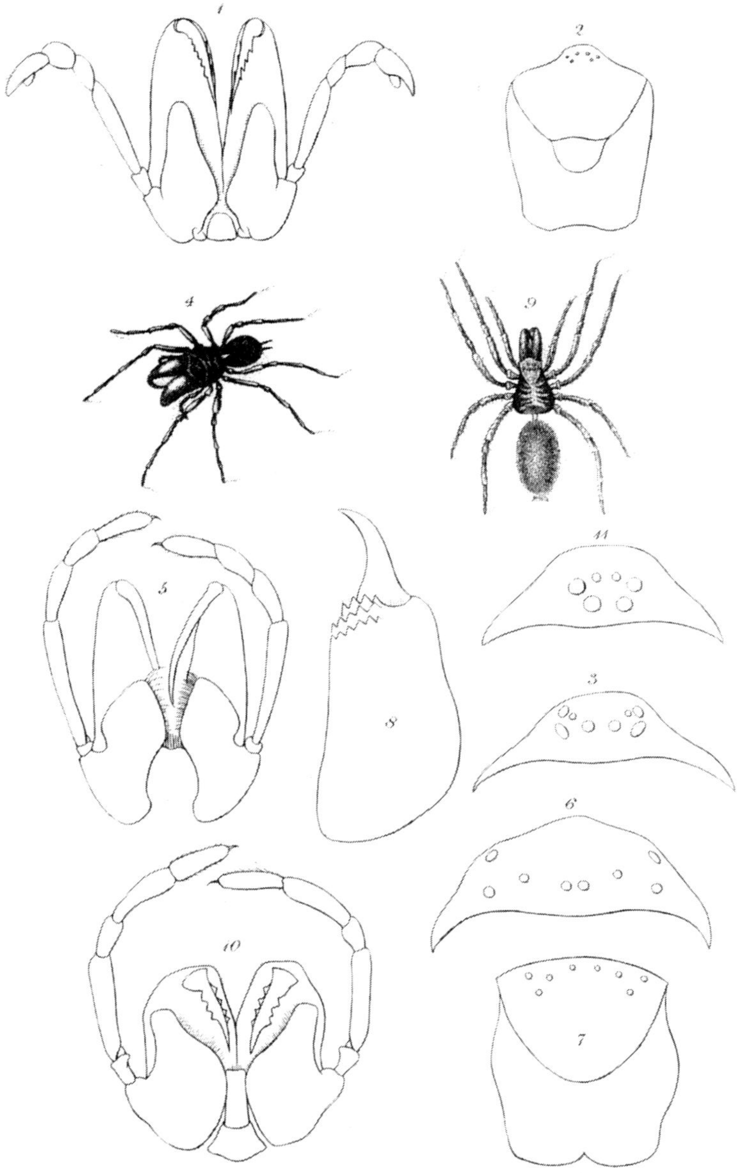

1.2.3.4. *Atypus subzeri* _ *Lat.*

5.6.7.8. *Eriodon occatarium* _ *Lat.*

9.10.11. *Disdera erythrina* _ *Walk.*

London Published by Whittaker & C.º Ave Maria Lane 1833.

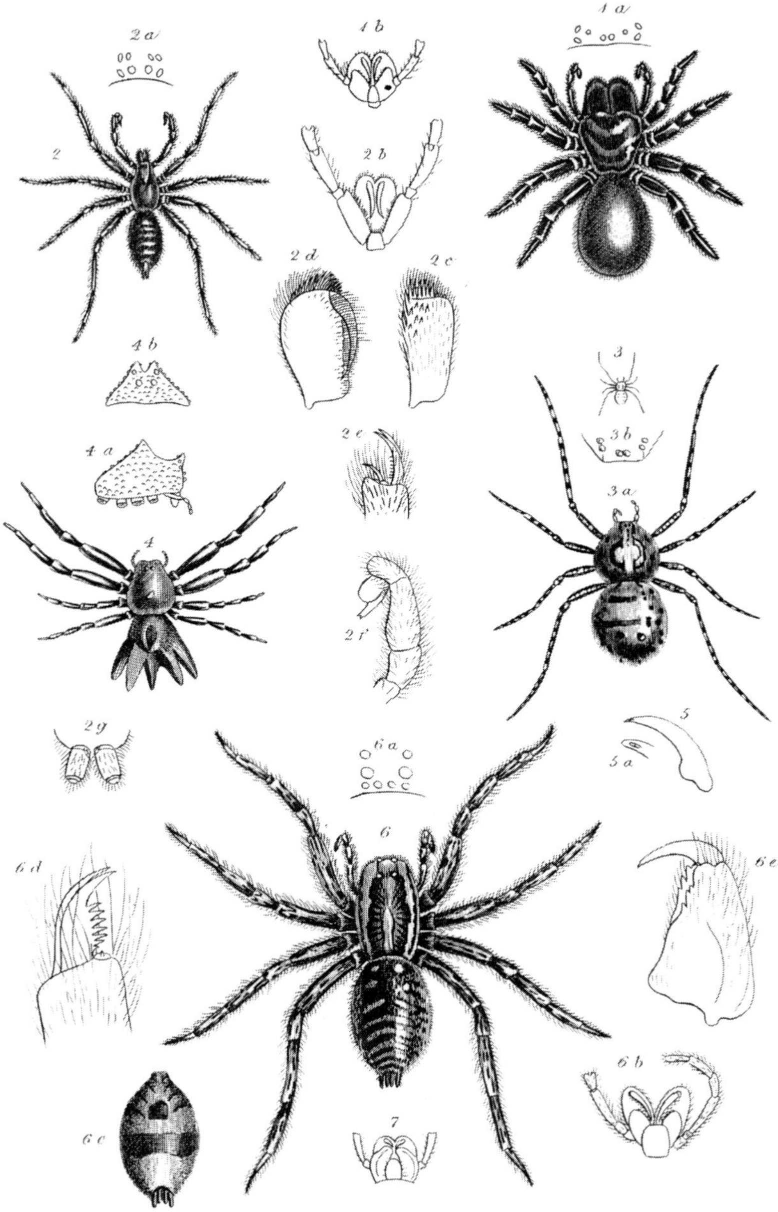

1 *Eriodon occatorius.* 4 *Thomisus heterogaster.*

2 *Mygale cæmentaria male* 5 *Hook of the mandible of Mygale avicularia.*

3 *Scythodes thoracica* 6 *Lycosa tarentula.*

7 *Mouth of Drassus melanogaster.*

London: Published by Whittaker & Co. Ave Maria Lane 1833.

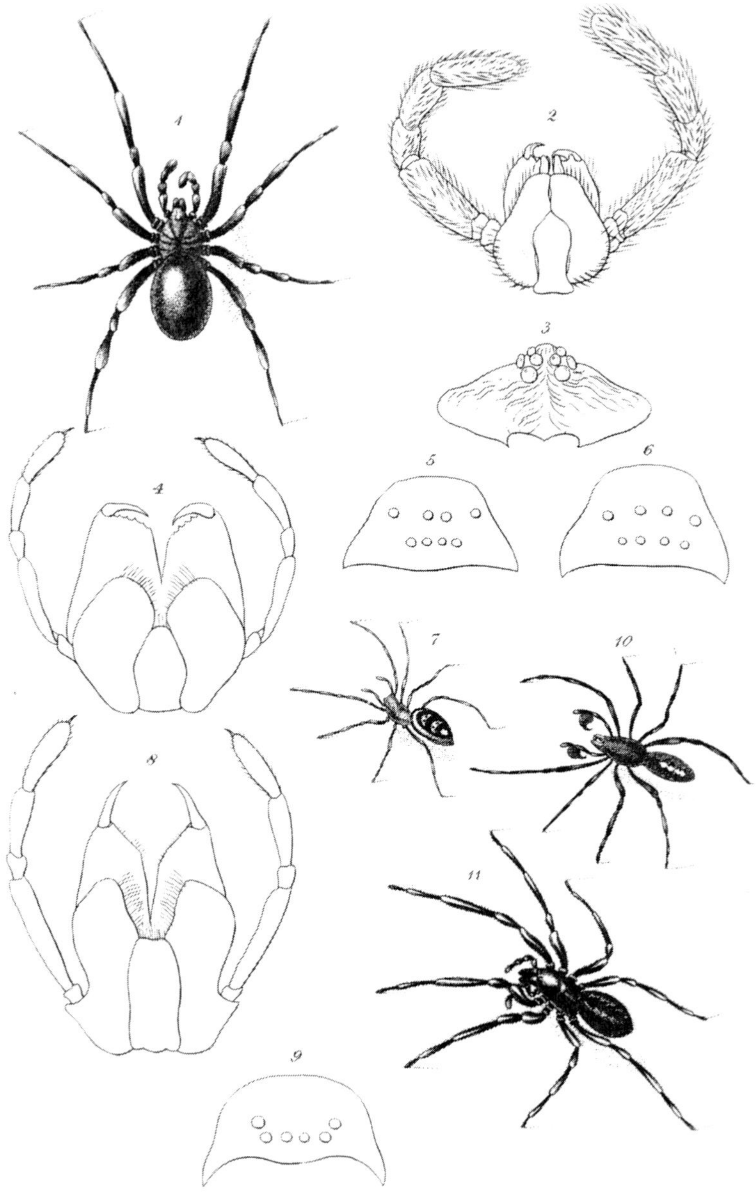

1.2.3. *Filistata bicolor.* 7. *Drassus viridissimus.*
4.5.6. *Drassus lucifagus.* 8.9.10. *Segestria senoculata.*
11. *Seg. perfida.*

London, Published by Whittaker & C.º Ave Maria Lane 1833.

and advanced, and whose jaws are straight, and dilated at the insertion of the palpi.

FILISTATA, *Latr.*,

Which have eight eyes grouped on a little elevation at the anterior extremity of the thorax, the forceps small, and the jaws arched at the external side, and surrounding the tongue as a centre.

We now pass to the araneïdes which have but a single pair of pulmonary sacs and stigmata. All present palpi with five articulations, inserted on the external side of the jaws, near their base, and most frequently in a sinus; a tongue advanced between them, either almost square, or triangular, or semicircular; and six nipples or spinnerets at the anus. The last articulation of the palpi of the males is more or less ovoïd, and encloses, most frequently, in an excavation, a complicated and very varied copulatory organ. Rarely (*Segestria*) it is uncovered.

With the exception of a small number of species, entering into the genus mygale, they compose that of

The SPIDER, ARANEA of *Linnæus*, or ARANEUS of some writers.

A first division will comprehend the SEDENTARY SPIDERS. They make webs, or at least send forth threads, to surprise their prey, and remain habitually within those snares, or close beside them, as well as near their eggs. Their eyes are approximated on the breadth of the forehead, sometimes to the number of eight, four or two of which are at the middle, and two or three on each side; sometimes their number is six.

Some, which in walking always proceed straight forward, and which we name on that account RECTIGRADES, weave webs, and are always stationary. Their feet are elevated in a state of repose; sometimes the first two and the last two, sometimes those of the two anterior pairs, or the fourth and the third, are the longest. The eyes do not form, by their general disposition, a segment of a circle or a crescent.

They may be divided into three sections: the first, that of TUBITELES, or WEAVERS, has cylindrical spinnerets, collected in a bundle, directed backwards, the feet robust, and the first two, or the last two, and *vice versâ*, longer in some, while in others the eight feet are nearly of equal length.

We shall commence with two subgenera, which, considered as to the jaws forming an arch round the tongue, approach the filistata, and depart from the following. The eyes are always eight in number, disposed by fours on two transverse lines. The first, that of

CLOTHO, *Walck.* UROCTEA, *Dufour*,

Is one of the most singular. Its forceps are very small, without denticulations, and but little capable of separation, which character approximates this subgenus to the last. The hooks are very small. In the short form of the body, and length of its feet, it resembles much the spider-crab or Thomisus. The relative length of these organs differs little. The fourth pair, and then the preceding, are only a little longer than the first four legs. The tarsi alone are furnished with prickles. The eyes are more remote from the anterior edge of the thorax than in the following subgenus, approximate, and disposed in the same manner as in the genus mygale of M. Walckenaer. Three on each side form an inverted triangle. The other two form a transverse line, in the space comprized between the two triangles. The jaws and tongue are proportionally smaller than those of the last-mentioned subgenus. The jaws have on the external side a short projection, or slight dilatation, serving as an insertion to the palpi, and terminating in a point. The tongue is triangular, and not almost oval, like that of drassus. The two upper spinnerets, or the most lateral, are long; but that which, according to M. Dufour, particularly characterizes his uroctea, or our clotho, is, that in place of the two intermediate spinnerets, we find two pectiniform valves, opening and closing at the will of the animal.

As yet but a single species is known, *Uroctea, 5—Maculata*, Dufour, Annal. des. Scienc Phys. V. lxxvi. 1; *Clotho Durandii*, Latr. Its body is five lines in length, of a chestnut-brown, with the abdomen black, having five round, yellowish spots above, four of which are disposed transversely in pairs, and the last or odd one is posterior; the feet are hairy. We see by the plates of the great work on Egypt, that M. de Savigny found it in that country, and that he proposed to form with it a new generic group. M. le Comte Dejean brought it from Dalmatia, and the Chevalier de Schreibers, director of the imperial cabinet of Vienna, has sent me some individuals collected in the same places. M. Dufour has also found it in the mountains of Narbonne, in the Pyrenees, and in the rocks of Catalonia. We are indebted to him, independently of the knowledge of the external characters of this araneïd, for some curious observations respecting its habits. "It makes," he tells us, "at the inferior surface of large stones, and in the clefts of rocks, a cocoon, in the form of a cap or little dish, a good inch in diameter. Its contour presents seven or eight emarginations, of which the angles alone are fixed upon the stone, by means of bundles of threads, while the edges are free. This singular tent is of an admirable texture: the exterior resembles the finest taffetas, composed, according to the age of the worker, of a greater or less number of doublings. Thus, when the uroctea, as yet young, commences to establish its retreat, it only fabricates two webs, between which it remains in shelter. Subsequently, and, I believe, at each moulting, it adds a certain number of doubles. Finally, when the period marked for reproduction arrives, it weaves a cell for this very purpose, more downy and soft, where the sacs of eggs, and the young ones newly disclosed, are to be shut up. Although the external cap or pavilion is designedly, without doubt, more or less soiled by foreign bodies, which serve to conceal its presence, the apartment of the industrious fabricator is always scrupulously clean. The pouches which enclose the eggs are four, five, or even six, for each habitation, which, nevertheless, forms but a single habitation. These pouches are of a lenticular form, and are more than four lines in diameter: they are formed of a sort of taffeta as white as snow, and furnished internally with a down of the finest kind. It is only at the end of December, or in the month of January, that the laying of the eggs takes place; it was therefore necessary, beforehand, to provide for the defence of their progeny against both the rigour of the season and hostile incursions. Every thing of this kind has been carefully done. The receptacle of this precious deposit is separated from the web, immediately applied upon the stone, by a soft down, and from the external cap, by the various stories of which I have spoken. Among the emarginations which border the tent, some are altogether closed by the continuity of the stuff; others have their edges simply lapped over, so that the uroctea, by raising them, may issue at will from the tent, and re-enter. When it quits its domicile to proceed to the chase, it has little cause to fear that its habitation should be invaded, for itself alone possesses the secret of the impenetrable emarginations, and the key to those by which it can introduce itself. When the young ones are in a state to do without maternal cares, they take their departure and proceed to establish elsewhere their particular habitations, while the mother dies in her own tent. Thus the last is at once the cradle and the tomb of the uroctea."

DRASSUS, *Walck.*,

Differ from Clotho in many characters. Their forceps are

robust, projecting, and denticulated underneath; their jaws are truncated obliquely at their extremity, and the tongue forms an oval, truncated inferiorly, or an elongated curvilinear triangle; the eyes are more approximated to the anterior edge of the thorax, and the line formed by the four posterior is longer than the anterior, or extends beyond it on the sides. The proportions of the external spinnerets differ but little, and we do not see, between them those two pectiniform valves which are proper to the clothos. Finally, the fourth feet, and next the first two, are very manifestly longer than the others. The legs, and the first articulation of the tarsi, are armed with prickles.

These araneïdes remain under stones, in the clefts of walls, and the interior of leaves, and there fabricate cells of a very white silk. The cocoons of some are orbicular, flatted, and composed of two valves applied one upon the other. M. Walckenaer distributes the Drassi into three families, according to the direction and approximation of the lines formed by the eyes, and the greater or less dilatation of the middle of the jaws.

The species which he names *viridissimus*, (Hist. des Aran. fasc. iv. 9.) and which alone composes his third division, constructs on the surface of leaves a fine, white, and transparent web, beneath which it establishes itself. One of the sides of the leaves of the pear-tree has sometimes presented to my observation a similar web, but angular at the sides, in the form of a tent, like that made by Clotho, and under which was the cocoon. It is, I presume, the work of this species of Drassus, and indicates the analogy of this subgenus with the preceding. M. Leon Dufour has given us, in the Annals of the Physical Sciences (*Drassus segestriformis*, VI. xcv. 1.) a very complete description of a species of Drassus, which he found under stones in the high mountains of the Pyrénées, and never below the Alpine Zone. It is one of the largest of this subgenus, and appears to me closely allied with that which I have named *melanogaster*, and which I believe to be the *Drassus lucifugus* of M. Walckenaer, (Schœff. Icon. ci. 7.)

One of the prettiest species, and which is very commonly found in the neighbourhood of Paris, running on the ground, is the *D. relucens.* It is small, almost cylindrical, with the thorax fawn-coloured, covered with a silky and purple down; the abdomen mixed with blue, green, and red, with metallic reflexions, and two transverse lines of a golden yellow, the anterior of which is arched. There are also seen there sometimes four golden points.

In the other tubiteles, the jaws do not form a kind of arch enclosing the tongue. Their external side is dilated inferiorly, below the origin of the palpi.

Some of them have but six eyes, four of which are anterior, forming a transverse line, and the other two posterior, situated one on each side, behind the two lateral ones of the preceding line. Such is the essential character of

SEGESTRIA, *Latr.*

Their tongue is almost square and elongated. The first pair of feet, and after them the second, are the longest; the third is the shortest. These araneïdes spin for themselves, in the clefts of old walls, silken, cylindrical, elongated tubes, in which they remain, having their first pair of feet directed forwards. Some diverging threads border externally the entrance of their habitation and form a small web proper for catching insects. The genital organ of the segestria perfida (*Aranea florentina*, Ross., Faun., Etrusc., xix. 3.), a tolerably large species, black, with green forceps, and not uncommon in France, is in the form of a drop, or ovoïdo-conical, very sharp at the end, entirely projecting, and red.

The other tubiteles have eight eyes. We may, by reason of the difference of the medium in which they live, divide them

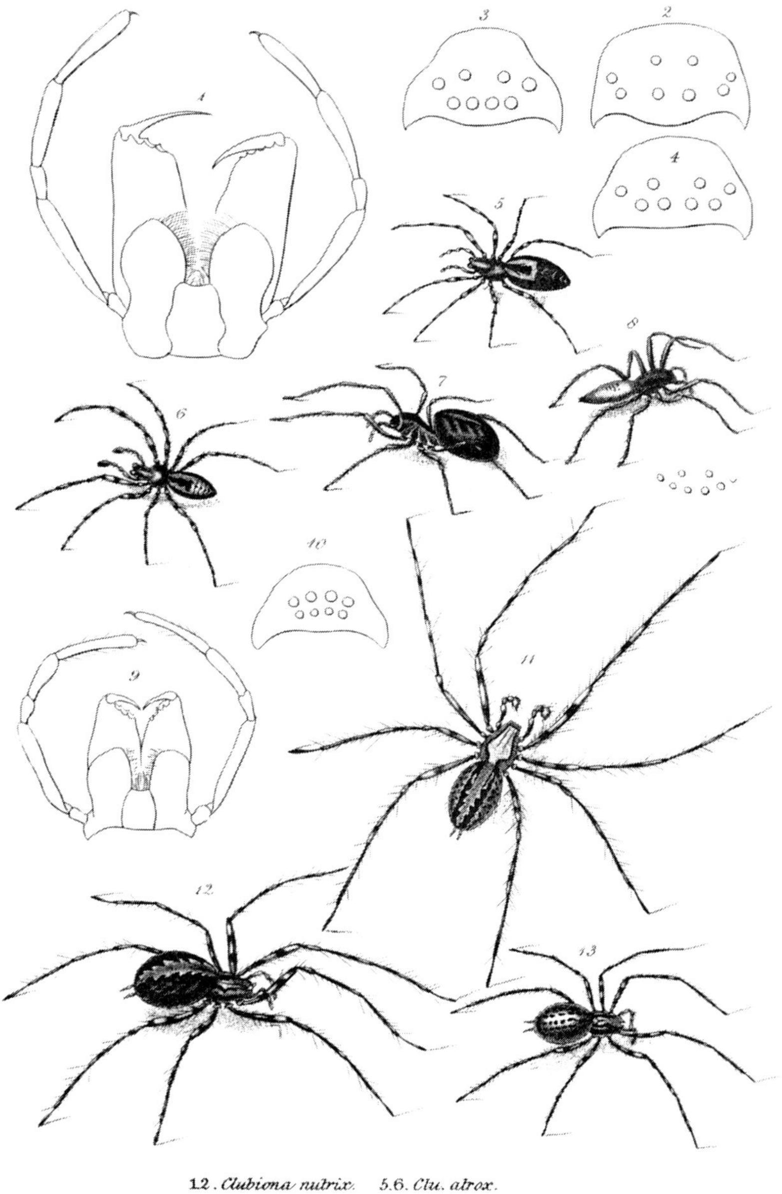

1.2. *Clubiona nutrix.* 5.6. *Clu. atrox.*

3. *Clu. accentuata.* 7. *Clu. ferox.*

4.8. *Clu holosericea.* 9.10.11.12 *Tegenaria domestica.*

13. *Teg. civilis.*

London, Published by Whittaker & C°. Ave Maria Lane. 1833.

into terrestrial and aquatic. Although M. Walckenaer has made of the last his final family of araneides, that of *nayades*, they have so much relation with the other tubiteles that, notwithstanding this disparity of habits, they must be placed with them. In those which are terrestrial, the tongue is almost square, or but little narrowed, very obtuse, or truncated at the summit. The jaws are straight, or almost straight, and more or less dilated towards their extremity. The two eyes of each lateral extremity of the ocular group, are in general tolerably distant from each other, or at least are not grouped in pairs, and borne on a particular eminence, like those of the aquatic tubiteles.

CLUBIONA, *Latr.*,

Are scarcely distinguished from the following subgenus, except in this, that the length of the external spinnerets are but little different, and the line formed by the four anterior eyes, is straight, or almost straight. They form silken tubes, serving them as an habitation, and which they place either under stones, in clefts of walls, or between leaves. The cocoons are globular.

THE SPIDERS, proper. (ARANEA),

Which we had at first designated under the generic name of *tegenaria*, preserved by M. Walckenaer, and to which we reunite his *agelena*, and his *nyssus*, have their two upper spinnerets remarkably longer than the others, and their four anterior eyes disposed in a line curved backwards.

They construct in the interior of our habitations, at the angles of walls, on plants, hedges, and often on the edges of roads, either in the earth, or under stones, a large web, nearly horizontal, and at the upper part of which is a tube, where they remain without making any motion.

Now come the *nayades* of M. Walckenaer, or our aquatic tubiteles, and which compose the genus

ARGYRONETA, *Latr.*

The jaws are inclined on the tongue, the form of which is triangular. The two eyes of each lateral extremity of the ocular group are very much approximated one to another, and placed upon a special eminence. The other four form a quadrilateral figure. *Argyroneta* (*Aranea aquatica*, Lin., Geoff., Deg.) is of a blackish brown, with the abdomen deeper, silky, and having on the back four deep points.

It lives in our stagnant waters, swims there, the abdomen being enclosed in an air-bubble, and forms there, as a retreat, an oval cocoon, filled with air, lined with silk, from which proceed threads, directed on all sides, and attached to the neighbouring plants. There it lies in wait for its prey, fixes its cocoon for the eggs, which it watches assiduously, and encloses itself in it to pass the winter.

The second section of the sedentary and rectigrade spiders, that of INEQUITELES, has the external spinnerets almost conical, projecting but little, convergent, disposed like a rosette, and the feet very slender. Their jaws are inclined on the labrum, and grow narrow, or at least do not sensibly widen, at their posterior extremity.

The majority have the first pair of feet, and then the fourth the longest. Their abdomen is more voluminous, more soft, and more coloured, than in the preceding tribes. They make webs with an irregular net-work, composed of threads, which cross each other in all directions, and on different planes. They bind their prey with cords, watch carefully over the preservation of their eggs, and do not abandon them until they disclose the young. They live but a short time.

Some have the first pair of feet, and after them the fourth, the longest. Such are

SEYTODES, *Latr.*,

Which have but six eyes, disposed in pairs. According to

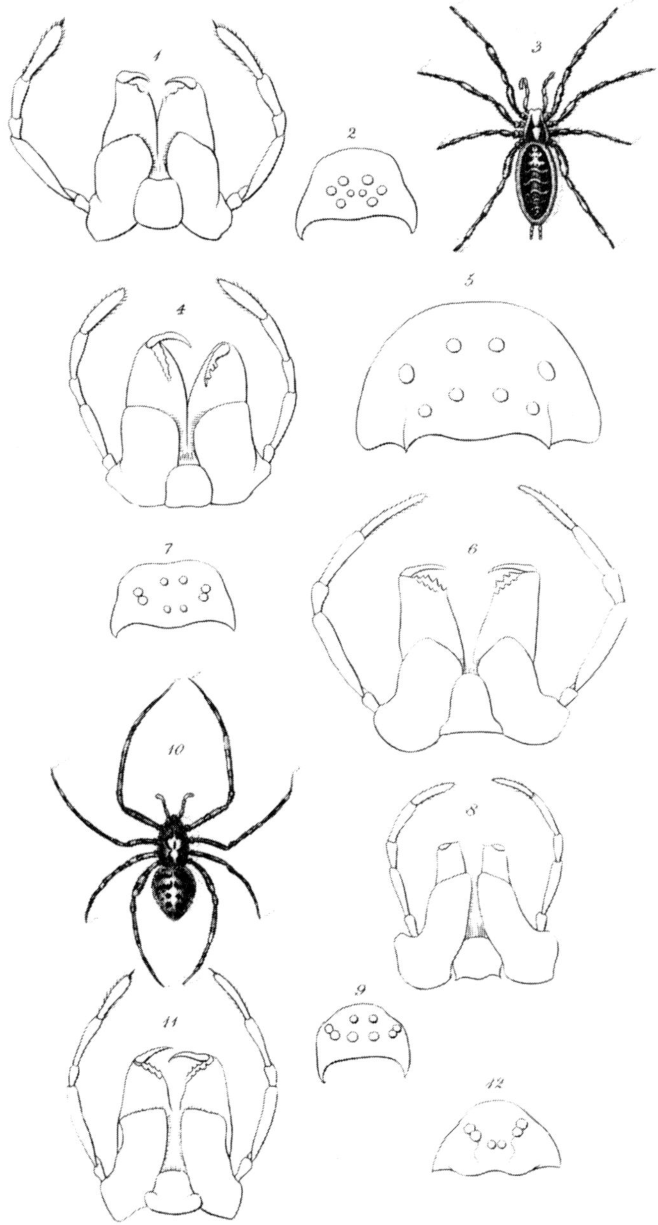

1.2.3. *Agelena labirinthica.* 6.7. *Aygyroneta aquatica.*
4.5. *Myssus coloripes.* 8.9.10. *Scytodes thoraca.*
11.12. *Theridion redimitum.*

London; Published by Whittaker & C° Ave Maria Lane. 1833.

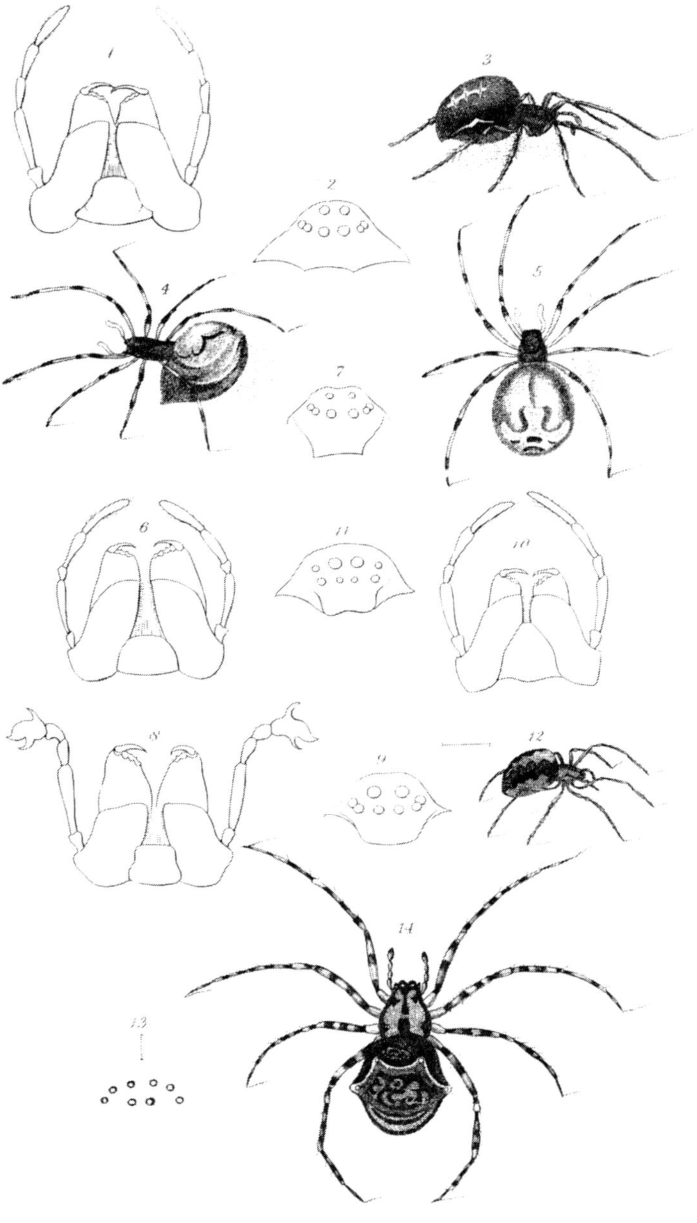

1.2. *Theridion 4 punctatum.* 8.9. *Ther. crypticoleus.*

3. *Ther. paykullianum.* 10.11.12. *Ther. triangulifer.*

4.5 6.7. *Ther. sisiphum.* 13.14. *Ther. aphane.*

London, Published by Whittaker & C.º Ave Maria Lane 1833.

M. Dufour, the hooks of the tarsi are inserted on a supplementary articulation.

Two species are known, one of which, *thoracica*, inhabits the interior of our apartments, and the other, *blonde*, (Ann. des Sc. Phys. V. lxxvi. 5.) has been found by this naturalist, under calcareous debris, in the mountains of the kingdom of Valencia. It weaves itself a tube, of no regular form, of slender fabric, and milky whiteness, pretty nearly like the *dysdera erythrina*.

THERIDION, *Walck.*

The eyes of which are eight in number, and thus disposed: four in the middle, in a square, of which the two anterior are placed upon a small eminence, and two on each side, also situated on a common elevation. The corslet is in the form of a reversed heart, or almost triangular. This subgenus is very numerous.

Aranea, 13—*guttata*, Fab., Ross., Faun., Etrusc. II. ix. 10. Eyes lateral, apart from each other; body black, with thirteen round spots, one of a blood-red, on the abdomen—Tuscany, Island of Corsica.

Its bite is believed to be very venomous, and even mortal.

The *A. mactans* of Fabricius, another species of Theridion, but from South America, inspires the same fears in that country. It would seem that these prejudices arise from the black colour, cut with sanguine spots, of these animals.

EPISINUS, *Walck.*,

Have also eight eyes, but approximated on a common elevation, and the corslet narrow, and almost cylindrical.

The other INEQUITELES have the first pair of feet, and then the second, the longest. Such are

PHOLCUS, *Walck.*,

Whose eyes, eight in number, are placed on a tubercle, and

divided into three groups; one on each side, formed of three eyes, disposed in a triangle, and the third in the middle, a little anterior, composed of two other eyes, and on a transverse line.

Pholcus phalangista, (Araignée domestique a longues pattes Geoff.) *ph. phalangioïdes,* Walck., Hist. des Aran. fasc. 5, tab. x. Body long and narrow, of a very pale, or livid yellow, pubescent; abdomen almost cylindrical, very soft, and marked above with blackish spots; feet very long, very fine, with a whitish ring at the extremity of the thighs and legs.

Common in houses, where it spins, at the angles of the walls, a web composed of loose threads, with but little adherence between them. The female agglutinates her eggs into a round naked body, which she carries between her mandibles.

M. Dufour has found a second species, *Pholque à queue* (Annal. des Scienc. Physique, V. lxxvi. 2.), in the clefts of rocks at Moxente, in the kingdom of Valencia. Its abdomen is terminated by a conical projection, thus forming a kind of tail, like that of the conical epeira. Like the preceding, it balances its body and feet. The palpi of the male have the genital organ extremely complicated.

The third section of the sedentary rectigrade spiders, that of ORBITELES, has the external spinnerets almost conical, but little projecting, convergent, and disposed like a rosette, and the feet slender, as in the preceding. They differ from those in the jaws, which are straight, and sensibly wider at their extremity.

The first pair of feet, and then the second, are always the longest. The eyes are eight in number, and thus disposed: four in the middle, forming a quadrilateral figure, and two on each side.

They approach the *Inequitelcs* in the size, the softness, and

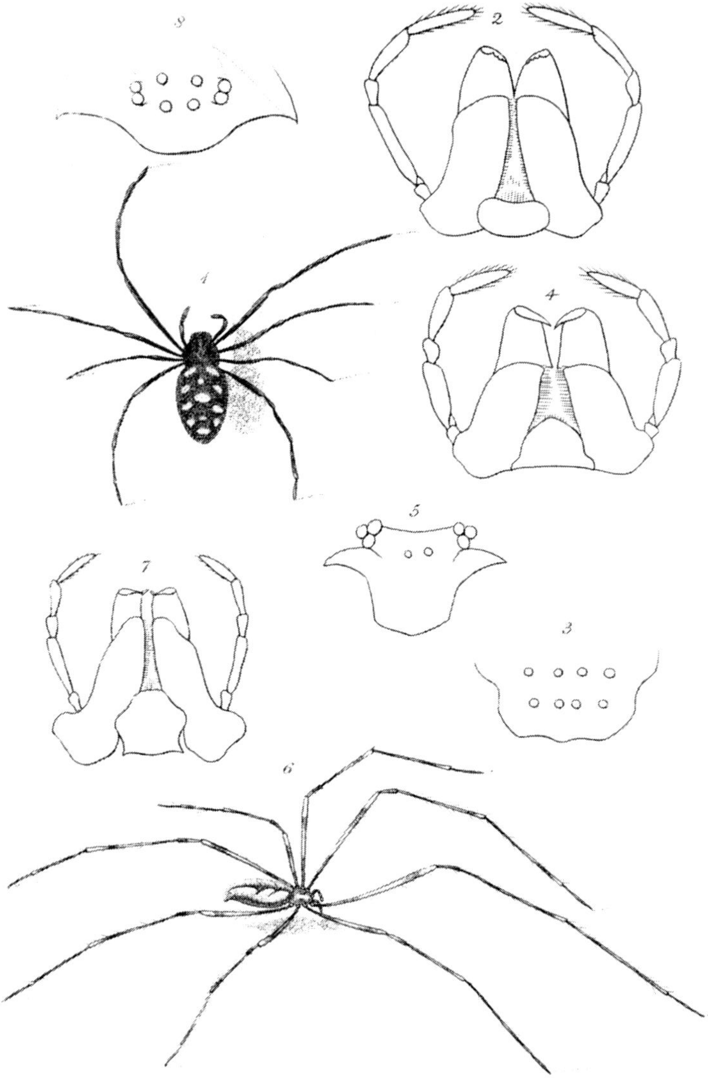

1.2.3 *Theridion (Latrodecta) 12-guttatum.*

4.5.6. *Pholcus phalangioides.*

7.8. *Theridion incertum.*

London, Published by Whittaker & Co. Ave Maria Lane, 1833.

the variety of colours of the abdomen, and the short duration of their life; but they make webs with a regular net-work, composed of concentric circles, crossed by straight radii, diverging from a centre, where these spiders always remain in an inverted position, towards the circumference. Some conceal themselves in a cavity, or a lodge which they have constructed near the edges of the web, which is sometimes horizontal, sometimes perpendicular. Their eggs are agglutinated, very numerous, and enclosed in a voluminous cocoon.

The threads which support the web, and which may be elongated to about one-fifth of their length, are used for the divisions of the micrometer. This observation has been communicated to us by M. Arago.

LINYPHIA, *Latr.*,

Well characterized by the disposition of their eyes: four in the middle, forming a trapezium, of which the posterior side is wider, and occupied by two eyes, much thicker and more apart, and the four others grouped by pairs, one on each side, and in an oblique direction. Their jaws are widened only at their superior extremity.

They construct on bushes, brooms, &c. a horizontal, thin web, of no great compactness, and spread above it, on several points, or in a very irregular manner, other threads. This web is thus a sort of mixture of those of the inequiteles and orbiteles. The animal remains at the lower part, and in an inverted posture.

ULOBORUS, *Latr.*,

Have the four posterior eyes placed at equal intervals on a straight line, and the two lateral ones of the first line more approximated to the anterior edge of the corslet than the two comprized between them, so that this line is arched backwards. The jaws, like those of the epeira, begin to widen a little above their base, and terminate in the form of a pallet or spatula. The tarsi of the last three pair of feet terminate in a single claw. The first articulation of the two posterior ones has a range of little hairs.

These weavers, as well as the species of the following sub-genus, have the body elongated, and almost cylindrical. Placed at the centre of their web, they carry forward, and in a straight line, the four anterior feet, and extend the last two in an opposite direction; those of the third pair are directed laterally.

These arachnides form webs similar to those of the other orbiteles, but more loose, and horizontal; they completely envelope, in less than three minutes, the body of a small coleopterous insect, which is caught in their net. Their cocoon is narrow, elongated, angular on the edges, and suspended vertically, by one of its ends, to a net-work; the other extremity is as it were forked, or terminated by two prolonged angles, one of which is shorter, and obtuse. Each side has two acute angles.

I am indebted for these interesting observations to my friend M. Leon Dufour.

Ul. Walckenarius, Lat. Nearly five lines in length, of a reddish yellow, covered with a silky down, forming on the upper part of the abdomen two series of little bundles; some rings, paler at the feet. Of the woods in the environs of Bordeaux, and the other southern departments.

TETRAGNATHA, *Latr.*,

Whose eyes are situated, four by four, on two lines almost parallel, and separated by intervals nearly equal, and which have jaws long, and narrow, and widened only at their superior extremity. Their forceps are also very long, particularly in the males. The web is vertical.

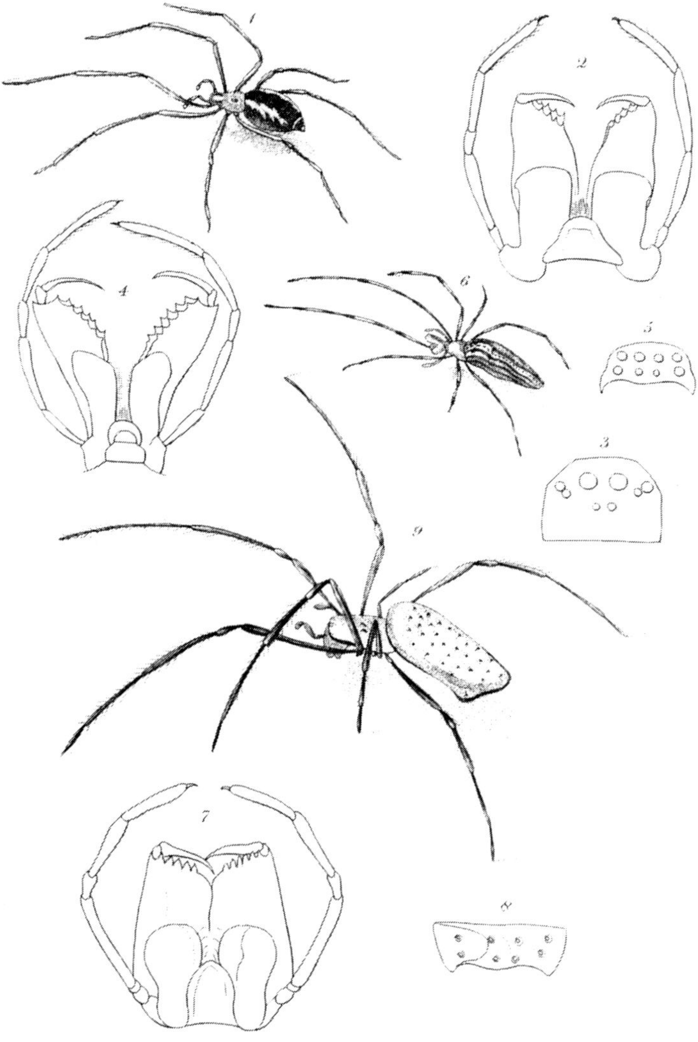

1. *Linyphia triangularis.* 4.5.6. *Tetragnatha extensa.*

2.3. *Lin. montana.* 7.8. *Epeira clavipes?*

9. *Epeira clavipes.*

London, Published by Whittaker & C.º Ave Maria Lane, 1833.

EPEIRA, *Walck.*,

Which have the two eyes of each side approximated by pairs, and almost contiguous, and the four others, forming in the middle a quadrilateral figure. The jaws are dilated from their base, and form a rounded palette.

The *epeïra curcubitina* is the only one known whose web is horizontal; that of the others is vertical, or sometimes inclined.

Some place themselves in the centre of the body inverted, or head downwards; the others make for themselves a dwelling close by, either arched on all sides, and sometimes in the form of a silken tube, sometimes composed of leaves brought together, and connected by threads, or open at the top, and imitating a cup or bird's nest. The web of some foreign species is composed of threads so strong, that it arrests little birds, and even embarrasses a man, who may happen to be engaged in it.

Their cocoon is most frequently globular, but that of some species has the figure of a truncated ovoïd, or of a very short cone.

The natives of New Holland (Voyage à la recherche de La Peyrouse, p. 239), and those of some islands of the South Sea, eat, for the want of other aliment, a species of epeira, very near the *Aranea esuriens* of Fabricius.

M. Walckenaer mentions, in his tabular view of the araneïdes, sixty-four species of epeïræ, generally remarkable for the variety of their colours, forms, and habits. He has distributed them into divers small and very natural families, and of which we have endeavoured, in the article Epeïra, in the second edition of the New Dictionary of Natural History, to simplify the study. Some important considerations, such as those of the sexual organs, have been neglected, or not sufficiently pursued. It is thus, for example, that the female of the epeïra diadema, and some others, present, at the part which characterizes their sex, a very singular appendage, which reminds us of the apron of the Hottentot women. These species should form a particular division. Others may probably be established, not less natural, by pursuing this examination.

We shall confine ourselves to citing some principal species, commencing with the indigenous.

E. diadema, Aran. diadema, Lin., Fab. Rœs. Insect. IV. xxxv.—xl. Large, reddish, and hairy; abdomen very voluminous in the females, especially when they are on the point of laying their eggs; of a deep brown or yellowish red, with a thick and rounded tubercle on each side of the back, near its base, and a triple cross, formed of small spots or white points; palpi and feet spotted with black.

Very common in Europe, in autumn. The eggs disclose in the spring of the following year.

E. scalaris, Aran. scalaris, Fab., Panz., Faun. IV. xxiv. has the corslet reddish, the upper part of the abdomen usually white, with a black spot in the form of an inverted triangle, oblong, and denticulated. It makes its web on the edge of ponds, streams, &c.

E. cicatrosa, Aran. cicatricosa, Deg., *A. impressa,* Fab., whose abdomen is flatted, of a greyish brown or obscure yellowish, with a black band, festooned, and bordered with grey along the middle of the back, and eight or ten thick sunken points, situated on two lines.

It weaves its web against walls or other bodies, and remains concealed in a nest of white silk, which it forms under some projecting part, or in some cavity, near its web.

It neither works, nor takes any nutriment, except at night, or when the light of day is feeble. It withdraws under the old bark of trees or stakes.

Epeïra sericea, Walck., Hist. des Aran. iii. 11, is covered

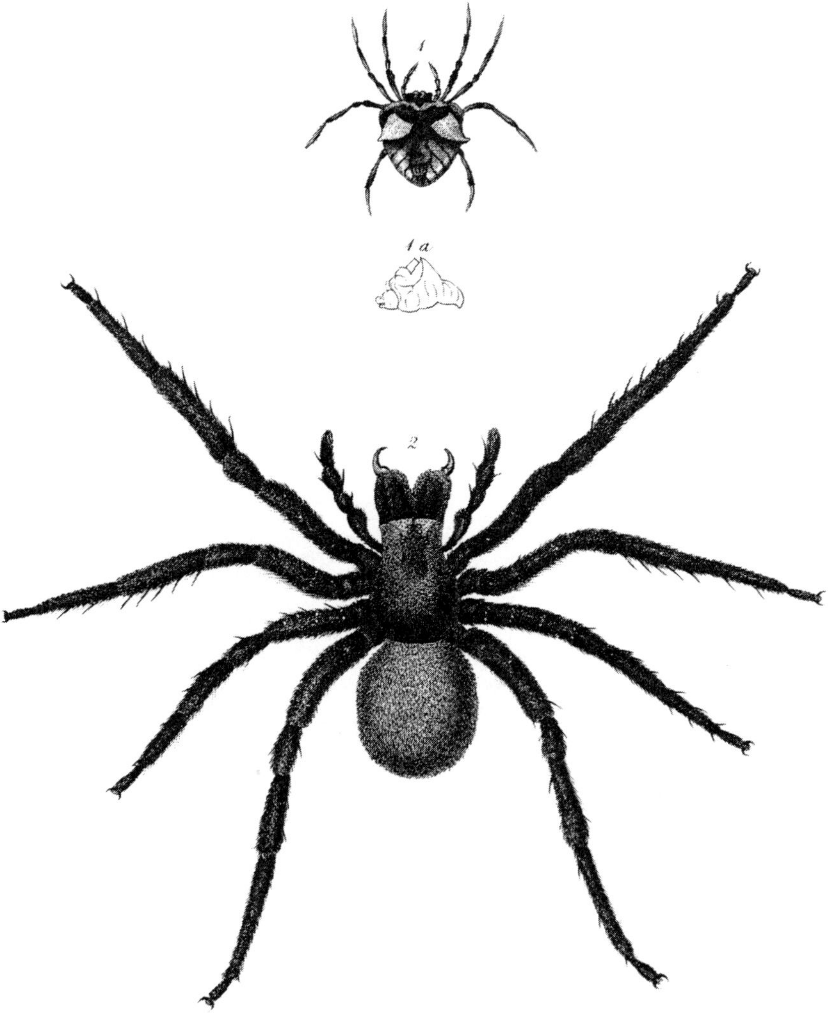

C. M. Curtis del.t

1 Epeira Australasia

2 Ctenus Walkenaerii.

London; Published by Whittaker & C.o Ave Maria Lane 1833.

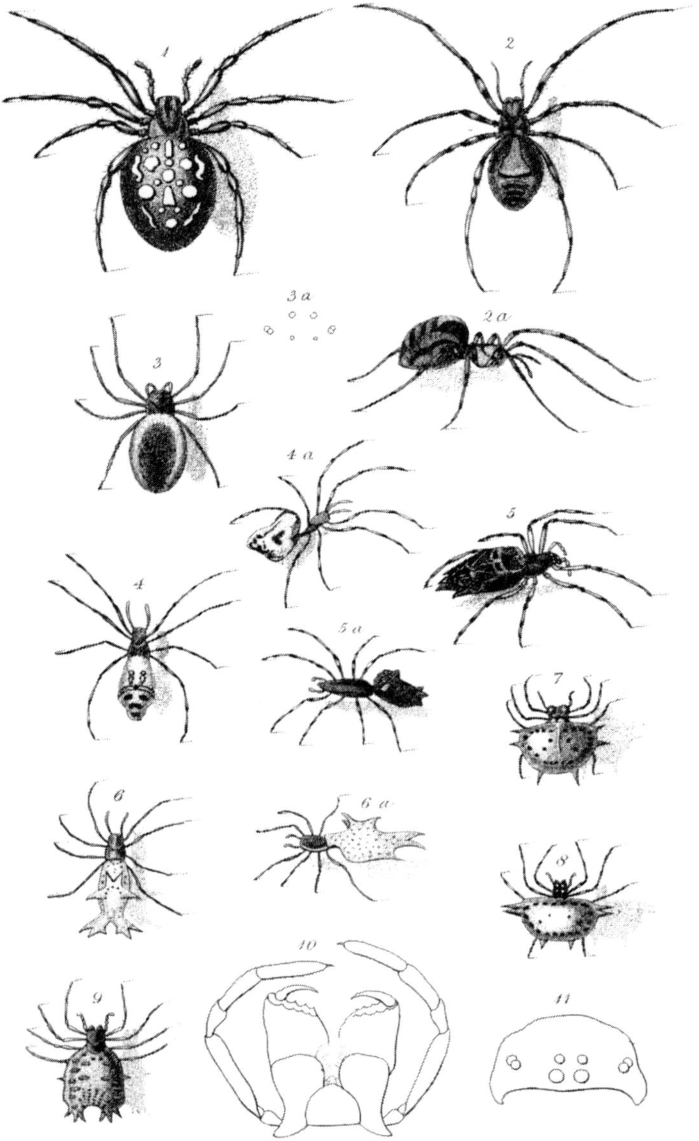

1 Epeira quadrata. 6 Ep. gracilis.
2 Ep. fusca. 7 Ep. cancriformis.
3 Ep. cucurbitina. 8 Ep. geminata.
4 Ep. conica. 9 Ep. clypeata.
5 Ep. oculata. 10. 11. Epeira ....?

London. Published by Whittaker & C.º Ave Maria Lane, 1833.

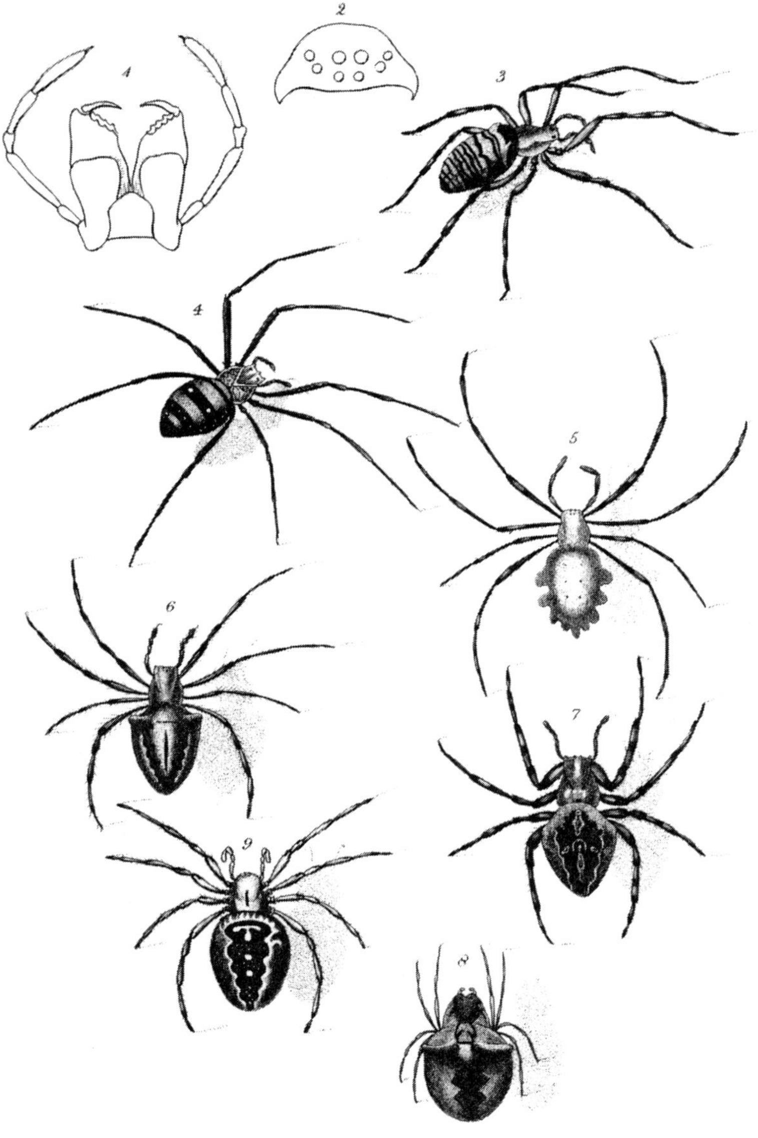

1.2. *Epeira?* — 6. *Ep. cornuta.*

3. *Ep. fasciata.* — 7. *Ep. angulata.*

4. *Ep. Latreillana.* — 8. *Ep. bicornis.*

5. *Ep. sericea.* — 9. *Ep. diadema.*

London, Published by Whittaker & C.º Ave Maria Lane, 1833.

above with a silvery silken down; its abdomen is flatted, without spots, and festooned upon the edges. It is found in the south of Europe and in Senegal.

Epeïra fusca, Walck., Hist. des Aran. ii. 1. fem., is very common in the cellars of the town of Angers. Its cocoon is white, almost globular, fixed by a pedicle, and composed of very fine threads, and as soft to the touch as wool.

That of the *Epeïra fasciata*, Walck., Hist. des Aran. iii. 1, fem., is about an inch in length, resembles a small balloon, of a grey colour, with some black longitudinal radii, and one of the extremities is truncated, and closed by a flat and silky opercle. The interior presents a very fine down, which envelopes the eggs. This species establishes itself on the banks of rivulets, and there spins a vertical web, not very regular, at the centre of which it remains. It is very common in the South of France. Its corslet is covered with a silken and silvery down. Its abdomen is of a fine yellow, interrupted at intervals by transverse lines, black, or of a blackish brown, arched, and a little waved.

M. Leon Dufour has given us, in the Annal. des Sc. Phys. (tom. VI. pl. xcv. 5), a detailed description of this species, of its habits, and was the first to make us acquainted with the male. He has figured the sexual organ: it is in the form of a twisted hair. *Epeïra cucurbitina*, (*Aran. cucurbitina*, Lin.; *A. senoculata*, Fab.), Walck., Hist. des Aran. III. iii., small; abdomen ovoïd, of a citron yellow, with black points, a red spot at the anus. It spins between the stems and the leaves of plants a horizontal web of no great extent.

Epeïra conica (*Aran. conica*, De G., Pall.), Walck. Hist. Nat. des Aran. II. iii., remarkable for its abdomen, humped in front, and terminated in the form of a cone, with the anus placed at the centre of an elevation.

It suspends to a thread the insect which it has sucked.

We may place after this species that which M. Dufour

names *Epeïre de l'opuntia* (Annal. des Scienc. Phys. V. lxix. 3), because it constantly remains at the middle of the leaves of the *agave* and *opuntia*, and there establishes its toils by means of a net-work, with loose threads, and irregularly interlaced. It is black, with white and inclined hairs, forming the appearance of scales. Its abdomen has on each side two pyramidical tubercles, and terminates posteriorly with two others, but obtuse, and separated by a broad emargination. The posterior face of each of these pyramidical tubercles presents a spot of a fine snow-white. These spots are connected with one another, and with one or two more which are posterior to them by white zig-zag lines. These tubercles do not exist in individuals which are but just born. The cocoons are oval, whitish, and formed of two tunics, the interior of which is a species of wadding enveloping the eggs. We often find seven, eight, and even ten of these cocoons in file, one after the other. This species inhabits Catalonia and the kingdom of Valencia.

Among the exotic species there are some very remarkable. Some have the abdomen invested with a very firm skin, with corneous points or spines; others have bundles of hairs to the feet.

We shall now pass to spiders, sedentary, like the preceding, but which can walk sideways, backwards, and forwards; in a word, in all directions. This is the section of LATERIGRADE SPIDERS. The four anterior feet are always longer than the others; sometimes the second pair exceeds the first, and sometimes they are nearly equal. The animal extends them in their whole length on the plane in which it is situated.

The forceps are usually small, and their hook is bent transversely, as in the four preceding tribes. Their eyes are always eight in number, often very unequal, and form, by their union, the segment of a circle or a crescent; the two lateral posterior ones are more removed backwards, or more approaching the lateral edges of the corslet than the others. The jaws are in a great number inclined upon the labium. The body is generally flatted, crab-formed, with the abdomen large, rounded, and triangular.

These arachnides remain tranquil, the feet extended, on vegetables; they form no web, but simply throw out some solitary threads, for the purpose of arresting their prey. Their cocoon is orbicular and flatted. They conceal themselves between leaves, the edges of which they draw together, and watch assiduously until the birth of the young.

MICROMMATA, *Latr.* SPARASSUS, *Walck.*,

Which have the jaws straight, parallel, and rounded at the edge, and the eyes disposed four by four on two transverse lines, the posterior of which is longer, and arched behind. The second feet, and then the first, are the longest of all. The tongue is semi-circular.

We find very commonly in the woods, in the neighbourhood of Paris, the

Micrommate Smaragdine (*Aranea Smaragdula*, Fab.; *A. viridissima*, De G.) Clerck. Aran. Suec. pl. 6. tab. iv., which is of middle size, of a grass-green, with the sides edged with a clear yellow, and the abdomen of a greenish yellow, cut on the middle of the back by a green line.

It collects three or four leaves together in a triangular packet, lines the interior with a thick silk, and places its cocoon in the middle, which is round, white, and allows the eggs to appear. These eggs are not agglutinated.

The *Micrommate argelas* (Dufour, Ann. des Scienc. Phys. vi. p. 306, xcv. 1; Walck. Hist. des Aran. IV. ii.) whose denomination reminds naturalists of one of our most zealous philosophers, whom I have already recommended to their esteem, as my preserver, in the storm of the revolution, is one of our largest species. M. Dufour has completed the description which I had given of it, and has observed its habits. Its body is seven or eight lines in length, of an ash-coloured blonde, furnished with down, and more or less spotted with black. The upper part of the abdomen presents from its middle, as far as the end, a band, formed by a series of small spots, in the form of a hatchet, of this last colour. There is a longitudinal band under the belly, equally black, but grey in the middle. The feet are ringed with black. This species was discovered in the environs of Bourdeaux by the naturalist to whom I have dedicated it. M. Dufour has since found it in the most arid mountains of the kingdom of Valencia; it runs with velocity, the feet being laterally extended. Its feet, cushioned, and furnished with claws, give it the facility of hooking itself on the smoothest surfaces, and in every position. It fixes, at the lower face of the fragments of rocks, a cocoon, which has much analogy, from its contexture, with that of the clotho of Durand. It retires there to shelter itself from bad weather, to escape from its enemies, and to lay its eggs. It is an oval tent, nearly two inches in diameter, applied on the stones, pretty much after the manner of the marine patellæ: it is composed of an external envelope, of a yellowish taffeta, as fine as the peel of an onion, but capable of resistance, and of an interior sheath, more supple, softer, and open at the two ends. It is through apertures, provided with valves, that the animal comes out. The cocoon for eggs is globular, placed underneath its dwelling, so that the spider can cover it, and it contains about sixty eggs.

The same naturalist has described and figured another species, the *M. à tarses Spongieux* (Ann. des Scienc. Phys. V. lxix. 6), which he found on a tree in a garden in Barcelona; but I presume, from its habits, and some descriptive characters, that this araneïd belongs to the genus *Philodromus* of M. Walckenaer.

1.2. *Micrometa.* 6. *Philodroma rhombifera.*
3. *Mic. ornata.* 7. *Phil. oblonga.*
4. *Mic. argelaria.* 8.9. *Thomisus cancerides.*
5. *Mic. rosea.* 10. *Thom. rotundatus.*

London, Published by Whittaker & C° Ave Maria Lane, 1833.

SENELOPS, *Dufour,*

Form the passage from the preceding subgenus to the following: the jaws are straight, or but very little inclined, without lateral sinus, and proceed into a point, being truncated obliquely at the internal side. The tongue is semi-circular, like that of the micrommata; but the eyes are differently disposed: there are six in front, forming a transverse line; the other two are posterior, and situated, one on each side, behind each extremity of the preceding line. The feet are long; the second, and then those of the two following pair, exceed the first two in length.

The species which serve as type, *Senelops omalosoma,* (Dufour, Ann. des Scienc. phys. V. lxix. 4.) has been found by M. Dufour in the kingdom of Valencia, but it is very rare there. Its body is about four lines in length, very flatted, of a reddish grey, with ash-coloured spots, and black rings to the feet; the abdomen appears to present behind some vestiges of rings, forming laterally the appearances of teeth. It inhabits rocks, and flies with the rapidity of an arrow. It is also found in Syria, (Collection of M. de Labillardière) and Egypt, Senegal, the Cape of Good Hope, and the Isle of France, furnish other species of the same subgenus.

PHILODROMUS, *Walck.,*

Differ from the two preceding subgenera, by having their jaws inclined upon the tongue. This part also is more high than broad. The eyes almost equal among themselves, always form a crescent, or a semi-circle; the lateral ones are never placed on tubercles, or eminences; the forceps are elongated and cylindrical; the last four, or the last two feet, do not differ remarkably in length from the preceding.

According to M. Walckenaer, these araneïdes run with rapidity, the feet extended laterally, watch for their prey, spread solitary threads to retain it, and conceal themselves in clefts, or in leaves which they draw together to lay their eggs in.

Some have the body flatted, broad, the abdomen short, widened posteriorly, and the four intermediate feet more elongated. Such is the *Philodrome tigré, (Thomise tigrée,* Lat., *Araneus margaritarius,* Clerck, VI. iii. Schœff., Icon. lxxi. 8; Frisch., Ins., 10. centur. II. xiv.; *Aranea levipes,* Lin.) This species is three lines in length. Its two intermediate anterior eyes, and the four lateral, are situated on a space a little more elevated, and the lateral, according to the same naturalist, are a little larger, or at least more apparent. The thorax is very broad, flatted, of a reddish fawn-colour, brown laterally, and posteriorly, and white in front; the abdomen, which appears to form a pentagon, is striped like a tiger, by means of the red, brown, and white hairs, with which it is invested. It is edged with brown on the sides, and has at the middle of the back, four or six sunken points; the belly is whitish; the feet are long, slender, reddish, with brown spots.

This species is very common on trees, wooden partitions, walls, &c., and sticks there with the feet extended, and as it were glued. As soon as touched, it takes to flight with extreme rapidity, or suffers itself to fall by unwinding a thread which supports it. Its cocoon is of a fine white, and contains about one hundred eggs, which are yellow, and not agglutinated. It places it in the clefts of trees, or of posts exposed to the north, and watches it assiduously.

The other philodromi, which in the method of M. Walckenaer, form many small groups, have the body, and sometimes the forceps, proportionally longer. The abdomen is sometimes pyriform or ovoïd, sometimes cylindrical; the second pair of feet, and then the first or fourth, are the longest.

We shall cite the *Philodromus rhombiferus* (Faun., Franc. araneide, vi. 8, male). Its body is three lines and a half in length, reddish; the second feet, and then the last two are the longest; the thorax is brown on the sides; the abdomen is ovoid, and has above a black or brown spot, like a lozenge, and bordered with white.

Philodromus oblongus (Walck., ibid. ead. fig. 9.) belongs to the same division, as regards the relative proportions of the feet, and the disposition of the eyes; but the abdomen is longer, almost cylindrical, or in an elongated cone, with three longitudinal stripes, and some brown points, on a yellowish ground, which is also the colour of the thorax. This part presents in the middle two brown stripes forming an elongated V.

These two species are found in the neighbourhood of Paris. *See*, with respect to the others, the French Fauna, from which we have extracted the preceding descriptions.

THOMISUS, *Walck.*,

Differ from philodromus in their forceps, proportionally smaller and cuneiform, and in their four hinder feet, very sensibly, or even suddenly shorter than the preceding. The lateral eyes are often situated on eminences, while those of the philodromi are constantly sessile. Here, again, the two lateral posterior eyes are more cast behind than the two intermediate ones of the same line, while in the thomisi, these four eyes are pretty nearly on the same level.

The species of this subgenus are those which have been more especially designated by the name of *crab*-spiders. The males are often very different in colour from the females, and sometimes much smaller.

Some, all exotic, have the eyes disposed four by four, on two transverse lines, almost parallel, and of which the hinder one is the longest.

In the others, which form the greater number, the assemblage of these eyes represents a crescent, the convexity of which is anterior and external.

T. globosus (Aranea globosa, Fab.) *A. irregularis,* Panz., Faun., Ins., Germ., Fasc., 74. tab. xx. fem.; Walck., Faun., Franç. araneïd. vi. 4; nearly three lines in length, black, with the abdomen globular, red or yellowish all around the back.

Thomisus cristatus, Clerck, Aran., Suecic. pl. 6, tab. vi.; size of the preceding; body of a reddish grey, sometimes brown, sprinkled with hairs, with small spines to the feet; the lateral eyes larger, and placed upon a tubercle; a transverse yellowish stripe on the front of the corslet, and others forming a V of the same colour on the back; the âbdomen rounded, with a yellowish band, having on each side three divisions, in the form of teeth, on the middle of its back. This species is common, and is usually found upon the ground.

T. citreus, Aranea citrea, Deg., Schœff., Icon., Insec., tab. xix. 13.; of a citron yellowish, with the abdomen large, broader behind, and having often on the back two red stripes or spots. Found on flowers.

A subgenus established by M. Walckenaer, under the name of Storena, but which is yet but imperfectly known, seems proper to terminate this section, and conduct to the oxyopes, which partake as much of the crab-spiders as of the wolf-spiders. The storenæ have the jaws inclined upon the tongue, which is almost as long as they are, and in the form of an elongated triangle. The forceps are conical; the two anterior feet, and then the second, are the longest of all; the two following exceed the last; the eyes are disposed on three transverse lines, 2. 4. 2.; the posterior two form, with the intermediate two of the second line, a small square, and the anterior two are apart.

Other spiders, whose eyes, always eight in number, are

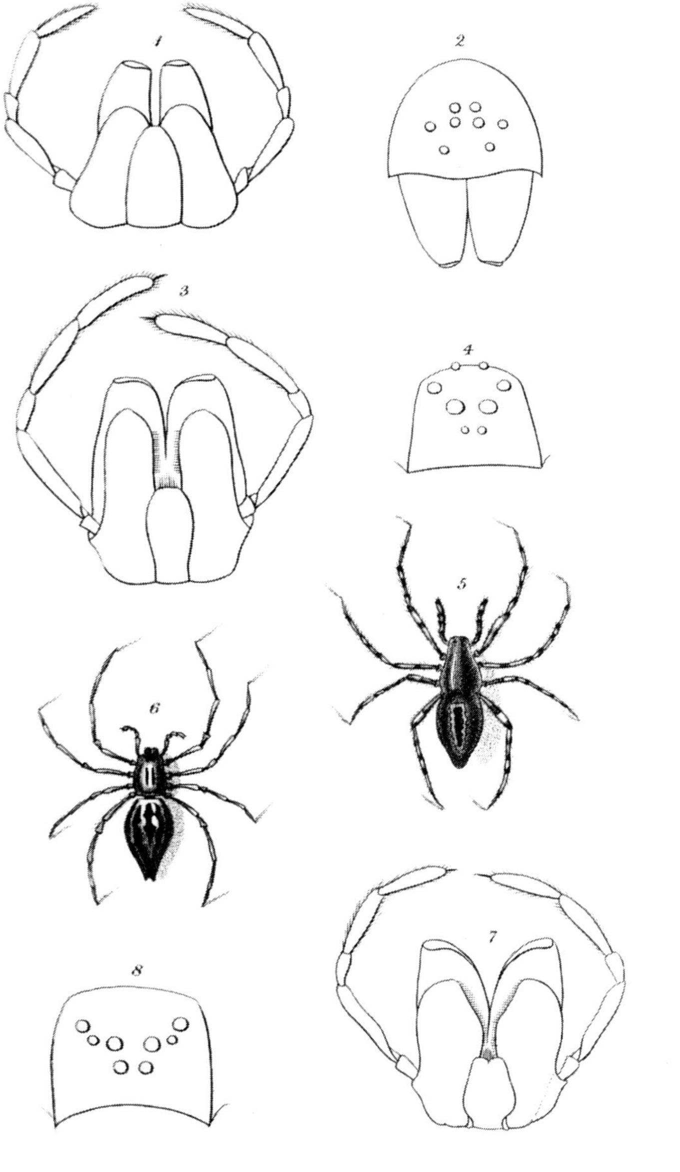

1.2. *Storene cyanea.* 5. *Ox. heterophtalmus.*

3.4. *Oxyopes Indicus.* 6. *Ox. Italicus.*

7.8. *Ctenus dubius.*

London, Published by Whittaker & C° Ave Maria Lane, 1833.

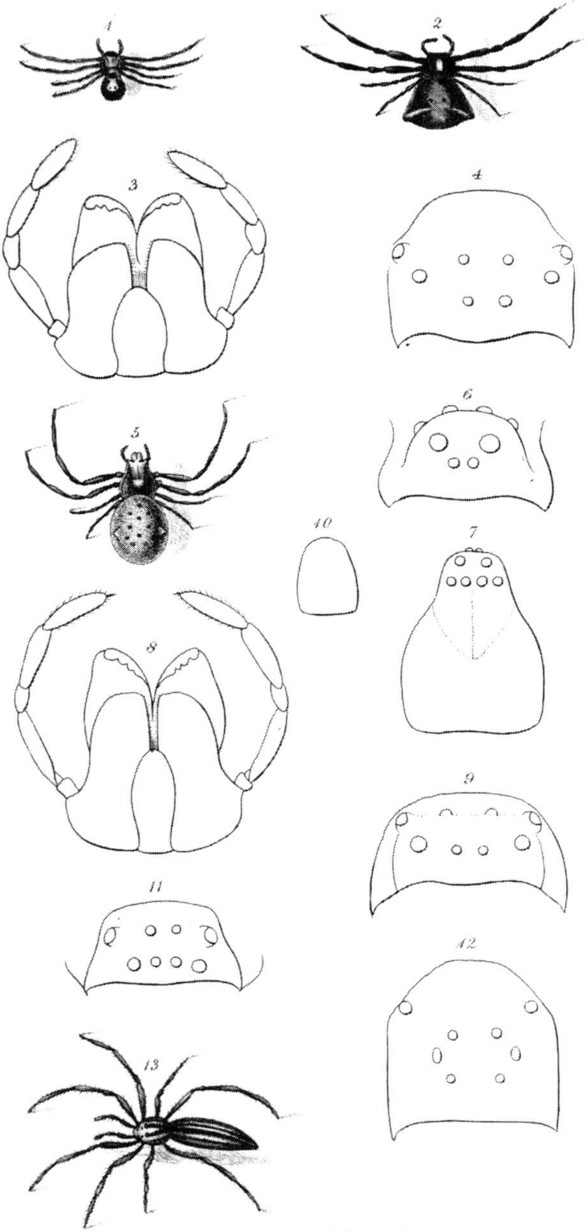

1. *Thomisus diana.* 6.7. *Tho. malacostraceus.*
2. *Tho. truncatus.* 8.9.10. *Tho. pigrus.*
3.4. *Th. citreus.* 11. *Tho. leucosia.*
5. *Th. onustus.* 12.13. *Tho. oblongus.*

London, Published by Whittaker & C.o Ave Maria Lane, 1833.

more extended in the direction of the length of the corslet, than in that of its breadth, or at least as much in one as the other, and which form by their union, either a curvilinear triangle, or an oval truncated, or a quadrilateral figure, compose a second general division, the ERRATIC SPIDERS, which I thus name in opposition to those of the first division, or the SEDENTARY.

Two, or four of their eyes, are often much larger than the others; the thorax is large, and the feet are robust; those of the fourth pair, the first two, or, after them, those of the second pair, usually exceed the others in length.

These spiders form no webs, watch their prey, seize it by running, or by jumping on it.

We shall divide them into two sections.

The first, that of CITIGRADES, is composed of the WOLF-SPIDERS of many writers. The eyes form, by their disposition, either a curvilinear triangle, or an oval, or quadrilateral figure, but the anterior side of which is much narrower than the thorax, measured in its greatest breadth. This part of the body is ovoid, narrowed in front, and keel-formed in the middle of its length; the feet in general are proper only for running; the jaws are always straight and rounded at the end.

The majority of the females remain upon the cocoon, or even carry it along with them, applied against the breast, and at the base of the belly, or suspended to the anus; they do not abandon it but in extreme necessity, and return to seek it when they have nothing more to fear; they also watch during some time over the preservation of their young.

OXYOPES, *Latr.*, SPHASUS, *Walck.*

Which have the eyes ranged two by two, on four transverse lines, and of which the two extreme are shorter; they design a sort of oval, truncated at both ends; the tongue is elongated, more narrowed at its base, dilated and rounded towards the end; the first pair of feet is the longest; the fourth and second are almost equal; the third is the shortest.

CTENUS, *Walck.*,

Have the eyes disposed on three transverse lines, 2, 4, 2, and forming a sort of curvilinear triangle, inverted, and truncated before or at its point; the tongue is square, and almost isometrical; the fourth pair of feet, and, next to them, the first, are the longest; the third is the shortest.

This genus has been established on a sort of arachnid, tolerably large, which is found at Cayenne. Some others have been since discovered, either at the same colony, or at Brazil, but all unpublished.

DOLOMEDES, *Latr.*,

Whose eyes, disposed on three transverse lines, 4, 2, 2, represent a quadrilateral, a little broader than long, with the two last, or posterior ones, situated on an eminence, and which have the second pair of feet as long, or longer than the first; those of the fourth are the longest; the tongue is square, and as broad as high, like that of Ctenus.

Some have the two lateral eyes of the anterior line larger than the two comprized between them, and the abdomen in an oblong oval, and terminating in a point.

The females construct at the summits of trees loaded with leaves, or in bushes, a silken nest, in the form of a funnel or bell; lay their eggs there, and when they proceed to the chase, or are forced to abandon their retreat, they always carry with them their cocoon, which is fixed upon the chest. Clerck tells us that he has seen individuals leap very promptly on flies which were hovering around them.

The others have the four front eyes equal, and the abdomen oval, or rounded at the end.

1.2. *Dolomedes marginatus.* 7. *Dol. limbatus.*

3.4.5.6. *Dol. mirabilis.* 8.9. *Lycosa vorax.*

London, Published by Whittaker & C°. Ave Maria Lane, 1833.

They inhabit the edges of the waters, run on their surface with astonishing swiftness, and even enter them a little without being wet. The females make, between the branches of plants, a thick irregular web in which they place their cocoon; they guard it until the young are disclosed.

LYCOSA, *Latr.*,

Which again have eyes disposed in a quadrilateral figure, but as long, or longer than broad, and the two posterior ones are not placed on an eminence. The first pair of feet is sensibly longer than the second, but shorter than the fourth, which, in this respect, exceeds all the others; the jaws are truncated obliquely at their internal extremity; the tongue is square, but longer than broad.

Almost all the lycosæ remain upon the ground, where they run very fast; they lodge in holes which they find formed, or which they have excavated, strengthening the walls with silk, and they enlarge them in proportion as they grow. Some establish themselves in the cavities or clefts of walls, forming those tubes of silk, which they cover at the exterior with parcels of earth or sand. It is in these retreats that they moult and pass the winter, after having closed, as it would appear, the aperture. It is there also that the females lay their eggs. When they proceed to the chase, they carry their cocoon, which is attached by threads to the anus. The little ones, on issuing from the egg, fasten themselves on the body of the mother, and remain attached there until they are strong enough to seek nutriment for themselves.

The lycosæ are very voracious, and defend courageously the possession of their domicile.

One species of this genus, the *Tarantula*, thus named from the town of *Tarentum* in Italy, in the environs of which it is common, possesses a great celebrity. In the opinion of the people, its poison produces very grievous accidents, often even followed by death, or the *Tarentismus*, as it is called, and which cannot be relieved but by the assistance of music and dancing. Enlightened and judicious persons think that it is more necessary to combat the terrors of the imagination, than the effects of this poison, and medicine, besides, offers remedies more to be relied on than those just mentioned.

M. Chabrier has published (Soc. Acad. de Lille. 4e cahier) some curious observations on the *lycosa, tarantula* of the south of France.

This genus is very numerous in species, but which are not as yet very well characterized.

Lycosa tarantula (*Aran. tarantula*, Linn., Fab.) Abbin. Aran., tab. xxxix. Senguerd, de Tarent. About an inch in length; red under the abdomen, traversed in the middle by a black band.

The *tarantula* of the south of France (*Lycose Narbonnaise*, Walck., Faun., Franç. Aran. I. 1—4.) is a little smaller, with the under part of the abdomen very black, bordered with red all round.

An analogous species is found in the neighbourhood of Paris, *Lycose ouvrière (Fabrilis*, Clerck., Aran., Suec., pl. 4. tab. ii., Walck., Faun., Franç. aran. ii. 5.)

L. Saccata, *(Aran. Saccata*, Lin., *Araneus Amentatus*, Clerck. iv. tab. viii. Lister. tit. 25. fig. 25.) small, blackish; keel of the corslet of an obscure reddish, with an ashy line; a small bundle of grey hairs at the upper base of the abdomen, feet of a livid red, intersected with blackish spots; cocoon flatted, and greenish. Very common in the neighbourhood of Paris.

We shall terminate this section with the subgenus

MYRMECIA, *Latr.*,

Which appears to conduct to the following, and of which we have exposed the characters in the Annals of the Natural Sciences, (vol. iii. p. 27.) The eyes form a short and broad

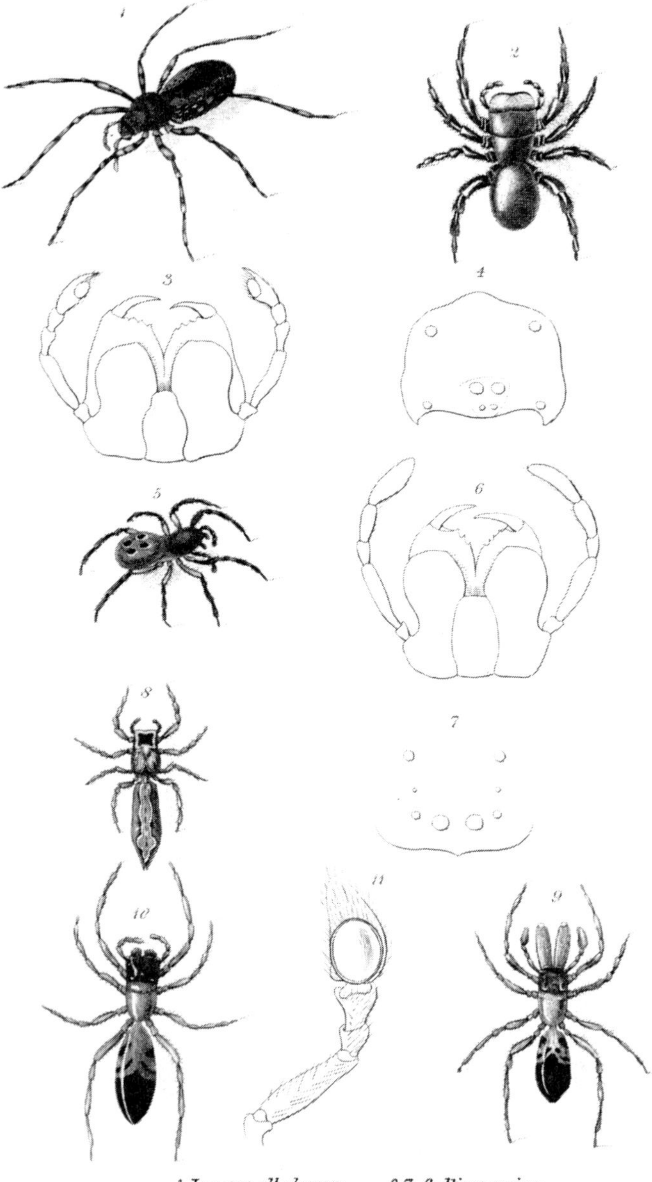

1. *Lycosa allodroma.* 6.7. *Salticus senicus.*
2. *Ly. fabrilis.* 8. *Sal. tardigradus.*
3.4.5. *Eresus cinnaberrinus* 9.10. *Sal. formicarius.*
11. *Palpus of male of same.*

London. Published by Whittaker & Co. Ave Maria Lane, 1833.

trapezium. There are four in front on a transverse line; two others, more interior than the two preceding extreme ones, compose a second transverse line; the last two are behind the two preceding; the forceps are strong; the jaws are rounded, and very hairy at the end; the tongue is almost square, a little longer than broad; the feet are long, almost filiform; those of the fourth, and the first pair, are the longest of all; the thorax seems divided into three parts, the anterior of which, much larger, is square, and the other two in the form of knots or bosses; the abdomen is much shorter than the thorax, and covered from its origin to its middle with a solid epidermis.

Myrmecia fulva, on which I have established this genus, is found in Brazil. But it appears that other species of it exist in American Georgia.

The second section of ERRATIC SPIDERS, that of SALTIGRADES, has the eyes disposed in a large quadrilateral figure, the anterior side of which, or the line formed by the first, extends through the whole breadth of the corslet; this part of the body is almost square, or semi-ovoïd, plane, or but little gibbous above, as broad in front as in the rest of its extent, and falls abruptly on the sides; the feet are proper both for running and leaping.

The thighs of the two fore-feet are generally remarkable for their size.

The spider *à chevrons blancs* of Geoffroy, a species of Salticus, very common in summer on walls or glass-windows exposed to the sun, walks as it were by jerks, stops short after having made a few steps, and raises itself on the anterior feet. If it discovers a fly, or a gnat more especially, it approaches quite softly within a distance which it can clear at a jump, and then instantly darts on the animal which it was watching. It does not hesitate to leap perpendicularly from a wall, in consequence of the thread of silk which attaches it, and which it unwinds in proportion as it advances. This also serves to suspend it in the air, to enable it to reascend to the point from which it had descended, and to waft it from one place to another, by the assistance of the wind. These habits are generally common with species of this division.

Several construct between leaves, under stones, &c., nests of silk, in the form of oval sacs, and open at both ends. These arachnides retire thither to repose themselves, to change skin, and to shelter themselves from the inclemency of the weather. If any danger menaces them, they issue forth immediately, and run off with great agility.

Some females form, with the same material, a sort of tent, which becomes the cradle of their posterity, and where the little ones live for some time in common with the mother.

Some species resemble ants, in raising their anterior feet and causing them to vibrate very rapidly.

The males sometimes betake themselves to combats very singular in their manœuvres, but which have no fatal issue.

One subgenus, established by M. Rafinesque, that of

TESSAROPS,

Appears to us very much to approach the following in most of its characters and habits, but to differ greatly, if there be no error, as to the number of eyes, which should be but four. (See the Annales Générales des Sciences Physiques, tom. viii. p. 88.)

Another subgenus, which is equally unknown to us, except by its description, is that of

PALPIMANUS,

Published by M. Dufour in the Annales des Sciences Physiques (V lxix. 5), and which appears to him to be intermediate between Eresus and Salticus. The disposition of the eyes is pretty nearly the same as in the first of these subgenera. The tongue is equally triangular and pointed, and the jaws again are dilated and rounded at the end; but, according to this naturalist, they would be inclined, and not straight, like those of eresus. The terminal articulation of the anterior tarsi would be inserted laterally, and destitute of hooks.

He describes but a single species (*Palpimane bossu*). It does not jump, walks rather slowly, and is found under stones in the kingdom of Valencia; but it is very rare there.

M. Lefevre has brought from Sicily a new species of araneïd, which appears to me to belong to this genus.

In the two following subgenera the number of the eyes is always eight, and the jaws are straight.

ERESUS, *Walck.*,

Which have, near the middle of the anterior extremity of the corslet, four eyes, approximated in a small trapezium, and the four others on its sides, forming also another quadrilateral figure, but much larger. Their tongue is triangular and pointed. Their tarsi are terminated by three hooks.

SALTICUS, *Latr.* ATTUS, *Walck.*,

Which have four eyes, placed in a transverse line along the front of the corslet, the centre two of which are the largest; the other four are situated near the lateral edges, two on each side; they thus form a large square, open posteriorly, or a parabola. The tongue is very obtuse, or truncated at the summit. The tarsi present at their extremity but two hooks. Many males have very large forceps.

Some have the corslet thick and sloping, very much inclined at its base.

Saltique de Sloane (*Aranea sanguinolenta*, Lin.) Black, a white line, formed by a sort of down, on each side of the corslet; the abdomen of a cinnabar red, with an elongated black spot at the middle of the back. South of France, on stones.

The others have the corslet very much flatted, and almost insensibly sloping at its base.

Sometimes their body is simply oval, furnished with hairs, or with thick down, and the feet are short and robust.

Saltique chevronné (*Aranea Scenica*, Lin.; *Araignée à chevrons*, Geoff.), *Araignée à bandes blanches*, De. G., Insect. VII. xvii. 8, 9. About two lines and a half in length, black above, with the edges of the corslet, and three lines in the form of chevrons on the upper part of the abdomen, white. Very common.

Sometimes their body is narrow, elongated, almost cylindrical, and smooth; the feet are long and slender.

Salticus formicarius; A. formicaria, De G., Insect. tom. VII. xviii. 1, 2; *Atte fourmi*, Walck. Faun. Franç. Aran. v. 1. 3. Red, front of the corslet black; some black bands and two white spots on the abdomen.

The second family of the PULMONARY ARACHNIDES, that of

The PEDIPALPI,

Presents to our observation very large palpi, in the form of advanced arms, terminating like a pincers or talon; forceps or antennæ-pincers, with two fingers, one of which is mobile; an abdomen composed of very distinct segments, without spinnerets at the end, and the sexual organs situated at the base of the belly. All the body is clothed with a tolerably solid dermis. The thorax is of a single piece, and presents, near the anterior angles, three or two simple eyes, approximated or grouped; and near the middle of its anterior extremity, or posteriorly, but in the middle line, two other simple eyes, equally approximate. The number of the pulmonary sacs is four or eight.

Some, which form the genus

TARANTULA, *Fabr.*,

Have the abdomen attached to the thorax by a pedicle, or by a portion of their transverse diameter, without laminæ, in the form of a comb, at its inferior base, or sting at the extremity. Their stigmata, four in number, are situated near the origin of the belly, and covered with a plate. Their antennæ-pincers, or forceps (mandibles of authors), are talon-like, or simply terminated by a mobile hook. Their tongue is elongated, very narrow, in the form of a dart, and concealed. They have but two jaws, which are formed by the first articulation of their palpi.

They all have eight eyes, three of which on each side, near the anterior angles, are disposed triangularly; and two near the middle, at the anterior edge, are placed on a common tubercle, or a small eminence, one on each side. The palpi are spiny. The tarsi of the two anterior feet differ from the others: they are composed of several articulations, in the form of a thread or bristle, and without a little claw at the end.

These arachnides inhabit only the very warm climates of Asia and of America. Their habits are unknown. At the present day we form two genera of them.

PHRYNUS,

Which have the palpi terminating in a talon, the body much flatted; the thorax large, almost in the form of a crescent; the abdomen without tail; and the two anterior tarsi very long, very slender, and similar to hair-like antennæ.

THELYPHONUS, *Latr.*,

Are distinguished from Phrynus by their palpi being shorter, thicker, terminating in a pincers, or in two united claws, by their long body, with the thorax oval, and the end of the abdomen provided with an articulated thread, forming a tail;

their two anterior tarsi are short, of the same size, and with but few articulations.

The others have the abdomen intimately united to the thorax in its whole breadth, presenting at its lower base two mobile laminæ in the form of a comb, and terminated by a knotted tail, armed with a sting at its extremity; their stigmata are eight in number, uncovered, and disposed four by four on each side, through the length of the belly; their *antennæ pincers* (forceps) are terminated by two claws, the exterior of which is mobile. They form the genus

SCORPIO,

Which have the body long, and terminated abruptly by a long and slender tail, composed of six knots, the last of which finishes in an arched and very sharp point, or in a dart, under the extremity of which are two small holes, serving for an issue to a poisonous fluid, contained in an interior reservoir. Their thorax, in the form of a long square, and generally marked in the middle with a longitudinal furrow, has at each side, near its anterior extremity, three or two simple eyes, forming a curved line, and towards the middle of the back two other simple eyes approximated; the palpi are very large, with a talon at the end in the form of a hand; their first articulation forms a concave and rounded jaw; at the origin of each of the four anterior feet, is a triangular appendage, and these pieces form, by their inter-approximation, the appearance of a labium with four divisions, the two lateral of which may be considered as kinds of jaws, while the two others form the tongue; the abdomen is composed of twelve rings, comprehending those of the tail; the first is divided into two parts, the anterior of which supports the sexual organs, and the other the two combs. These appendages are composed of one principal piece, narrow, elongated, articulated, mobile at its base, and furnished along its lower side with a series

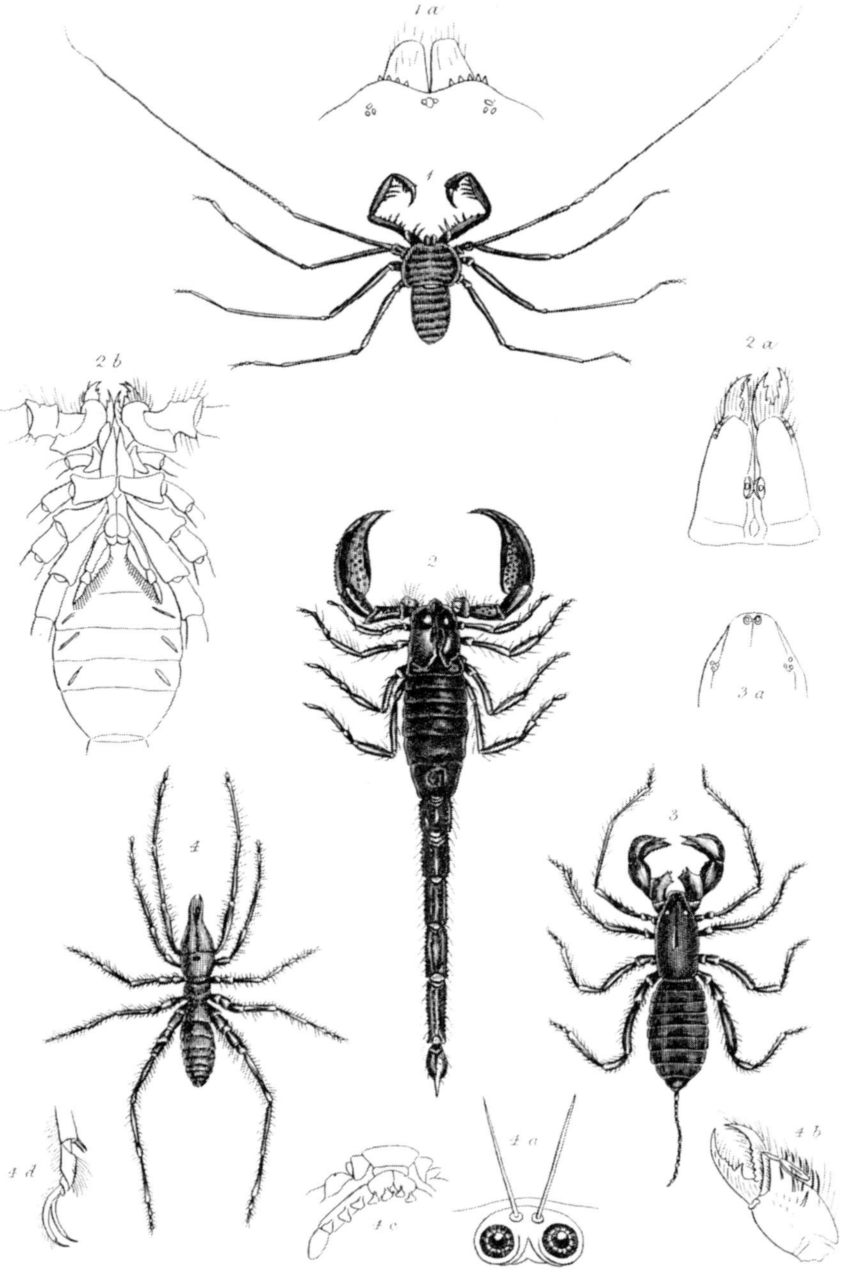

1 *Phrynus reniformis* 3 *Theliphonus caudatus.*

2 *Scorpio afer* 4 *Galeodes spinipalpis.*

London, Published by Whittaker & C.° Ave Maria Lane 1832.

of small laminæ, united with it by an articulation, narrow, elongated, hollow internally, parallel, and resembling the teeth of a comb. Their number is more or less considerable, according to the species; sometimes it varies within a certain quantity, and perhaps according to age in the same species. The use of these appendages has not yet been ascertained by any positive experiments. The four following rings have each a pair of pulmonary sacs and stigmata; immediately after the sixth, the abdomen suddenly grows narrow, and the other six rings, in the form of knots, compose the tail. All the tarsi are similar, having three articulations, with two hooks to the end of the last. The last four feet have a common base, and the first articulation of their haunches is soldered; the last two are even partly set back to the abdomen.

The two nervous cords proceeding from the brain, are united at intervals, and form seven ganglia, the last of which belongs to the tail. In all the other arachnides, the number of ganglia is but three at most.

The eight stigmata open into as many white pouches, each enclosing a great number of small and very slender laminæ, between which it is probable that the air is filtrated. A muscular vessel runs along the back, and communicates with each pouch by two vessels: other branches proceed from it to all parts. The intestinal canal is straight and narrow; the liver is composed of four pairs of glandular clusters, which pour their fluid into four points of the intestine. The male has two penes issuing forth near the combs, and the female two vulvæ. These last open into a matrix composed of several canals, which communicate one with another, and which we find at the time of parturition filled with living young. The testicles are also formed of certain vessels anastomosed together.

These arachnides inhabit the hot countries of both hemispheres, live on the ground, conceal themselves under stones or other bodies, most frequently in ruins, or in sombre and moist places, and even in the interior of houses. They run fast; curving their tail in the form of an arch over the back, they turn it in all directions, and make use of it as an offensive and defensive weapon; they seize with their talons wood-lice and different insects, such as carabi, weevils, orthoptera, &c., on which they feed, wound them with the sting of their tail, pushing it in front, and then cause their prey to pass between their forceps and jaws. They are fond of the eggs of araneïdes and those of insects.

The sting of the *European scorpion*, as far as it appears, is not in general dangerous. That of the scorpion of Souvignargues, of Maupertius, or the species which I name *reddish* (occitanus), and which is stronger than the preceding, produces, according to the experiments which Dr. Maccary had the courage to make upon himself, accidents more serious and alarming. The poison appears to be more active in proportion to the age of the scorpion. Volatile alkali is employed, either internally or externally, to arrest its effects.

Some naturalists have advanced that our indigenous species produce two generations in the year. The one which appears to me to be the best authenticated, takes place in the month of August. According to M. Maccary, the female changes skin before she brings forth the young. The male does so likewise at the same period.

The female brings forth the young in succession. She carries them on her back during the first days, does not then come forth from her retreat, and watches for their preservation for the space of a month, at the end of which time they are sufficiently strong to establish themselves elsewhere, and provide for their own subsistence; after two years they become capable of engendering.

Some have eight eyes, and form the genus *Buthus* of M. Leach.

The African Scorpion, (*Afer*. Lin., Fab.) Rœs., Insect. 3. lxv.; Herbst., Monog., Scorp. 1.; five or six inches long; of a blackish brown, with the talons large, heart-formed, very much shagreened, and a little hairy. Anterior edge of the corslet strongly emarginated; thirteen teeth on each comb. Of the East Indies, Ceylon, &c.

The reddish Scorpion, (*Occitanus*, Amor.) *Tunitanus*, Herbst, Monog. Scorp. iii. 3.; *Buthus occitanus*, Leach. Zool. Misc. cxliii. Yellowish or reddish; tail a little longer than the body, with raised and finely crenulated lines. Twenty-eight teeth and upwards (52—65. Maccary) in each comb. South of Europe, Barbary, and very common in Spain.

The others have but six eyes, and compose the genus *Scorpio*, proper, of the same naturalist.

The *Scorpion of Europe*, (*Europæus*, Lin., Fab.) Brown, more or less deep; feet, and last articulation of the tail, of a more clear or yellowish brown; claws heart-formed, and angular; nine teeth in each comb. The most southern and eastern departments of France.

SUPPLEMENT

ON THE

CLASS ARACHNIDA*.

THE animals of this class, though approximating to the insects, and though always considered as such in popular language, differ from them, as may be seen in the text, by certain marked characters which it is unnecessary to repeat here. They were first separated under the above denomination by M. de Lamarck. He, however, comprehends in this class all the apterous insects of Linnæus, with the exception of the genera *cancer*, *monoculus*, *oniscus*, which compose his class of the crustacea, and those of *termes* and *pulex*, which he reunites to the insects properly so called.

The Arachnida, thus named from the principal and most numerous genus, that of the spiders, (*arachne* in Greek) have at the external surface of the body apertures for the admission of air, or stigmata, in this respect approaching the insects, and departing from the crustacea. But they have not, like the first, antennæ and a distinct head. Their mandibles, or rather the pieces which replace them, are contiguous and advanced parallel to one another, in the direction of the length of the body; the jaws, or the parts analogous thereto, are but an expansion of the first articulation of the haunches of the an-

* We adopt the word Arachnida in the Supplement as the term most used by English naturalists, though it will be observed that the equivalent Arachnides is the word made use of in the text.—ED.

terior feet, or the articulation itself. The eyes are invariably simple, and the feet almost always eight in number.

The abdomen of the arachnides, like that of the insects, is the seat of the vital functions. The external organs of respiration, however, occupy there a more circumscribed space, being exclusively situated on the sides of the belly, or those of the chest; and not along the whole lateral portions of the body, as in insects. The stigmata lead either to pouches, or sacs enclosing bodies analogous to gills, but performing the office of lungs, or to two trunks of tracheæ, which are divided almost from their origin, and in all directions, into a great number of branches.

The myriapods, the thysanoura, and the parasites, are alone, of all the numerous class of insects, like the arachnida, truly apterous. Like them also, they undergo no metamorphosis, properly speaking, can engender several times, and have their growth not limited to the term of the development of their several organs, or of their aptitude for reproduction. But these insects are, nevertheless, remote from the arachnida in the characters which we have already noticed, or those which are proper to their own peculiar class.

Among the arachnida, some have two articulated mandibles, terminating in a talon or pincers, similar to small feet; two palpi still more analogous to locomotive organs; two, or several jaws, formed by the dilatation of those palpi, or of the anterior pair of feet; and a lip without palpi. The other arachnida have a mouth after the manner of a sucker, but the pieces of which, though otherwise modified, appear to correspond with the preceding. It is also, most frequently, accompanied with two palpi. The number of simple eyes varies from two to eight. Their situation, their symmetrical disposition, their relative sizes, and their forms, often furnish to the naturalist the appropriate means of distinguishing the principal divisions.

The number of feet is usually eight. Some, however, have but six, and the females of some others have two more, which only, however, answer the purpose of carrying the eggs.

Most of the arachnida subsist on living prey, or suck the blood, or other fluids of several animals.

The sexual organs of the females, and even in many of the males, are situated at the base of the belly, or near its junction with the thorax. They are double in most of the pulmonary arachnida, and perhaps in all. The males of the araneides, or spinning arachnida, have theirs at the extremity of the palpi; a character which indicates the natural affinity between these animals and the last order of the crustacea.

The araneida do not undergo any essential change of form, and are only subject to moulting. In some, however, two of the feet are not developed, until some days have elapsed from their birth. It is only after the fourth or fifth change of skin, that these animals become adapted for the purposes of generation.

Most of the pulmonary arachnida are suspected, or dreaded, as being venomous. The bite or incision made by some of them may produce, under some circumstances, and more especially in warm climates, accidents of a serious nature.

The body of the araneides, or spinning arachnida, is composed of an inarticulate trunk, with which the head is confounded, and the abdomen fixed to the posterior extremity of the trunk by a small thread. It is usually soft, without rings, or with nothing but folds, and having from four to six external nipples placed under the anus. These nipples are a sort of spinnerets, and exist in both sexes.

The trunk is crustaceous, and presents, at its anterior part, a triangular space, which appears to correspond with the head, and on which the simple eyes are situated. These organs evidently replace the composite eyes of the insects, and are always six or eight in number. Their size and disposition vary according to the manner in which these animals hold themselves in a state of repose, and according to some peculiar habits. They are very brilliant, and in some species offer the appearance of a pupil and an iris.

The organs of manducation occupy the anterior and lower extremity of the trunk. They consist of two mandibles, two palpi, a lip, and a kind of epiglottis, or interior tongue. The mandibles advance in parallel directions, and are composed of two tubular articulations, the first of which is the largest, and the terminal one more solid and scaly, in the form of a very sharp crook, and having at its extremity a small cleft, destined for the passage of a poisonous fluid, which is conducted thither by an interior canal, from the base of the first articulation, where its reservoir, or the poisonous receptacle exists. The palpi, like little feet, especially in *mygale,* are of the same thickness, or filiform, in the females, thicker at their extremity in the males, and composed of five, or even six, articulations. The jaws are composed of a single piece, in the form of a lamina, more or less oval and triangular. The palpi articulate with their summit in the mygale, so that the jaws in reality form the first articulation. However, in the other araneides, it is at the base of their internal side, that the palpi are inserted. The lip is also of a single piece, the figure of which most usually approaches a square, or an oval, truncated at its base, and is but an appendage of the anterior extremity of the breast. The interior of the mouth, or palate, presents a fleshy piece, hairy, in the form of a tongue, which, in most of the species, is applied against the internal surface of the lip. There is, probably, on each of its sides, an aperture for the passage of the alimentary fluids. The mandibles, no doubt, contribute to manducation; but though hollow and pierced at their extremity, they do not perform the office of a sucker. Their use is to retain the insect seized by the araneid, and facilitate the compression made upon it by the jaws, so that the nutritive fluids may be extracted. Besides, we know that the scorpions, though without perforated mandibles, nevertheless suck their prey in the same manner.

The feet, in their relative size, vary, according to the different positions, habitual to these animals, and sometimes even according to the sexes. They are inserted all around the sides of the chest, and composed of seven articulations. The last two constitute the tarsi, which are terminated by two hooks. It is easy to conceive that these two hooks answer the purpose to the arachnides of holding by their web, and that they must also be useful in its construction.

The spinning epeiræ are the only animals of this family that have the abdomen covered with a crustaceous or solid dermis, folded in the form of a ring. In all the other species this part of the body is soft, and without any apparent division. Its envelope is only a sort of sac, in which the organs of circulation, of respiration, the intestines, the vessels that secrete the silk, and in the females the ovaries, and the other sexual parts, are enclosed. Immediately under the dermis, the mucous tissue is perceptible, composed of a soft matter, divided into an infinite number of small grains. The heart is a large vessel along the back, throwing out branches on each side. The respiratory organs are two in number, and composed of small laminæ, adherent to the interior parietes of two pouches situated one at each side, near the base of the belly, and covered by a membranaceous lid, leaving a transverse cleft, for the passage of air. The place of these organs is usually indicated by two yellow or whitish spots.

The intestinal canal is straight. There is a first stomach, composed of several sacs; then, towards the middle of the abdomen, is a second stomachal dilatation, surrounded by the liver. The silk-vessels, usually six in number, extend, on each side, along the whole interior: they resemble tortuous intestines, filled with a yellowish matter; contracted rather abruptly towards their extremity, they terminate in a straight thread, leading to the nipples, which are cylindrical or conical parts, and membranaceous, serving as a conduit to the threads of silk, which are named spinnerets.

The mygalæ have but four apparent nipples, the upper two of which form a small forked tail; but in the others they are six. In these the nipples have but two articulations, the last of which, in the form of a head, is bordered all round, like a crown, with several conical pieces, which give issue to the silken threads, and are in fact the spinnerets properly so called.

The proper spinnerets of each nipple are carried by some authors to the number of a thousand, so that, when all the nipples are at work, the quantity of threads which proceed from them should be six thousand. But these animals manage with economy a substance which constitutes a portion of their means of existence, and which is also necessary to the preservation of their posterity. Moreover, this calculation is not applicable to all the species, since many of them form no web, and employ their silk only in the construction of the cocoon, which is to envelope their eggs.

It has been attempted, by weaving, to derive some advantages from the silk of certain of the araneides of the genus epeira, and gloves and stockings have been made of this silk, of a greyish colour, almost as strong as those fabricated with the ordinary silk. Lebon employed that of the cocoon of these animals, and thirteen ounces of these cocoons yielded four ounces of silk. To put it into a state for being wove, he caused it to be beaten with the hand, and with a small stick, for the purpose of expelling the dust; he afterwards washed it in tepid water, and put it then into soap-suds, in which some saltpetre and gum-arabic were dissolved. The whole was kept boiling over a slow fire for two or three hours, and the cocoons after this operation were washed in tepid water, until they yielded to the water the soap with which they were impregnated. They were then suffered to dry, and softened a little with the fingers, to fit them for being carded the more easily. This carded silk was easily spun upon a spindle. The thread drawn from it was finer and stronger than that of common silk, and easily assumed, in dyeing, all the colours which were desired. The Academy of Sciences, however, to which Lebon imparted the result of his experiments, considered that this branch of industry presented but little expectations. Reaumur, attempted, but to no purpose, to rear araneides on vegetable substances. These were by no means to their taste,—nothing would do but insects, and therefore an education of this kind would prove more embarrassing than useful, if attempted to be executed on a great scale, as it would be necessary to rear flies for the purpose of supporting the spiders. It was calculated that seven hundred thousand spiders would produce but a pound of silk. Moreover, these animals devour each other, and the threads of their silk are so extremely fine, that, as it is said, ninety of them would be necessary to equal the thickness of the thread of a silk-worm, and eighteen thousand to produce a thread proper for the purpose of manufacture. The thread of a young spider is still finer. Notwithstanding this, these threads can support, without breaking, a weight six times as heavy as that of the body of these animals. Wilhelm informs us, that a fabricator of stuffs at Paris also made silk stockings with the cocoons of the epeira diadema. He reared eight hundred individuals in one room, the ceiling of which was covered with a great number of pack-threads crossing each other. These animals were so tame, under his management, that when he entered the room with a plate full of flies, they immediately used to descend to take their food, and reascend after they had feasted. This they would also perform when he entered without bringing them anything.

It is well known that poor Pelisson, confined in the prison

of the Bastile, had so completely familiarized a spider, which had taken up its abode on the edge of a chink giving light to his gloomy dwelling, that it would come at the sound of an instrument, and at a certain signal, to take a fly, even on the knees of its teacher. It is truly painful and disgusting to remember, that the inhuman governor of that castle deprived the prisoner of this trifling consolation, by crushing the animal, even at the very moment in which it was exhibiting proofs of its docility.

Astronomers alone derive some advantages from the threads of the spider: they employ the one which sustains its web, and which is the strongest, for the divisions of the micronometer. It acquires, through its ductility, about a fifth more than its ordinary length.

Although these animals inspire a kind of horror in a great number of persons, which is founded on the opinion that they are generally venomous, they do not the less merit to be known, either for this very reason itself, or in consequence of the interesting facts which their economy presents to our contemplation. If they live, like so many other beings, on the fruit of their rapine, they achieve this end by means very different from those adopted by other races: to remain in ambush, to dart like an arrow on their prey, or to catch it by speed, are the ordinary means employed by carnivorous animals to satisfy the wants of nature. There is nothing in all this but a simple exercise of that superiority which nature has gifted them with above the beings on which they subsist. All their stratagems, all their instinctive combinations, are reduced to the concealment of the irresistible arm of strength, so that they may surprise the feeble with greater success; but the majority of the araneides have been provided by nature with peculiar resources, and highly worthy of our utmost attention: she has instructed them in the art of laying snares, and that with a substance drawn from their own entrails. Let us observe the delicate and ingenious manner in which this aerial web is woven, which is suspended over our heads; the regularity of the numerous concentric circles, and the radii which cut them, and give to this web the form of a circular net-work. Examine its points of attachment: it is wonderful to conceive how the animal which has constructed it has been able to fix them, and that too at such considerable distances from each other. Our garrets and apartments, long left in a state of neglect, serve for an habitation to many species whose mode of labour is not the same. Some of them give to their web a more compact and thicker tissue, which allows no meshes to appear, and which they place in a horizontal situation. Another spider, having established its sojourn in cellars, exhibits a tapestry, whose whiteness rivals that of snow. Some form a species of cylinder in a hole between leaves, and there remain in ambush. The argyronetes form, in the midst of the waters, an oval cocoon, filled with air, carpeted with silk, from which proceed threads, in all directions, and attached to the plants in the neighbourhood; there they watch their prey, and when the necessity of respiration, or other motives, force them to issue from their domicile, they envelope their abdomen with a bubble of air, which presents to the eyes of the astonished observer the spectacle of a silver globe rolling rapidly in the midst of the waves. Some mining araneïdes, or those which dig subterraneous galleries, know how to close the entrance of their habitation, with a door of clay, fixed by means of a sort of hinge, opening at the will of the animal, and falling by its own proper force and position.

Endowed with an instinct less surprizing, the araneïdes (*inequiteles*, Latreille), attach on trees, to the corners of walls, in garrets, &c. some threads, the union of which has no determined figure, but forms loose and irregular webs.

These animals remain tranquil at the centre of the snare, or in the cell which they have constructed near it. Woe to the imprudent insect which should fall into their net!—the slightest impression is sufficient to advertise the spider of the presence of its prey. It proceeds with the rapidity of lightning to the spot where the insect appears. If this, for instance, should prove to be a large fly, it envelopes it with a tolerably strong layer of silk, which it draws from its spinnerets; it then attaches it to its own hinder part, and drags it within its den, that it may suck and devour it at leisure. If the fly be small, the spider carries it off without any envelope. But if, on the contrary, an insect which is larger than itself should fall into the net, it assists to disembarrass and disengage it, by breaking some threads of the web, which it mends afterwards; or, if the efforts which it has made have broken the web too much, it abandons it and forms a new one. Some species simply suck flies; others devour them altogether, leaving only the hardest parts. As the araneïdes have not always as many flies as they can eat, they are so organized as to be able to support a very long fast; but when opportunity offers, they make full amends for this by gormandizing. They pass the winter in a sort of lethargy, and take no nourishment during that season. In every other, they can still remain many months without eating. It appears, from the observations of M. Amédée le Pelletier, that they have the faculty of reproducing the feet which they have lost.

When one of these animals wishes to commence its web, it causes to issue from its nipples a drop of the silky fluid. It applies it against a wall or tree, and then removes from it, spinning. In proportion as it proceeds, this fluid, which at first was soft, assumes a consistence, grows thick, and forms a thread, which the spider glues to the opposite end of the wall, or to another branch of the tree. It is thus that all the araneïdes commence their web; but they do not all finish it in the same manner. The domestic spider returns along the line to fix another to the place from whence it set out, returns on its path, to do the same at the other end, and continues the same manœuvre until it has fixed a great number in this direction; after this, it places more in an opposite line; and as all these threads are gluey, they become cemented one to the other, and soon form a tolerably solid web.

The *Epeira diadema,* which makes a perpendicular web with radii, and the threads of which lead to a common centre, proceeds in another fashion. According to the majority of authors, it suffers itself to hang to its thread, and the wind carries it to a tree different from that to which it was at first attached; it there applies one end of its thread. This done, it returns to the middle of this thread, along which it walks, where it attaches a second, the extremity of which it fastens to some branch near the first, and so on. The opinion of Lister is, that the araneïdes can shoot their threads to a very great distance, as the porcupine shoots his quills, with this difference, however, that the quills of the porcupine are detached from its body, whereas the threads of the araneïdes remain attached. This opinion has been combated. People have been unable to conceive that the silk, which hardens in the air, can be syringed in this manner, like a fluid. Besides, it has been contended, that so weak a thread could not be shot any distance without being forced by the resistance of the air to fold upon itself, and envelope the body of the animal. Be this, however, as it may, M. Latreille has very distinctly observed the *Aranea cancer* turning on itself, and darting in all directions, in a horizontal line, a thread proceeding from the anus.

We shall now explain how the Epeira diadema makes its web between two branches, or two trees, separated from each other by a ditch, or by a stream which it cannot cross. In calm weather, placed at the end of some branch, it remains firm on its front feet, and with its two hinder feet it draws from its nipples a thread, tolerably long, which it suffers to float in the air. This thread is pushed by the wind against some solid body, where it is quickly cemented by its natural gluten. The epeira draws it to itself from time to time, to ascertain if it be attached. When it is assured of this by the resistance which it experiences, it binds and fastens the thread to the place where it is itself: the first thread serves as a point of communication for placing the others. The spider imparts to it more solidity. Afterwards, it spins others, perpendicular and oblique, which it attaches to different branches, and the ends of which all repair to a common centre. When this labour is finished, it spins others, which it fastens above. They are apart from each other, and it places them circularly round the centre. The web being finished, the epeira constructs at one of the upper extremities, between two approximating leaves, a little lodge, which serves it as a retreat. It usually remains there all day, and does not issue forth but in the morning and evening. It chooses the top of its web for a sheltering place, because insects ascend better than they descend.

We have now exhibited the most general and interesting points in the economy of the *sedentary* araneides. Those which M. Latreille has designated under the name of *wandering* araneides, seize their prey by running or leaping upon it.

These animals being carnivorous, and devouring each other when they meet, the sexual intercourse does not take place without great precautions on the part of the male, who is obliged to make the advances. The coupling which has been most observed by naturalists is that of the Epeira diadema, so common in gardens, towards the commencement of autumn, which is the season of its amours. The female remains tranquil in the midst of its web, the head down, and the belly upwards. The male rambles around the web, and finally ventures to mount upwards; but he takes care first to attach a thread to some place, at no great distance, so that he may make use of it to save himself if the female should not be disposed to grant him a favourable reception. As soon as he has ascended, he walks gently over the web, approaches the female by little and little. If she remain quiet, he touches her gently with one of his fore-feet, and quickly draws back, and then the female makes some slight movements to touch him in turn. During this, which appears to be the prelude of the coupling, the antennulæ of the male half open at their extremity, the buds which enclose the generative organs become humid, and the sexual part of the female also opens a little; then the male, emboldened, inserts there one of his antennulæ, and retires. A moment after, he returns, and inserts the other antennula. He touches the female several times in succession in the same manner, using the two antennulæ alternately. During the act, which appears to consist only in these simple touchings, the male introduces into the female organs a part which appears to be the organ of generation, issuing from the bottom of the antennulæ during the act, and re-entering there immediately after. The same precautions are used by the males of those species which do not spin. Audebert has observed, that in a species commonly found in houses, a single act suffices for the fecundation of all the eggs which a female can lay at different times, for many years in succession. In general there is but one brood of eggs in the year, and which, in our climates, takes place towards the end of summer, or at the commencement of autumn.

Soon after the females are fecundated, their belly, always much larger than that of the male, increases very much. All are oviparous, and lay a great number of eggs. The spinners, and those which form no web, envelope them with a thick bed of white silk, in the form of a cocoon; some place them on a tree or wall. Some species carry their's enveloped in a round cocoon, very close, and they are often seen dragging this cocoon after them, by means of a thread which keeps it attached to the hinder part of their body: it has the form of a truncated ovoïd, of a grey or whitish colour, and divided longitudinally by blackish bands. Its aperture is hermetically sealed by a silken plug. This envelope contains a second, the tissue of which is still softer: it is a true down, which preserves the eggs from any accident. Almost all the cocoons of the araneïdes are equally composed of two kinds of silk, the interior of which is finer, and softer to the touch.

The insects which do not live in society, as soon as their eggs are laid, give themselves no further care touching their posterity. Many of the araneïdes, on the contrary, guard with the utmost vigilance the fruit of their amours; some even carry between their feet their eggs, shut up in the cocoon. When the young ones are first disclosed, they attach themselves on the back of their mother. All the eggs are of a round form, white, or yellowish white. Those of many species exclude fifteen or twenty days after they have been laid; others pass the winter, and do not exclude until spring. Some days before the young issues forth from them, the pellicle, which is very slender, changes form, and suffers all the parts of the insect to appear.

As soon as the young araneïdes which make webs have left the egg, they begin to spin directly. The female lycosæ tear open the cocoon, which encloses their young, to give them a greater facility of coming forth, at the moment in which they should quit it. These mount on the back of the mother, who carries them along with her; and when she finds an insect, she shares it with them. The araneïdes, in general, exhibit a very great attachment to their young. All the young araneïdes live, as it were, in society, until the first moulting: they then separate, and become mutual enemies. They grow very much in their youth, and in augmenting in volume, they change their skin. It is believed that they quit it three times before they are in a state for reproduction. Their life is more or less long. Audebert brought one up, and preserved it for several years.

The araneides, that cause so great a destruction of flies and other insects, are not without their enemies. Birds, and some insects feed their young with them. Many species of wasps, the *sphenges*, carry them off from the middle of their webs to bring them to their larvæ. The slightest wound which a spider receives, effectually puts it out of all condition for battle, and it dies in a short time after having received it.

According to the observations of Homberg, the domestic spiders are subject to a malady which makes them appear hideous. Their body becomes covered with scales, bristling one above the other, and among which species of mites are discovered. When the spider walks, it shakes itself, and throws off part of the scales and of the insects. This malady seldom occurs to the spiders of cold countries. The author whom we have cited, says, that he has never observed it but in those which are found in the kingdom of Naples.

The body of the araneïdes is in general hairy, with colours most frequently sombre, and forms far from agreeable; women, children, and even men, have an insurmountable repugnance to them. This aversion is not founded merely on the ugliness of these animals; it is also caused by the opinion that their bites are dangerous. Many authors, in fact, relate, that different persons have died after having been bitten by them. Other testimonies, however, combat the preceding. Clerk and Lebon, who were often bitten by araneides, assure us that they never felt any other inconvenience from their wounds than what might be occasioned by gnats, and some insects, whose stings produce upon the skin a trifling inflammation and itching. Degeer is also of opinion, that the araneides of Europe are formidable only to flies and other insects. With regard to the pretended mortal bite of the tarantula, a species of lycosa, which is found in the most southern parts of France, and in Italy, of which so many authors have made mention, and on which Baglivi has more especially written;

people are pretty well recovered from the fright which it occasioned in his time, and no longer believe that it can be the cause of the malady which was imputed to it. The bite of the three species of tarantula which he has described, occasioned maladies, the symptoms of which were different. Those which followed the bite of one of them, were, according to him, very terrific, and sometimes assumed all the characters of a malignant fever. The patient often died of this malady, or if the symptoms were mitigated, he fell into a melancholy of a peculiar kind, and of which music alone could cure him. But we are at present aware, that the tarantula never occasioned any such disease, which was merely feigned; accordingly the bite of this animal is no longer dreaded.

Nevertheless, we cannot avoid noticing some observations in the *Encyclopedie Methodique*, which prove that sometimes the bite of the araneides is followed by accidents more or less grievous. In the southern part of Provence, a young peasant girl, when sitting, found herself stung in the right thigh; when she was about to rise, on shaking herself, there fell out a very large spider, which the pressure of her hand had killed. She had crushed it at the moment on the wound, and experienced only a trifling swelling round the place which was bitten, and slight cramps in the thigh and leg, which were removed by time, and a sudorific drink. A farmer of the isles of Hyères, aged more than sixty years, according to the report of his children, was bitten by a large spider, in gathering up a sheaf of corn. This bite at first occasioned only a slight inflammation, to which the man paid but little attention. But the inflammation soon very considerably increased, and terminated some time after in mortification and death, the remedies which were employed being found altogether inadequate to stop the progress of the disorder. From these facts, we may conclude, that under certain circumstances, the bite of the araneïdes may be dangerous. The more or less serious effects resulting from it, also depend upon the state of health and constitution of the bitten person; but, speaking generally, the spiders of cold countries are not formidable.

Travellers tell us of some species which are reputed to be venomous. The *avicularia* of Linnæus, which is found at Cayenne and Surinam, is considered dangerous to men, and its bite is always followed by grievous accidents, according to those writers. It is frequently fatal to the humming-birds, and colibris, on which it feeds. The least wound which it inflicts instantly kills them, which is not surprizing, considering the strength of its hooks, and the extreme delicacy of those little birds. Swammerdam, and other naturalists, have endeavoured to discover if the araneïdes have really a poison which they insinuate into the wound, after they have bitten. Nothing, however, was discovered which could indicate that they poison the wound which they inflict. Poultry, and birds in general, eat these animals without suffering any inconvenience. It also sometimes happens to men to swallow small araneïdes in eating fruit, without experiencing any accident in consequence; and some persons have even eaten very large ones to prove that they are not venomous. The astronomer Lalande, swallowed four of them in the presence of M. Latreille, without any evil result *. The latter gentleman is, however, of opinion that they do possess a venom, though it ordinarily produces no sensible effect upon us. It is very certain, though contrary to the testimony of some naturalists, that the talons of their mandibles are pierced at the extremity. It is equally easy to be convinced that insects which have been bitten by a tolerably strong spider, die almost immediately. Rossi informs us that a species of *theridion* inflicts wounds which are mortal, even to man.

* This, however, does not negative the existence of venom, as we know that the poison of a viper may be taken into the stomach without injury.

It would therefore appear, that some caution is requisite, respecting the larger species. In endeavouring to avoid the Charybdis of credulity, we should take care not to stumble on the Scylla of imprudence.

The araneides are very generally spread through all countries, and found, in fact, in every habitable portion of the globe. Those of warm climates are larger than the spiders of temperate regions. The males and females live separately. The latter are more frequently met than the former, which do not approach the females but at the time of coupling, for fear of being devoured. Nevertheless, in some small species, both inhabit the same web; the male, however, remains a little apart. All are extremely carnivorous, and live only by rapine. They seize flies and other insects, which fall into their nets. Those which construct no webs, such as the wandering araneïdes, catch their prey by running, or darting upon it from above; others watch for it concealed under a leaf. The males often fall victims to the females, and the latter carry on a cruel war against each other when they meet. If one spider should happen to fall into the web of another, a desperate and mortal combat immediately takes place. When the two combatants are of equal strength, they wound each other reciprocally. The proprietor of the web is almost always the aggressor. The stranger remains on the defensive; but when the first finds itself the weaker of the two, it flies, and yields its web to the other, which never pursues, but remains, and profits by the labour of its adversary. It often occurs, according to Geoffroy, that some old araneïdes take possession by force of the web of a younger spider, because with the advance of age, the reservoir of the fluid which furnishes their threads becomes exhausted, and they can no longer construct a web, of which they, nevertheless, have need, for the purpose of catching their prey; they therefore compel a young one to give up its own. Nature, according to the same author, has accorded to each spider a sufficient quantity of the silky matter to make six or seven webs during its life. When no more remains, they must either die, or appropriate to themselves the production of others.

The life of many species scarcely extends beyond the term of eight or twelve months. But the mygales, the spiders properly so called, and the lycosæ, can live many years. Many pass the winter shut up in holes concealed under stones. Some even construct for themselves, in that season, a cocoon of silk, which serves them as a retreat.

In the fine days in autumn, we may see floating in the air a tolerable quantity of threads of silk, which are often carried by the wind to a considerable height. Many of these threads are the work of some young araneides; of this we may convince ourselves by examining them closely. We shall find at one of the ends, little spiders occupied in producing new threads, or elongating those which have been already spun, until they are fixed at a distance to some solid place, whither they can transport themselves.

Quatremèe d'Isjonval believed that he discovered in the epeiræ, a natural barometer; but it does not appear that this opinion has been followed up, or has led to any results.

The spiders of the genus MYGALE appear to be nocturnal animals. Their sombre colours, and some observations which have been made upon them, seem to authorize this conjecture. They establish their domicile in cavities usually subterraneous, which they either prepare for themselves or find by chance, and whose aperture they line after the manner of the tubicolæ, which are likewise nocturnal animals.

In this genus are found those monstrous araneïdes which can occupy a circular space of from seven to eight inches in diameter, and which sometimes even seize small birds.

These species are in general peculiar to the equatorial countries, and those which border on the tropics. They are extremely dreaded in Antilles, and in South America, where they are called *crab*-spiders.

The habits of all these mygales are probably the same, and we shall therefore present here what authors have collected on the subject, though as far as actual observation is concerned, they principally apply to the species called *avicularia*.

That which Pison, in his Natural history of Brazil, names *nhamdu*, or *nhamdu-guaçu* (great spider) is a species very much akin to *avicularia*. According to him, it nidificates, after the manner of birds, in the rubbish and cavities of old and decaying trees. It lives a very long time, and can support an extreme degree of abstinence. Some individuals, which the author had shut up in boxes, have lived there some months without any sort of nourishment. This species constructs, though but seldom, with the two projecting spinnerets, which it carries at the anus, webs, similar, as he says, in disposition to those made by the other spiders. But the generality of this assertion, and the description which this author gives of the webs of these mygales, would seem to prove, that he does not speak *ex visu*, but abandons himself to argument and conjecture. Such, again, is the case when on the subject of the coupling of these animals, he advances that their bodies are opposed one to the other *(aversis clunibus)*. The females carry their eggs under the belly. The talons of their mandibles are enchased in gold, to serve as tooth-picks, and are even supposed to be an excellent odontalgic. Not only the pricking of these animals, but the liquor which is distilled from their mouths, and even, it is reported, their hairs, are reputed venomous. The part of the body which the animal has wounded, grows benumbed, livid, and blackish, swells considerably, and the malady proceeds, according to Pison, to such an extent as to prove incurable. The wound is cauterized; but the best antidote, as the same author tells us, is furnished by that preparation of the crab, which he names

Aratu (grapsus pictus). It is pounded, and a potion made of it with wine: it acts as an emetic.

The ancients, in like manner, boast of the anti-poisonous qualities of the crustacea, and the employment of them may certainly prove salutary, in circumstances where the use of alkalies is required. M. Arthaud has caused death to chickens, by causing them to be stung by the large crab-spider of the Cape of Good Hope *(mygale cancerides,* Lat.). This species, according to him, frequents humid places, kills and sucks large insects, kakerlaks, and often its own congeners. He maintains that this mygale often perishes, by the sting of a sort of gadfly under the belly, most probably in the organs of respiration. The touching of this last araneïde, or rather its hairs, produce severe smartings, similar to those caused by the introduction of the hairs of certain caterpillars, into the epidermis.

Though sound criticism may authorize us to call in doubt, or suspect of exaggeration or partiality, the testimony of some travellers, or of some historians respecting the venomous powers of these araneïdes, prudence, enlightened by observation, forbids us to deny the existence of their poison, or to lull ourselves in a fancied security against the dangers which it may occasion. Here, as in so many other cases of uncertainty, further experiments will be necessary to decide our judgment.

The hairs of these mygales, as we have already hinted, make the same impression upon the skin as the hairs of certain caterpillars. " One morning, as I was getting up, one of the Spanish travellers made an exclamation on seeing on my garments, from the feet even to the shoulders, a brown trace, occasioned by the passage of one of these crab-spiders, and by an acrid and caustic liquor which distils continually from its mouth and feet. Happily it passed innoxiously while I was sleeping profoundly, and contented itself with thus leaving its visiting card." (Lescallier, Notes sur la traduct. Franç. du Voyage du Capitaíne Stedman.)

Pison relates that the mygale of which we have spoken a little farther back, sheds its hairs with age, and that then, the skin of its belly is of a pale carnation colour.

Madlle. Merian informs us that she had found many individuals of the *mygale avicularia* on the tree named *guajave*, there making their domicile, and remaining in ambush in the cocoon, which is formed by a caterpillar of the same tree, for its change into the form of a chrysalis. She assures us explicitly, that this mygale does not spin long cocoons, as some travellers would have us to believe. The majority of the other testimonies which we could allege here, do not appear to us of great authority, either because they were not ocular, or because it is difficult to ascertain to what species of araneides they should be applied. The author of the Natural History of Equinoxial France, places the habitation of the avicularia, or that of some other species, in the clefts of rocks. In Stedman's Voyage to Guyana, this animal is called the *bush-spider*, and its web is said to be of small extent, but strong. The *mygale avicularia* is provided with two long spinnerets; thus there can be no doubt of its capacity for spinning. But when we examine the form of the hooks of its tarsi, when we find them so small, and almost without denticulations, and thus so different from those of the industrious araneïdes, we must feel inclined to refuse to this mygale the faculties which the majority of the araneïdes possess, and to suppose that its strength may suffice for all the purposes of its existence. It lives, according to Madlle. Merian, on ants, which escape with difficulty from its vigilance and pursuit. In failure of these it endeavours to surprise small birds in their nests, whose blood it sucks with avidity. This change of nutriment is rather different, but the appetite of the animal is equally voracious and accommodating. The ants occasionally avenge themselves for the evils they endure from their formidable adversary, and fall upon him in such numbers, that he is unable to defend himself, and is finally devoured.

M. Moreau de Jonnès, who, during a residence of many years in Martinique, had especially devoted himself to the study of the natural productions of that island, communicated to M. Latreille a digest of the observations which he had made on the subject of a species of mygale very common in that country. We shall present our readers with the short, but interesting statement of facts which he has adduced.

"The *mygale avicularia* bears in the Antilles the name of *crab-spider*. It also preserves that of *Matoutou*, given to it originally by the Caribs. This species is the largest of two hundred which are known to naturalists. Its length is an inch and a half, and when its feet are extended, it covers a surface of six or seven inches. It avoids inhabited places, and I have never found it in the towns, where the *hunting-spider* of Linnæus, and six other species of the same genus are, on the contrary, extremely multiplied.

"As M. Latreille had recognized by the mere inspection of the organization of the animal, it spins no web, to serve it as a dwelling. It burrows, and lies in ambush in the clefts of hollow ravines, in volcanic tufas, or in decomposed lava. It often hunts to a considerable distance, and conceals itself under leaves to surprise its prey, or it climbs on the branches of trees to devour the young of the colibris and the *certhia flaveola*. It usually takes advantage of the night to attack its enemies, and it is commonly on its return towards its burrow that one may meet it in the morning, and catch it when the dew, with which the plants are charged, slackens its walk.

"The muscular force of the mygale is very great, and it is particularly difficult to make it let go the objects which it has seized, even when their surface affords no purchase, either to the hooks with which its tarsi are armed, or to the claws which it employs to kill the birds, and the anolis. The obstinacy and bitterness which it exhibits in combat cease only with its life. I have seen some which pierced twenty times through and through the corslet, still continued to assail their adversaries, without showing the least desire of escaping them by flight. In the moment of danger, this spider usually seeks a support against which it can rise itself, and mark the opportunity of casting itself upon its enemies. Its four posterior feet are then fixed upon the ground; but the others half extended, are ready to seize the animal which it is about to attack. When it darts upon it, it fastens itself upon its body with all the double hooks that terminate its feet, and stretches to obtain the superior base of the head, that it may sink its talons between the cranium and the first vertebra. In some other American insects, I have recognized the same instinct of destruction.

"When the mygale applies its claws on a hard and polished body, we see there immediately the traces of a liquid, which must be the poison that it ejects, and that renders its sting, or bite, dangerous. Nevertheless I have been unable to discover the issue through which the emission of this fluid is made, the effects of which are considered to be very formidable in the West Indian islands. Never, either, have I seen the mygale employ, as has been strongly asserted, another fluid secreted by glands situated at the extremity of the abdomen, and which is said to be darted by it against its adversaries, to blind them by its corrosive qualities. The individuals of this species, which I have preserved for a long time, and in great numbers, never had recourse to this means, in the combats which they carried on for the possession of their prey; but I have recognized the existence of this liquor, which is lactescent, and singularly abundant in proportion to the size of the animal.

"The mygale carries its eggs enclosed in a cocoon of white silk, of a very close tissue, forming two rounded pieces, united at their border. It supports this cocoon under its corslet, by means of its antennulæ, and transports it along with itself. When very much pressed by its enemies, it abandons it for an instant; but it returns to take it up as soon as the combat is concluded.

"The little ones are disclosed in a rapid succession. They are entirely white; the first change which they undergo is the appearance of a triangular and hairy spot, which forms on the centre of the upper part of the abdomen.

"I had preserved from 1800 to 2000 of these, all which proceeded from the same cocoon. They were all devoured in a single night by some red ants, which, guided by an instinct that set at defiance all my cares, discovered the box in which I had inclosed the spiders, and insinuated themselves into it by means of an almost imperceptible aperture, through which myriads of them passed, one by one, in the space of a few hours. It is owing, in all probability, to the destructive war waged upon the aviculariæ by these insects, that the number of these arachnides is confined within such narrow limits, which by no means correspond with their prodigious capability of reproduction."

According to M. Palisot de Beauvois, the *mygale avicularia*, inhabits the open country, and establishes itself in the cavities presented to it by the soil. It closes the aperture of its dwelling with a web, as do many other congeneric araneïdes.

Another mygale, called *recluse* by M. Latreille, constructs its nest in the same manner as the *mason*-mygale, of which more anon. It fixes its abode in stony places, and its bite, according to Brown, causes a very severe pain, which lasts for many hours, and is sometimes even accompanied with fever and delirium. The ordinary sudorifics, spirituous

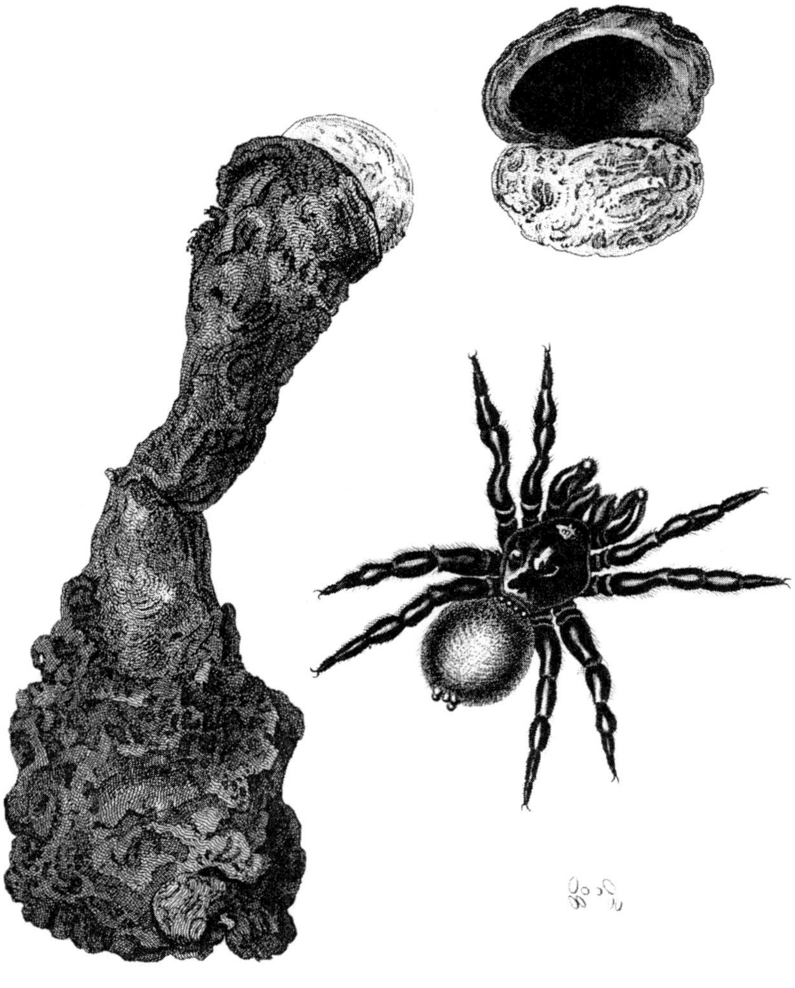

Mygale nitida.

C.M. Curtis del.^t

London, Published by Whittaker & Co Ave Maria Lane: 1833.

liquors, such as tafia, rum, soon relieve such symptoms, by inducing sleep and perspiration. Badier has often met with the same species in the island of Guadaloupe; but he has always found it in argillaceous soils which present a gentle declivity. Retired within its nest, this animal scarcely gives any sign of life.

Olivier has observed, in the environs of Saint Tropes, and in the isles of Hyères, the nest of a mygale, which, from its position and construction, would seem to be very distinct from the others, and to announce peculiar manners. This nest was situated in a horizontal soil. Its door, although of clay, and closing of itself by a sort of spring, resembled a circle, from which a small portion had been cut off. It was attached to one of the sides of the aperture, and the entrance was free. The inhabitant was absent, and this naturalist conjectured that it does not close this entrance but at the time when it is actually occupying the nest. M. Latreille believes this species to be his *mygale carminaus.* M. Boyer de Foulolombe has observed this same nest with more attention than Olivier, although he was never able to surprise the animal in its dwelling. It is formed of a sort of tube of silk, sunk vertically in the ground, and covered at its orifice by two shutters placed in a horizontal position, at the suface of the soil. A solid partition cuts this external door, a little above it. Some persons informed this naturalist that they had seen the animal come out, and re-enter, shutting the door after it.

A very curious species is the *mason-spider (mygale cementaria)* which is found in the neighbourhood of Montpellier. Almost all the araneïdes have the two upper hooks of their tarsi pectinated, or formed like a comb, we may easily conceive that from this arrangement of these parts, they find the means proper for the execution of their labours. But the hooks of this mygale, from their simplicity, seem but little adapted for working, though its industry yields in nothing to that of the other araneides, but even exceeds it. It was necessary, therefore, that nature should supply it with other instruments. These reflections led M. Latreille to a very attentive examination of the organs of these animals, and he discovered, above their mandibles, some hard, corneous points, the anterior of which, ranged in a transverse series, resemble a sort of rake. Without seeing these animals in the performance of their operations, it could scarcely admit of doubt that this peculiar instrument must be very useful to them for the formation of their nest.

In so carefully concealing their retreat, in preparing and constructing it with so much art, these araneïdes have less in view their own preservation than that of their offspring. Rossi has found in the nest of that species, which he names *Aranea sauvagesii*, its numerous family. These two species (*cementaria* and *sauvagesii*) excavate, in argillaceous soils, a burrow, or cylindrical trench, having the same diameter throughout. Its relative dimensions may vary according to the species and the age of the animal. It usually chooses soils in declivity, or cut vertically, so that it may not be stopped by the rains, and which besides are arid, and composed of a strong earth, without any mixture of pebbles or small stones. It takes care to unite the interior walls of its habitation, and to line them with a silken pellicle, so as to consolidate them, and prevent any fallings in. This web may also contribute to the facility of its movements, and advertise it, by the motions which it undergoes, of what is passing at the entrance. A door, or sort of flat trap, but tolerably thick, circular, composed of different beds of earth, moistened and bound together with silk, smooth, a little convex, covered with very strong threads, forming a very close tissue underneath, closes the aperture of this burrow. The threads with which the interior surface of this door is lined, are prolonged from the side of the most elevated edge of the entrance, fasten there, and attach the coverlid, forming a sort of hinge, so that being inclined in the direction of the soil, it falls back by its own proper weight, and the entrance of the habitation is always naturally closed. The contour of the door corresponds so well with that of the aperture, that it does not out-edge it in any place, that there is not the least vacancy in the joinings, and that the proportions could not have been better observed had they been taken by the compass. When this door, therefore, falls, it seals the entrance hermetically. The posterior convexity of the door also contributes to the precision of the closure.

The Abbé Sauvages, from whom these observations are taken, was unable to discover the manner in which this animal proceeds in the formation of this nest, or its mode of subsistence and propagation. The individuals which he took alive, all perished, in spite of the cares which he employed for their preservation.

This spider employs a singular degree of strength and address, when an attempt is made to open the door of its domicile. The observer just quoted, being desirous to raise it by means of a pin, experienced a resistance which he by no means expected. He saw the animal in a reversed attitude, hooked by the legs, on one side against the walls of the entrance of the hole, on the other at the web, which covers the hinder part of its door, dragging the door to itself, so that in this struggle it opened and closed alternately. The mygale did not give way until the trap was entirely raised. It then precipitated itself to the bottom of the hole. Every time, when similar attempts were made, even at the slightest movement, the animal runs forward immediately, to hinder its door from being opened, and never ceases to keep guard there. If it be closed, one may work at the clay all about, and excavate it to carry off the habitation, without the peril with which it is menaced causing it to desert its post. But as soon as it has been expelled from its dwelling, one would believe that it had lost all its vigour. It appears languid, benumbed, and if it makes some steps, it is only in a tottering manner. It is never seen of itself to issue forth from its habitation, and the light of day appears to be injurious to it. Olivier tells us that the *mygale ariana,* which belongs to the island of Naxos, remains constantly in its nest during the day, and never leaves it but at night.

The Abbé Sauvages had discovered the mason-spider in the neighbourhood of Montpellier, on the edges of the roads, and the high banks of the small river of Lez. But the description which he had given of it was very insufficient. This defect, however, was supplied by Dorthès, in a memoir in the second volume of the Linnæan Transactions.

Dorthès has added some observations to those of Sauvages. If the lid which closes the entrance be fixed with a pin, or if it be taken away, a new one is found to the aperture on the following day. It appears certain, that it is only by night that this animal plunders, and works at the construction of its abode. The bottom of this often contains debris of various insects, and even of tolerably large coleoptera. In August this spider attains its full growth, is disposed for coupling, and is most timid. Fecundity appears to change the character of the female; become a mother in September, she no longer flees, but grows fierce and more voracious. The threads which she extends over the inequalities of the ground near her dwelling, procure her different insects for nutriment, and more especially diptera. She then lives in society with the male, and Dorthès has found thirty little ones in the nest.

The SPIDERS proper *(aranea)* are almost all of them domestic. In the corners of neglected apartments, garrets, stables, &c., they spin large horizontal webs, which exactly occupy the angles. Their surface forms a triangular plane, but which becomes a little concave in consequence of the natural sinking in of the web. Its threads are very close, cross each other, and being strongly bound together by their viscosity, give it some resemblance to a very fine stuff, but which can, nevertheless, retain the various insects which come there, on which these animals feed, and frequently even coleoptera of tolerable size. Many loose, and as it were, floating threads, compared by Lister to cordages, or sail-yards of vessels, are placed on the upper side of the web, and become sorts of snares for the insects which get embarrassed there. Immediately at the angle formed by the union of the two walls, the spider weaves a cylindrical tube, having one of its apertures above, the other underneath. It there remains constantly in ambush, the head being turned forwards; as soon as a fly, or any other little animal is arrested in the web, it runs forward, promptly seizes its prey, and drags it to the bottom of its lodge, for the purpose of sucking it with the greater facility. If any pressing danger should frighten the spider, it takes to flight, and speedily escapes through the lower aperture of its habitation. Homberg has also described, in the memoirs of the Academy of Sciences, 1707, the manner in which these animals manage in spreading their webs. The cavities which are found under stones, in rubbish, also serve them for the purpose of retreat. Lister has occasionally encountered them in the woods. He has seen the male and female on the same web, in the commencement of June, the season of their amours. Having thrown some flies to them, each individual took one. The laying takes place towards the end of the following month. The cocoon has a double envelope of very white silk. It is placed near the anterior aperture of the tube where the spider remains, and it seems to form a portion of the web. The eggs are whitish, and have no adherence. Audebert brought up, and kept for some years, many individuals of the domestic spider. Some females which he had isolated, produced in succession several generations equally fruitful. The domestic spider grows considerably with age, and some individuals are found extremely large; but according to Lister the feet alone increase in volume, and become more hairy.

The *Aranea labyrinthica* constructs its web on the same model; but it places it on hedges, bushes, at the lower part of trees, or on different tufted vegetables. There may be seen in the fields, towards the end of summer, a great quantity of these webs. This species, however, is less common towards the north of Europe. Lister observes, that it establishes itself preferably in the neighbourhood of the habitations of the great ants. It also appears that this spider lays its threads in succession, to embarrass these insects while they are running; that it stings them when they stop, and returns to seek them some seconds after, when the poison has produced its effect. It also feeds on bees; but in failure of them, on other insects, even on coleoptera. The feet of the male are larger than those of the female, as well as in the common spider. Coupling takes place, at least in the south of England, towards the end of July. The cocoon is placed like that of the domestic spider, and contains about sixty eggs of a whitish colour, which Lister observes were the largest he had ever seen.

A female which he reared, suspended her cocoon to the middle of a glass in which she was enclosed. She enveloped it with different webs, divided by partitions, forming kinds of chambers or alleys, but which conducted to the depôt of the eggs. The cocoon had the figure of a star.

In the genus ARGYRONETA, the habits of the only species, *Aquatica*, have, from their singularity, attracted the attention of several naturalists. It lives in the dormant or very slow waters of marshes and dykes, which are not dried up, at least entirely. It is in the interior of these waters, and not at the surface that it inhabits, differing in this respect from some other species of aquatic spiders. We begin to find it from the first warm days in spring. It swims in a reversed position, having the under part of the body turned upwards; its abdomen is then enveloped with a ball of air, and appears like a little silvery and very brilliant globe. Degeer tells us that the entire body, feet excepted, is involved in a globe of air. This has not been verified by other observers to the same extent. It often comes to the surface of the water, keeping its body suspended there, the lower extremity upwards, apparently for the purpose of respiration. We at present know the place of the organs of respiration, and that they are not spinnerets, as Clerk supposed. But how does it manage to envelope a large portion of its body with this mass of air? what is the cause of its adhesion? These are problems which observation has not yet solved.

A property of these araneïdes, not less singular, is the capacity of constructing for themselves, in the bosom of the water, a kind of aerial mansion, a true diving bell, where they can respire freely, live in safety, and which serves as a cradle for their family. This may be compared to a diving bell, not only because it has the same destination, but the same form, namely, that of a cap, or that of one half of the shell of a pigeon's egg. It is entirely filled with air, perfectly close, the under part excepted, where there is a tolerably large aperture, giving an entrance and exit to the animal. Its walls are slender, and composed of a tissue of white silk, strong, and close, a great number of irregular threads fix it to the stems of plants, or other bodies. Sometimes the upper part is out of the water, but most generally it remains entirely immersed. Its inhabitant is thus environed with air; she remains there quietly, the head usually down, a situation which permits her to see more easily what passes, to watch her prey, and to escape from the least danger. Degeer has seen her with her head upwards, and the feet applied against the body.

It is easy to conceive the mode in which the argyroneta introduces the air into her ball, and how she totally fills it. At first the water may be supposed to occupy its interior capacity. To empty it, and substitute another fluid, the animal goes continually to the surface of the water, charges itself with a bubble of air, transports it into its habitation, unburthens itself of its aerial provision, and displaces an equal mass of water, which issues through the lower aperture. By repeating this operation several times, it succeeds in expelling all the water out of its cell, and introducing the same volume of air. The males, as well as the females, construct for themselves, at all the favourable seasons of the year, similar habitations, which demonstrates the analogy existing between these and the other araneïdes of the same tribe. Degeer found, in the month of December, one of these balls, closed in all parts, and in which the animal was, as it were, imprisoned. It issued forth through a rent made by this observer, and immediately proceeded to suck a fresh-water *assellus,* which he presented to it. It is probable that these araneïdes shut themselves up in this manner to pass the winter. The same naturalist, as well as Clerk, has preserved, in the same vessel, many individuals of both sexes, without one devouring the other; and though they had been deprived, for several days, of every kind of nourishment, all that passed in the rencontres between male and male, and female and female, was simply touching each other, or attacks which led to no murderous conclusion. Thus some writers have falsely attributed to the aquatic spiders, probably presuming from analogy, a cruel and voracious character, with regard to their own species.

The eggs are round, of a yellow sulphur colour, and shut up

in a globular silky cocoon, the volume of which occupies one-fourth of the internal capacity of the cell. The female remains constantly by them, having the abdomen in the interior of the habitation, and the trunk in the water. Clerk has seen many little ones swimming in the month of July. which leads to the supposition that the eggs are all laid in the course of the preceding month. It takes place a little sooner in France.

Some species of THERIDION remain under stones; others inhabit the parts of houses which are little frequented or seldom visited, and make their webs either at the angles of walls, or in closets, and amongst furniture. But most part of the others choose as a domicile, trees or flowers. Such is particularly the species which M. Walckenaer has named *benignum*, and whose manners he has studied with so much attention. He has given us in the fifth fasciculus of his History of the Animals of this family, some extremely curious and complete observations respecting the coupling of this species, of which the following is an abridgment.

This species is frequently seen, especially in autumn, in gardens, and kitchen gardens. Its irregular web, notwithstanding its extreme tenuity, often protects grapes from the bite of insects. It is seldom that this fruit is used without finding the animal there. It is also fond of spreading its threads over the surface of the leaves, between the flowers, and at the extremity of different vegetables. The female deposits her eggs three times a-year; her cocoon is lenticular, flatted, of a close tissue, and a very brilliant white.

The business of reproduction so completely absorbs both sexes, that when the coupling is commenced, the leaf on which they are may be detached, and the union observed with a microscope, without in the least disturbing them.

The coupling most usually takes place on the shrubs of our gardens, such as lilacs, rose-bushes, &c., towards the middle of May, and particularly in the morning of those days in which the weather is disposed to be stormy. The two sexes cover themselves with a thin and very delicate tissue, which they construct in common. The male, after having spread some threads on that side of the tent where the female is placed, advances towards her, touches her gently a minute or two on the back, and finally determines her to quit the motionless and contracted posture which she held. After this the coupling takes place. When the amours of these animals are terminated, the two sexes live together in tranquillity, forming, in this respect, a rare exception in the spider race. This good understanding appears to be general among all the Theridia.

Of all the known theridia, the species most deserving of our attention is that which Rossi calls 13-*guttata*, and which is the spider known in the island of Corsica, under the name *marmignatto*, or *marmagnato*. According to him its bite is mortal, even to man himself. It produces the most serious symptoms, which scarcely disappear by the operation of sudorifics and scarifications. This animal spreads along the furrows of the fields different threads, so as to stop or impede the progress of locusts, of which it makes its prey. Having the body reversed, and suspended by the fore-feet, it draws, by the help of the hinder feet, some new threads, which it shoots very quick, and by an undulatory movement, on the feet of the locust, until it has sufficiently entangled it to approach without any fear. It first bites it near the neck, and then sucks it at its leisure, as the animal, on being bitten, falls into a convulsion immediately, and perishes. If enclosed in a vessel with a locust, the theridion, by trying to envelope it, soon exhausts all its silky matter and dies itself, having lost all its strength. It will not attack the European scorpion, or different spiders which may share its captivity; but it is not so with individuals of its own species, with which it fights to the death. It has been observed, that where this theridion is most common, a hymenopterous insect, known by the trivial name of *St. John's fly*, in all probability a sphex, or pompilus, destroys a great number of them. The cocoon is about the size of a hazel-nut, and the mother guards it with the greatest assiduity.

Degeer has observed another Theridion *(aranea bipunctuata* of Linnæus) spin round the insect arrested in its snare, as well as all round the neighbourhood, fresh threads, drawing them with its hinder feet, so as to hinder its prey from breaking its fetters, and then attack it with open force, kill it, and drag it within its domicile; this is usually in the cleft of a casement. The angles, the corners of the walls which are near it, are curtained with its web, which is loose and spreading.

The species of the genus LINYPHIA, the most common *(triangularis)* spins on bushes, &c., particularly towards the end of summer, a horizontal web, hung between the branches, slender, not compact, and the extent of which, often considerable, varies in proportion to its proximity to, or distance from, the points of attachment. To maintain it in the same situation, and prevent its sinking, it spreads above and on all sides, perpendicular and oblique threads, which it fixes to the branches in the neighbourhood. They are sometimes even so drawn that the web becomes convex. It is suspended in the midst of this very irregular assemblage of threads, being directed and crossing on all sides. The animal remains in an inverted position, having the belly upwards and usually at the centre of the web; as soon as an insect is caught there, it runs out quickly, pierces it with its mandibles through the web, makes a rent there to allow it to pass, and without enveloping it with silk, sucks the insect which is dead, or excessively enfeebled by the effects of the poison. When many individuals of this species are placed together, they slaughter each other without mercy. There is scarcely any resemblance between the males and females. It appears, according to the observations of Degeer, that at the season of reproduction, the female receives the male without the least movement, or giving him the slightest cause of apprehension. In this respect, the males of this species are far more fortunate than most of their confréres, who, in their amorous essays, are in a state of perpetual alarm, lest they should be devoured by their mistresses.

The abdomen of the females enlarges considerably, as the time of laying the eggs approaches. The cocoon, composed of a loose silk, is placed close by the web. Lister has sometimes seen two, one by the side of the other, but unequal in size, one of which contained the young, and the other the eggs. These eggs, tolerably numerous, are of a reddish colour, bordering on brown, and not agglutinated together. The same observer has found some cocoons in the middle of June; but he has also seen at the commencement of September, a great number of females with their males, in the same webs, and ready to lay. He presumes that the latter conceal their cocoons under the moss, and at the roots of old trees, to preserve them from the rigours of winter. It is clearly certain, that many of these eggs, those probably which have been first laid, exclude the young before winter, Lister having met, in the month of November, a great number of young ones of this species casting threads, suspended with them, hovering in the air, and repeating these manœuvres, until they had escaped from his hands.

The EPEIRÆ most generally remain at the centre of their snare, the body being immersed, or the head downwards. But others construct for themselves a dwelling near the web, either entirely arched, and sometimes in the form of a silken tube, sometimes composed of leaves connected together by threads, or open at the top, and like a cup, or a bird's nest.

The cocoon is usually globular, or ovoïd, and presents at the interior, a wad of silk, tolerably thick, and often differently coloured from the silk which forms the exterior envelope. The eggs are very numerous, agglutinated, and placed in the middle of this sort of down.

Many of these araneïdes lay eggs but once a year, which happens at the end of summer, or the commencement of autumn. Some epeiræ compose webs of very strong threads, capable of arresting the flight of birds even as large as a wild pigeon. The epeiræ described by M. Labillardiere, in his Voyage in search of La Peyrouse, is a viand greatly in estimation among the inhabitants of New Caledonia. They first kill these animals in earthen vessels, which they cause to be heated, and then grill them on the coals. The naturalist just mentioned saw two children swallow about one hundred of them. This species inhabits the woods, and its web also opposes much resistance.

The *Epeira cicatrosa* spins its web against walls or other bodies, and remains concealed in a nest of white silk, which it forms under some projecting part, or in some cavity in the neighbourhood of its web. It gives no sign of life when it is taken, and never comes forth except at night; it is then, or at all events when the light is weak, that it spins. Its web is often loaded, but without any order, with the carcases of the different insects which have served it as food; even scolopendræ have been found there. Clerk, however, tells us that this species prefers phalenæ, and other nocturnal lepidoptera to flies. It is also in the darkness of night that it devotes itself to the pleasures of love. The female lays in spring, and conceals its eggs in its habitation, or near it. According to Clerk, the cocoon is the size of an ordinary pea. Lister tells us that the eggs are very crowded, and placed one upon the other in several strata, so that they form a firm, flatted, and orbicular body, like in figure and bulk to a lupin seed. They are covered with a loose wad, and the animal often cements over its cocoon a tolerable quantity of detritus of various kinds. The young are excluded towards the end of spring or the beginning of summer. Arrived to a considerable degree of their growth, when the cold weather comes on, they support its rigours, by remaining concealed under the old barks of trees, and stakes.

The *Epeira apolisca,* is found in woods, near ponds, and in humid places. Its nest is composed of a very close silky matter, which Lister compares to the substance yielded by prepared flax. There is but one little aperture placed underneath, which the animal closes with its feet, when one attempts to seize it. On the approach of winter, it consolidates this habitation with grains, or pieces of vegetables, which it attaches there. It remains there entirely closed up, not to come forth until the return of fine weather. But it appears, however, that, according to Lister, this epeïra sometimes selects for itself, for hybernation, a *locale* of a different kind, and one more sheltered. It proportions the extent of its web to that of the soil, so that the number of concentric circles of its net-work varies from fifteen to eight and thirty. The same naturalist has seen males confine themselves to spreading simple threads, and without much order, upon the summits of gramineous plants. He has ascertained that the same female, in the space of about two months, laid in succession, three broods of eggs, which were indicated by so many cocoons, and even four in a little longer time. The first took place towards the end of May. He amused himself for nearly a month and a half, in undoing every day the web of an individual of this sex, which he had transported from the country into his garden, and which had established its nest between the green leaves of the rose-tree. The animal was never tired in reconstructing its work, and never abandoned the cradle of its offspring. It appears, that under such circumstances, the mothers do not change place, not even for the purpose of seeking food. With respect to the other species of this genus, we could add nothing, except details pretty nearly similar to the foregoing.

In the genus micrommata, the manners of the species have been but little observed, with the exception of one (*smaragdina*). It is found in spring on plants, and particularly on the yoke-elm, and on trees, of which it even gains the summit. Clerk says that it leaps uncommonly well, a fact also remarked by M. Walckenaer. It is also extremely agile in running. A female individual which the first of these naturalists had reared, gave him an opportunity of observing the mode in which this spider performs the business of manducation. As soon as it had seized a fly, it pierced it with the talons of its mandibles, then compressed it, and chewed it with its jaws. It seemed to move their denticulations, or rather the hairs with which their internal side is provided, then held it, and turned it over with its palpi, and then withdrew one of its talons and plunged it into another part. In the interval of these jaws, or what this naturalist terms the throat, a frothy matter was observed, which absorbed the nutritious juices expressed from the carcase. The action of these different organs was more easily observed when the body of the fly was reduced one-third. All its soft, or liquid substance being exhausted, the animal threw away the remains. It then cleaned the extremity of its palpi, using the claws of its mandibles, its jaws, and particularly by the assistance of a liquid matter which it caused to flow from the œsophagus.

The female lays her eggs in June or July. She draws together, and binds with a great number of threads, three or four leaves, of which she makes a packet, which has something of a triangular form. Its interior is lined with a thick silk, and in the middle of this nest is placed the cocoon itself, composed of the same material. It is round, white, formed of a single stratum, and the tenuity of its walls allows the egg to be perfectly well distinguished. Clerk has counted about one hundred and forty of them. They are the size of a grain of beet-root, spherical, of a clear shining white, with white circles on one of the sides. They are exceedingly smooth, and when placed on paper they roll like little drops of mercury. They are seen to go equally from one side to the other in the cocoon, not being agglutinated. The female fixes herself in the midst of the packet of leaves, to watch for the preservation of her posterity. Clerk tells us that the young are born towards the end of July.

It is more particularly to the species of the genus THOMISUS, that Europeans give the name of *crab-spiders*. They are seen running on the ground, climbing on the bushes, on plants, and even on elevated trees, from which they often descend by means of a thread which they unwind, and by means of which they can reascend. Accordingly Lister has compared them to rope-dancers. Contracting their feet against their body, they balance themselves in some sort, in the air, impress a movement on their thread, and direct it as if nature had given them wings and oars. Degeer also tells us, that these araneïdes unwind always in walking, a thread which is attached to the place where they were seated. They are again to be met with in the corollæ of flowers, where they seize the little insects which come to settle. The *Thomisus tigrinus* is very common on the stems of trees. Clerk has seen the *Thomisus cristatus*, which he preserved in a box, make at one of its angles a little web as thin as paper. It appears, however, certain, according to the observations of other naturalists, that the Thomisi do not weave nets for the purpose of surprizing their prey, which they take by running, or they wait patiently until it imprudently approaches within their reach. M. Walckenaer tells us that they introduce themselves into the webs abandoned by other araneïdes, and profit by the fruits of their labours. We speak here only of species indigenous to Europe. It would seem, according to the observations of some travellers, that the other exotic thomisi are more industrious, and approach in that respect to the epeiræ, and that they even live in houses. It may, nevertheless, be possible that they take possession of the webs of other araneïdes, as M. Walckenaer observes concerning the thomisi of Europe. The latter are sometimes embarrassed in the threads of the epeiræ, and serve them for food, as Lister has observed.

Degeer has witnessed the coupling of the *Thomisus citreus*. Having found, in the month of May, many individuals on a willow bough, he put them into the same sand-box; the weakest soon became the prey of the strongest, and he was obliged to separate them. He discovered among them an individual differently coloured, which he imagined to belong to another species. But he soon was perfectly convinced that it was a male, by seeing it couple. There is nothing different from the other genera in this particular.

The cocoon is composed of a white silk, very close, and forming a papyraceous or membranous tissue. It is usually orbicular or very flatted. We shall have an idea of its form, by imagining two caps a little gibbous, applied one against the other in an opposite direction, and united at their edges. Lister describes that of one species, which is snow-white, angular, and of a radiated form. It was found attached to a little branch of the *ulex Europæus*, at the commencement of June. He saw, on the same shrub, at the same period, the cocoon of another species of the same genus, and which was attached to one of the summits of the branches; the female was clinging to the cocoon. Lister having detached the branch which bore it, and having placed it along with the spider in a box, this tender mother did not attempt to ascend the branch, but remained constantly on her cocoon, placing it under her breast. The *Thomisus citratus*, of which we have spoken above, bends a willow leaf in two, and fills its interior capacity with a web of white silk, in the midst of which it encloses its cocoon, which is oval, and about the size of a cherry stone. The leaf is then closed on all sides by a web similar to that of the interior, strong, and tolerably thick. The female places herself at its external surface, watches assiduously over her depot, and never lets it go, even when an attempt is made to drive her from it. Other species place their cocoons in the clefts of old stakes, &c. The eggs of the thomisi are round, more or less yellow, and forty or fifty in number in some cocoons, a hundred in others; they are not coherent. The young are born in June or July. To pass the winter, they, as well as their mothers, conceal themselves under heaps of dry leaves, under different bodies, and sometimes even in the nests of small birds. They re-appear in the earliest fine days of spring. When the thomisi are seized, they contract their feet towards the body, and roll themselves into a ball, as do some other species of araneïdes.

The Lycosæ remain almost continually on the ground, where they run very fast; the holes which they find there, or those which they make, and enlarge as they grow older, preventing them from tumbling in, by strengthening the interior walls with a web of silk, serve them as an abode. The *Lycosa perita* raises, above the hole which she inhabits, a small cylindrical tube, formed of earth. Some others establish themselves in cavities and clefts of walls. The species *Allodroma* even constructs a tube there, composed of a fine web, covered at the exterior with parcels of earth or sand, a few lines in length; it closes it at the time of laying. Placed near the entrance of their dwellings, they there watch their prey. It is there also, or at least in similar retreats, that they hybernate. The *tarantula*, according to Olivier, takes the precaution of closing the aperture exactly, and it is probable that many other species are similarly prudent.

The lycosæ, especially such as frequent the neighbourhood of aquatic places, begin to be found from the earliest fine days of spring. Coupling takes place, according to the species and the temperature of the weather, from the month of May to the middle of July. The eggs are free, usually spherical, and their number varies according to the species, being sometimes twenty, seventy, eighty, and even one hundred and eighty. They are enclosed in a sac or cocoon, sometimes globular, sometimes flatted, circular, and formed by two caps united at their edges. It is membranaceous, and composed of a compact silk. Clerk has observed some which were whitish above, and blackish underneath. The cocoon of the *Lycosa littoralis*, one of those whose figure is lenticular, is grey externally, with a white circle, and formed of a less compact silk; its interior parietes are whitish; the egg-sac is always attached to the hinder part of the female by means of a small pellet, or silken tie, issuing from the spinnerets; she applies the threads over its surface, causing to act upon it with rapidity, the nipples, which are the conduits of the silk. If we detach this sac, we unwind, at the same time, a thread of silk which issues from the spinneret. Lister has even maintained that the animal can withdraw this sac into the interior of its spinnerets, which Degeer conceives to be impossible. The female always carries this precious deposit along with her, and in spite of her burthen, runs with celerity. If she is separated from it, she testifies her uneasiness by running to and fro on all sides, and as soon as she has found it, she seizes and runs off with it. Clerk tells us, with respect to the species which he names *Amentatus*, that when she has recovered her cocoon, she carries it at first, putting it under the belly, and approached a little to one side of the breast, where it may be in safety, and that after having attached it, as before, she betakes herself to flight.

Degeer having enclosed in a box, a female of the species *ruricola,* she spun there, against its walls, a stratum of white silk, to which she attached her cocoon; she then removed to a certain distance, but would return to it from time to time, sitting herself down upon it in an affectionate manner. This cocoon contained more than one hundred and eighty eggs.

This observer presumes that the mother assists the young in issuing from their prison, by piercing the cocoon, and that this assistance is even necessary for them. The eggs disclose the young in June or July. The little ones remain for some time longer, or until their first change of skin, in the cradle where they were born. Less feeble, after this transformation, they abandon their dwelling, mount on the body of the mother, cling all around her abdomen, more particularly on the back, and arrange themselves there in a large sort of ball or cushion, so that the mother becomes hideous, and not to be recognized. She walks about every where, thus loaded with her progeny, which do not abandon her, and with which no doubt she shares her booty. Towards the end of June, or the commencement of July, the *lycosa littoralis* is frequently observed in this situation.

Lister has observed, in the middle of October, when the weather was serene, a great number of young lycosæ hovering in the air. He tells us that he has sometimes seen them ejaculate from their spinnerets, several simple threads, like the rays of a comet, and which exhibited all the splendour of a brilliant purple. Sometimes they broke the threads, and sometimes collected them into a little ball of snowy whiteness, moving their feet with rapidity all around, and above their heads. They abandoned themselves to the impulse of the air, and were transported to very considerable elevations. These long aerial threads united in the form of unequal cords, and intermixed, often answer the purposes of a net for catching flies.

Among the different species of lycosæ, there is one which enjoys a very great celebrity, the far-famed *Tarantula*, so named from the town of Tarentum, in Italy, in the neighbourhood of which it is very common. The effects which have been attributed to the poison resulting from its bite, or that singular malady, called by some authors *Tarentismus*, and the cure of which, as was believed, could only be obtained by the assistance of music and dancing, have rendered this spider greatly renowned. But since these marvellous facts have been submitted to judicious investigation, and to the lights of experience, they have lost, at least in the opinion of educated and unprejudiced persons, this reputation, the unhappy fruit of the terrors of a credulous imagination. It is ascertained, at the present day, that the poison of the tarantula is very little, or rather not at all dangerous to man, and that it is very easy, through the means which medicine affords, to prevent the least ill consequence from its reception.

In the most southern departments of France, there is a species of lycosa, which differs very little from the tarantula of Italy, and which has even been confounded with it by Olivier. He has studied its habits, and has published the result of his observations in the fourth volume of the Natural History of the *Encyclopedie Methodique*. It strengthens with a fine and compact web, the interior surface of its cell, and its eggs are in a silken cocoon. This cell consists of a perpendicular cylindrical cavity, which it hollows in dry and uncultivated soils. Its dimensions augment progressively with the age, and often according to the bulk of the individual. It usually places itself at the entrance of this, and as soon as it perceives an insect, it darts upon it with prodigious swiftness, seizes it with its forceps, carries it to the bottom of its dwelling, and devours it almost entirely, or leaves nothing but the hardest parts. It often takes its course through the fields to carry on its trade of rapine, but invariably returns to its nest. Coupling takes place during the hottest part of summer, or from the month of June to the middle of July. Towards the end of the month of August, the female lays a very considerable quantity of eggs, perfectly similar to the seeds of the white poppy. She encloses them in a cocoon of white silk, of a very compact tissue, which is strongly attached to her anus, and which she always carries with her. When the little ones are excluded, the mother tears the envelope, to enable them to get out. She carries them on her back, and nurses them until the first moulting, or until they become sufficiently strong to form an habitation for themselves, and provide for their own necessities. "The tarantula," says Olivier, "which we have always observed, dies at the end of summer, or passes the winter in a lethargic state, shut up in its nest, after having closed it exactly, to protect itself. It does not come forth until the heat of spring is sufficiently strong to re-animate it." The death, however, of this spider is not owing to cold, but to many fortuitous circumstances which kill other spiders that are naturally long lived. The mother usually passes the winter with her family under the same roof, and dispersion does not take place until the return of the fine season. Occasional inclemency, or variations of temperature, cause the destruction of a great number of these young individuals.

They may be seen, in the first fine days, at the end of March, issuing from their dwelling, to enjoy the gentle heat of the sun, and making excursions, but of short duration. The slightest zephyr is sufficient to cause the family to re-enter its habitation. At the end of the second winter, the tarantula has acquired about one-third of its proper size, and it is not until the third year that its growth is terminated. The duration of their existence may be very considerable; but the heavy rains of autumn, and a large species of scolopendra, are enemies from which few of them escape. This scolopendra attacks the largest tarantulæ, and, after an obstinate combat, succeeds in killing them, and taking possession of their habitation. The two sexes live separately, and, except during the season of reproduction, carry on a mortal warfare. This tarantula is susceptible of anger to an extreme degree, especially when forced to quit its habitation, which it never does without a desperate struggle.

In the genus SALTICUS, the species named *Aranea scenica* by Linnæus is very common, and remains generally on walls exposed to the sun, or on the glasses of casements, where it parades at all hours during the summer. It walks, as it were, by jerks, stopping short altogether after having gone some paces. It rises on its first feet, elevates the anterior part of its body, to consider on which side it shall leap; and it is thus that it seizes small insects, especially gnats, which it appears to prefer. When it has discovered the object of its prey, it approaches it softly, with gentle steps, until it comes to a distance which it can cross by a single jump, and fall upon the little animal which it has been watching. It does not fear to leap perpendicularly from a wall, because it always finds itself attached by a thread of silk, which it continually unwinds in walking, and which, under these circumstances, holds it suspended. The other species of Saltici also use the same precaution when they fall, either of their own accord or from some sudden impulse; and this thread serves the purpose, being moved by the wind, to transport them with facility from one place to another. They can also re-ascend to the point from whence they had descended.

Some individuals of the species above mentioned, and which Degeer kept in a box, spun against the walls little nests in the form of oval or rounded sacs, composed of white silk, and pierced on both sides with an aperture. Lister says that this spider passes the winter in a thick web, which it has made for itself, and from which it does not issue forth until the middle of February. But it results from observations collected respecting other species, that they also form a cocoon with another view, namely, that of preserving their posterity, and sheltering themselves at those critical periods when they change skin.

Degeer found, at the end of July, on a branch of pine, a large oval cocoon, of white silk, placed around the branch, and intertwisted with its leaves. It was the dwelling of one of those leaping spiders, and of its young, which were living along with it, in a state of good intelligence, and appeared to subsist in common on the prey taken by it. On the middle of one of the sides of the cocoon was a cylindrical aperture, a sort of door, where the mother used to remain in ambush. The same observer found under stones, on the shores of the Baltic sea, many individuals of another species, resembling an ant, which M. Walckenaer has placed in a particular family. All the individuals were lodged separately in small oval cocoons of white silk, having an aperture at each end, and which they had spun against the under-side of the stones. If he touched their cocoons ever so slightly, they would issue forth from one of the apertures, and betake themselves to flight with great rapidity. When he wanted to take them, they easily escaped, by suffering themselves to descend on a thread of silk. They abandoned their nests without any difficulty, as they were able very speedily to construct new ones. Degeer has witnessed their changing skin. When they walk, they stop short at intervals, raise the two anterior feet in the air, agitate them up and down like antennæ, and feel the ground with them, as they would with true antennæ. They would then appear to have but six feet. The individuals of this species, which this naturalist kept in a sand-box, seemed to dread each other extremely. When they met, they first placed themselves in a posture of defence, face to face, covering the body, lowering the abdomen, contracting the feet, making a few steps on one side, and then forward, approaching nearer to each other. They opened their mandibles, and seemed desirous to fight; but the combat ended, either by the flight of one or both. Another species does not fear the approach of the human hand, but will present to it its large forceps.

Degeer has remarked, relatively to another species (*Grossipes*), that it can run equally sideways and backwards, as well as forwards; and that it often makes leaps in its walking.

M. Walckenaer, with equal exactitude and conciseness, has given us a definitive recapitulation of what is most general and certain in the history of these araneïdes, in the following terms:—"Araneïdes watching their prey, seizing it by running or leaping, enclosing themselves in a sac of fine and white silk, between leaves which they draw together, or in empty shells, receptacles of fruits, clefts, and cavities."

We shall conclude this supplement on the pulmonary arachnida with a few general observations on the genus SCORPIO.

The scorpions live exclusively in the warm climates of both hemispheres, and are so multiplied in certain districts, that they become a subject of continual terror to the inhabitants, to such an extent, according to some accounts, as to oblige them to abandon the soil. The zodiacal constellation of the scorpion proves that the knowledge of this animal is of the very highest degree of antiquity. Its effigy became the symbol of Typhon, the maleficent genius of the Egyptian mythology. On the antique engraved stones which present us with the traces of this mythology, Anubis is represented facing the scorpion, as if he intended to conjure and annihilate the influence of this evil principle. All the fables which superstition and ignorance have brought forth, during a series of ages, respecting this animal, are exhibited at length in the natural history of Pliny. The ancients, however, did observe that it coupled, was viviparous; that its sting was pierced, so as to give passage to the poison, and that this poison was white. They further remarked, that the females carried their young; but they supposed that there was but one to each mother; that this had escaped by stratagem from the general slaughter which she had made of her posterity, and that it finally revenged its brethren, by devouring the author of its life. According to others, the mother became the prey of her own family; but at all events the voracity of these animals was fully recognized. We cannot rank among the number of these fables the existence of scorpions, since there are specimens of such to be found. It is likewise possible that individuals may have been found whose tail was composed of seven knots instead of six, which is the common number. It is probable that the winged scorpions, which excited astonishment from their size, such as those which Megasthenes informs us were to be found in India, are orthoptera of the genus *Phasma* or spectrum, or hemiptera of that of *nepa* of Linnæus. Indeed, the name of *aquatic scorpion* has been given to an insect of the latter genus, extremely different from the arachnida so designated. Pliny informs us that the Psylli endeavoured to naturalize in Italy the scorpions of Africa, but that their attempts proved wholly unsuccessful. He distinguished, on the authority of Appollodorus, nine species. Nicander, who reckons one less, gives some particular details on the subject, but guided by views purely medical. It is from modern writers, and more especially from Dr. Maccary, that we must look to obtain more certain information respecting the habits of these curious animals. The scorpions live on the ground, concealing themselves under stones, most frequently in ruined buildings, in sombre and humid situations, and even sometimes in the interior of houses; they have even been found in beds. They run swiftly, curving their tail arch-like over the back. They are able to turn it in all directions, and employ it as an offensive and defensive weapon. They seize with their claws wood-lice and different insects, such as carabi, weevils, orthoptera, &c. wound them with the sting of their tail, pushing it forward, and then devour them, placing them between their mandibles and jaws. They are fond of the eggs of araneïdes and insects; they even attack araneïdes much larger than themselves, and appear to carry on a special war against them.

They vary considerably in size: those of Europe are scarcely more than an inch in length, whereas in India there are some five inches long. It is supposed that these are very venomous, and that the wound which they make with their sting very frequently causes death, from the introduction of the poisonous fluid.

It is an error, however, to believe that all these animals are venomous to us. It has been proved that those of Tuscany are not so, for the peasants of that country touch them, and allow themselves to be stung by them, without suffering any serious inconvenience. This observation, however, must not be too generally extended, as we find from the experiments of Redi and Maupertuis. These writers, who have made many experiments on the effect of the poison of another species of scorpion, larger than the common (*occitanus*), and which is found in Languedoc, at Tunis, in Spain, &c. have seen young pigeons die in convulsions, and vertigoes, five hours after they were stung, and others which evinced no sign of pain from the wounds which they had received. Redi attributes this difference to the exhaustion of the scorpion, which, in his opinion, has need of time to recruit its forces for the purpose of reproducing venom. Of this he had a proof in a fresh experiment which he made after having allowed the scorpion to repose for a single night.

In his experiments, Maupertuis caused many dogs and pullets to be stung by the scorpions of Languedoc; but of all these animals only a single dog died, which had received on that part of the belly without hair three or four wounds from the sting of a scorpion which had been previously much irritated. All the other dogs, and even the pullets, in spite of the fury and repeated blows of scorpions recently taken in the country, suffered nothing.

The author of this last experiment tells us, that in an hour after the dog which fell a victim was stung, it became very much swelled and staggering. It ejected all that it had in its stomach and intestines, and continued for three hours to vomit up from time to time a sort of viscous slaver. Its belly, which was very tense, diminished after each vomiting, and then swelled anew. These alternations of swellings and vomitings lasted about three hours, at the end of which the dog had convulsions, bit the ground, dragged himself along on his fore-paws, and died at last six hours after he was stung.

Dr. Maccary had the courage to try upon himself, and with the same species of scorpion, some experiments, which prove that its poison may produce very serious accidents, and that it is more active in proportion as the animal is older. He told M. Latreille that several of the French soldiers died in Spain from the sting of this scorpion. Accidental circumstances, as a morbid habit of body, for example, may augment the danger.

D'Opsonville, in his *Essais Philosoph. sur les Mœurs des divers Animaux Etrangers,* says, "the bite of marsh or field-snakes, such as those we see in Europe, is commonly as little dangerous in Asia. A slight scarification, and the application of a little quick lime, or of a piece of copper rusted with verdigris, which is fixed on the wound, will suffice to effect a cure. These two receipts are also employed against the sting of the scorpion called *agrab* in Persian, *gargonali* in Hindostanee, and *australis* by Linnæus, which in different parts of Asia is almost as common as the spider.

On this subject, therefore, we may refer to that part of our work which treats of the serpent tribe. Olivier informs us, that the sting of the scorpion, which he terms *crassicauda*, and which is very common in the Levant, is never dangerous to life, and that the effects of its poison are easily dissipated by analogous remedies.

According to the observations of M. Maccary, the scorpions couple very nearly in the manner of crabs. The female changes her skin before she brings forth the young, and the male does the like at the same epoch. The species indigenous to Europe produce two generations every year. M. Dufour has found in summer some females whose eggs were at their full growth, and others in autumn with but very small germs, the perfect development of which would not take place until the following spring. These facts, and those ascertained by Dr. Maccary, seem to establish that there are in fact two generations, one in this last season, and the other in summer. The female brings forth the young one by one. She carries them on her back during the first days, never issues then from her retreat, and watches over their preservation for the space of about a month, a period, at the end of which they become strong enough to establish themselves elsewhere, and provide for their subsistence. They are not in a state to reproduce their species until the end of two years.

It has been asserted that the scorpion, when enclosed in a circle of lighted coals, and when it finds that it is impossible to escape from the action of the heat, will sting itself to death. Maupertuis, after some experiments, has combated this opinion. Other observations, however, are in its favour; and M. Latreille informs us, that Count de Senneville made several experiments on this subject, and in the presence of a great number of persons, the result of which went to confirm this popular notion respecting the scorpion.

The scorpions, at least under certain circumstances, kill and devour their young. Maupertuis, having shut up about one hundred of them, found at the end of a few days, no more than fourteen alive. Another curious instance of this kind occurred about a dozen years ago to the Baron Cuvier. A collection of more than four hundred living scorpions, which he had received from Italy, was reduced, in the course of a very little time, to a few individuals.

The *scorpio occitanus* remains under stones, in the mountains of southern countries, exposed to a strong heat. M. Dufour has described this species at full length, and in a very exact manner, and a part of his description is common to all the species of the genus. It is extremely common in the kingdom of Valentia, and Lower Catalonia provinces, in which M. Dufour was unable to discover any individual of the European scorpion. These two species appear to exclude each other reciprocally from the same localities. Thus we should look in vain for the second, or the European scorpion, in the mountains, or arid hills of the environs of Narbonne, on those of schistose nature, or the deserts which form, from north to south, a maritime border of eight or ten leagues in breadth, between Barcelona and St. Philippe, as well as on the confines of Lower Catalonia and Arragon, all localities where the *occitanus* or *reddish* scorpion is found, and often in very great abundance. Its country in Spain is the same as that of the carob-tree (*ceratonia Siligua*, Linn). Thus, for example, a little beyond Barcelona, where we meet the first plantations of this tree, we also begin to find the first individuals of this species of scorpion. This concomitance is entirely referrible to the identity of the temperature and of the soil. The carob-tree, as well as this arachnid, can prosper only in dry soils, exposed to a tolerably strong heat, and situated at a little distance from the sea. M. Dufour presumes this scorpion never advances inland beyond the limits above indicated, and does not think it is to be met with at a height of more than 150 fathoms above the level of the sea, since the mountains of *Porta-Cœli*, situated six leagues to the west of Valentia, although within the Zone, or locale of this scorpion, of an elevation favourable to the propagation of subalpine plants, have not presented to him, in spite of the most careful research, any trace of this animal. The habitat of the scorpion of Europe, is, in like manner, subjected to the influence of the soil and the temperature of the climate.

This observer was unable to discover any individual of either of these species in the plain of Madrid, the two Castilles, Guipuscoa, the environs of Tudela, and those of Tafalla, in Lower Navarre, although he pursued his investigations during the fine season. But in France the European scorpion begins to make its appearance at a higher latitude, towards the forty-fourth degree, or under the zone which is proper for the culture of the almond and the pomegranate, and approaching within the northern limits of that of the olive-tree. M. Latreille presumes that if it does not inhabit the provinces of Spain, the reason is that the winters there are longer or more rigorous than in the part of France just mentioned. It is also to be observed that the habitation of the reddish scorpion (*occitanus*), is also determined by the nature of the insects on which it feeds, and which are proper only to certain localities. M. Dufour has never met with more than two of these under the same shelter. Most usually they are solitary, and dig in the soil a conchoïd cavity, where they squat down. When they quit their retreat to seek their food, which is usually in the evening, or during the night, they put forward their palpi, and keep the tail dragging along the ground. But when irritated, or menaced by any danger, they throw the palpi back, and curve the tail over the body, so that the sting protects the head, and becomes an essential weapon, which the animal directs on all sides, for the purposes of attack or defence. These scorpions fight together to the last extremity, and finish by the one devouring the other. Divers insects, either in the perfect or the larva-state, which they seize with their forceps, and mash completely, constitute their food. But they can support very long fasts, and M. Dufour has preserved some during six months, deprived of all kind of aliment, without their having appeared to have suffered any thing. Redi had previously made a similar observation. They moult several times, like the other arachnida. The females carry the young on their back; the male does not differ from the other sex, except in being a little smaller and less bulky in the abdomen.

The gestation of the scorpions is considerably longer than that of insects. From the commencement of autumn, all the adult females are fecundated. Their eggs are then lateral, small, and pedicled. They augment in volume during the winter, so that in spring their bulk is four times greater than it was in autumn. They are at this period entirely in the matrix. The gestation of the scorpion thus continues for nearly a year, which is very extraordinary, even when compared with that of red-blooded animals.

The poisonous fluid which the scorpion distils through the two pores of its sting is of a whitish colour, analogous to the serosity of milk, and when spread on white paper, this fluid produces a spot such as oil or grease would make, and the stained paper, in drying becomes more consistent and transparent.

THE SECOND ORDER OF ARACHNIDES.

THE TRACHEAN (TRACHEARIÆ),

DIFFER from the preceding, in the respiratory organs consisting of radiated or ramified tracheæ, and not receiving the air but through two apertures or stigmata; in the absence of a circulatory organ; and in the number of eyes, which is but from two to four. From the want of anatomical observations sufficiently general, the limits of this order are not yet rigorously traced. Even some of these arachnides, such as the pycnogonides, present no stigmata, and their mode of respiration is unknown.

The trachean arachnides are very naturally divided into those which are provided with forceps terminated by two claws, one of which is mobile, or by one alone, likewise mobile, in the form of a talon, or hook; and those in which these organs are replaced by simple laminæ, or lancets, and which, with the tongue, constitute a sucker. But most part of these animals being very small, this examination involves great difficulties, and we feel that such characters should not be employed but when we find it impossible to do without them.

The first family of TRACHEAN ARACHNIDES, that of

PSEUDO SCORPIONES,

Has the thorax articulated, with the anterior segment much

more spacious, in the form of a corslet. An abdomen very distinct and annulated; palpi very large, in the form of feet or claws; eight feet in both sexes, with two equal hooks at the end of the tarsi, the anterior two, at most, excepted; two antennæ-pincers or forceps apparent, terminated by two claws, and two jaws formed by the first articulation of the palpi. They are all terrestrial, and have the body oval or oblong. This family comprehends but two genera.

GALEODES, *Oliv.* SOLPUGA, *Licht.*, *Fab.*,

Have two very large forceps, with vertical claws, strongly denticulated, one upper, fixed, and often provided at its base with a slender elongated appendage, terminating in a point, and the others mobile; the palpi large, advanced in the form of feet, or antennæ, terminated by a short articulation, in the form of a button, vesicular, and without a hook at the end. The two anterior feet are of a figure almost similar, equally devoid of hooks, but smaller; the others are terminated by a tarsus, the last articulation of which is provided at the end with two small cushions, and two long claws, with a hook at their extremity. There are five scales in the form of a half-funnel, and pedicled, on each of the hinder feet, disposed in a range along their first articulations; and two eyes, very much approximated on an anterior eminence of the first thoracic segment, which represents a large head, bearing, besides the parts of the mouth, the two anterior feet.

Their body is oblong, generally soft, and provided with long hairs. The last articulation of the palpi, or their button, encloses, according to M. Dufour, an especial organ, in the form of a disk, of a mother-of-pearl white, and which never presents itself externally, but when the animal is irritated. The two anterior feet may be considered as second palpi. The labrum has the form of a very small and compressed beak, curved, pointed, and hairy at the end. The tongue is

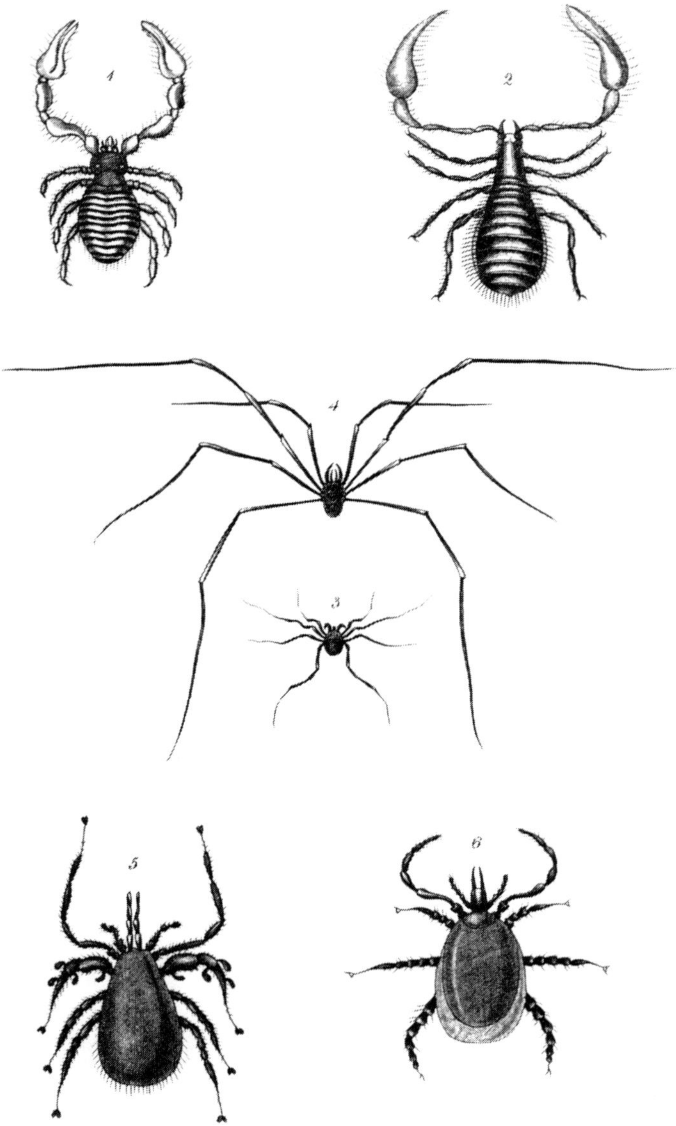

1 *Chelifer parasita.* 4 *Ph. annulatum.*

2 *Che. carcinoides.* 5 *Siro crassipes.*

3 *Phalanginum spinosulum.* 6 *Macrocheles marginatus.*

London, Published by Whittaker &C.o Ave Maria Lane. 1833.

small, in the form of a keel, and is terminated by two barbed threads, divergent, and placed each on a small articulation; the other pairs of feet are annexed to as many segments. I have perceived a large stigma on each side of the body, between the first and second feet, as well as a cleft at the base of the belly. The abdomen is ovaliform, and composed of nine rings.

We suspect that the ancients have designated these arachnides under the names of *Phalangium*, *Solifuga*, *tetragnatha*, &c. M. Poë has discovered a species in the environs of Havannah; but the others are proper to the hot and sandy countries of the old continent. These animals run with extreme swiftness, throw up their head, and seem to wish to defend themselves, when they are attacked. They are reputed venomous.

CHELIFER, *Geoff.* OBISIUM, *Ilig.*,

Have the palpi elongated, in the form of arms, with pincers like a hand, and didactylous at the end. All the feet are equal, terminated by two hooks, and the eyes placed on the sides of the thorax.

These animals resemble little scorpions deprived of a tail. Their body is flatted, with the thorax almost square, and having on each side one or two eyes.

They run fast, and often backwards or sideways, like the crabs. Rœsel has seen a female lay its eggs, and assemble them in a heap. Hermann, the father, says that these individuals carry them collected in a cushion under their belly. He even thinks, according to another observation, that these arachnides can spin.

His son (*Mém. apterol.*) divides this genus into two sections. Some (*Chelifer*, Leach) have the first segment of the trunk or thorax, divided into two by an impressed and transverse line; the tarsi with a single articulation; a species of stylet at the end of the mobile claw of the forceps, and the hairs of the body in the form of a spatula.

Phalangium cancroïdes, Lin.; *Scorpio cancroides,* Fab., Rœs., Ins., iii. supp. lxiv. vulgarly, *book scorpion* (*Scorpion des livres*) is found in herbals, old books, &c., where it lives on the little insects which destroy them.

Another, *Scorpio cimicoides,* Fab., Herm.; Mem., Apter. vii. 9. lives under the barks of trees, stones, &c.

Others (*Obisium,* Leach) have the thorax without division, the forceps without stylet, the hairs of the body in the form of threads; but the number of the eyes furnish us with a more important character. It is four in obisium, and two in *chelifer* proper.

The second family of the TRACHEAN ARACHNIDES, that of

PYCNOGONIDES,

Has the trunk composed of four segments, occupying almost all the length of the body, terminated at each extremity by a tubular articulation, the anterior of which, larger, sometimes simple, sometimes accompanied with forceps and palpi, or with a single species of these organs, constitutes the mouth. The two sexes have eight feet proper for running; but the females have, besides, two false feet, situated near the two anterior, and serving only to carry the eggs.

The *Pycnogonides* are marine animals, having some analogy with *cyami* and *caprellæ,* or with the arachnides of the genus *Phalangium,* to which Linnæus has united them. Their body is generally linear, with the feet very long, with eight or nine articulations, and terminated by two unequal hooks, appearing to form but one, and the smallest of which is cleft. The first articulation of the body, and which holds the place of head and mouth, forms an advanced tube, almost cylindrical, or in a truncated cone, having at its extremity a triangular aperture, or like a trefoil. It carries at its base the forceps and the palpi. The forceps are cylindrical or linear, simply prehensile, composed of two pieces, the last of which is pincer-like, with the lower claw, or that which is immoveable, sometimes shorter. The palpi are in the form of a thread, of five, or nine articulations, with a hook at the end. Each following segment, with the exception of the last, serves as an attachment to a pair of feet; but the first, or that with which the mouth is articulated, has on the back a tubercle, carrying on each side two simple eyes, and underneath, in the females only, two other small feet, folded back upon themselves, and carrying the eggs which are assembled all around them, in one or two pellets. The last segment is small, cylindrical, and pierced with a little hole at its extremity. No vestiges of stigmata are discoverable.

These animals are found among marine plants, sometimes under stones, near banks or shores, and sometimes also on cetaceous animals.

PYCNOGONUM, *Brun.*, *Müll.*, *Fab.*,

Are without forceps or palpi, and the length of their feet but little exceeds that of the body, which is proportionably shorter and thicker than in the following genera. They live on the cetacea.

PHOXICHILUS, *Latr.*,

Present no palpi, as in the preceding, but have very long feet, and two antennæ-pincers.

NYMPHON, *Fab.*,

Resemble the *Phoxichili*, in the very narrow and oblong form of their body, the length of their feet, and the presence of antennæ-pincers, or forceps; but they have, besides, two palpi.

The third family of the TRACHEAN ARACHNIDES, that of

HOLETRA, *Hermann*,

Has the thorax and abdomen united in a mass, under a common epidermis; the thorax is mostly divided into two, by a strangulation, and the abdomen presents only in some the appearances of rings, formed by some folds of the epidermis.

The anterior extremity of their body is often advanced in the form of a muzzle, or bill. The majority have eight feet, and the others six.

This family is composed of two tribes.

The first tribe, that of PHALANGITA, *Lat.*, has very apparent forceps, either projecting in front of the trunk, or inferior, and always terminated by a didactylous pincer, preceded by one or two articulations.

They have two palpi in the form of a thread, of five articulations, the last of which is terminated by a small claw; two distinct eyes; two jaws formed by the elongation of the radical articulation of the palpi, and often four additional ones, which are likewise nothing but a dilatation of the haunch of the first two pair of feet; the body oval, or rounded, covered, at least on the trunk, with a more solid skin; some appearances of rings, or folds, on the abdomen. The feet, always eight in number, are long, and distinctly divided after the manner of those of insects. Many at least (*phalangium*) have, at the origin of the two posterior feet, two stigmata, one at each side, but concealed by the haunches.

The majority live on the ground, on plants, at the bottom of trees, and are very agile; others conceal themselves under stones, and in moss. Their sexual organs are placed under the mouth, and internal.

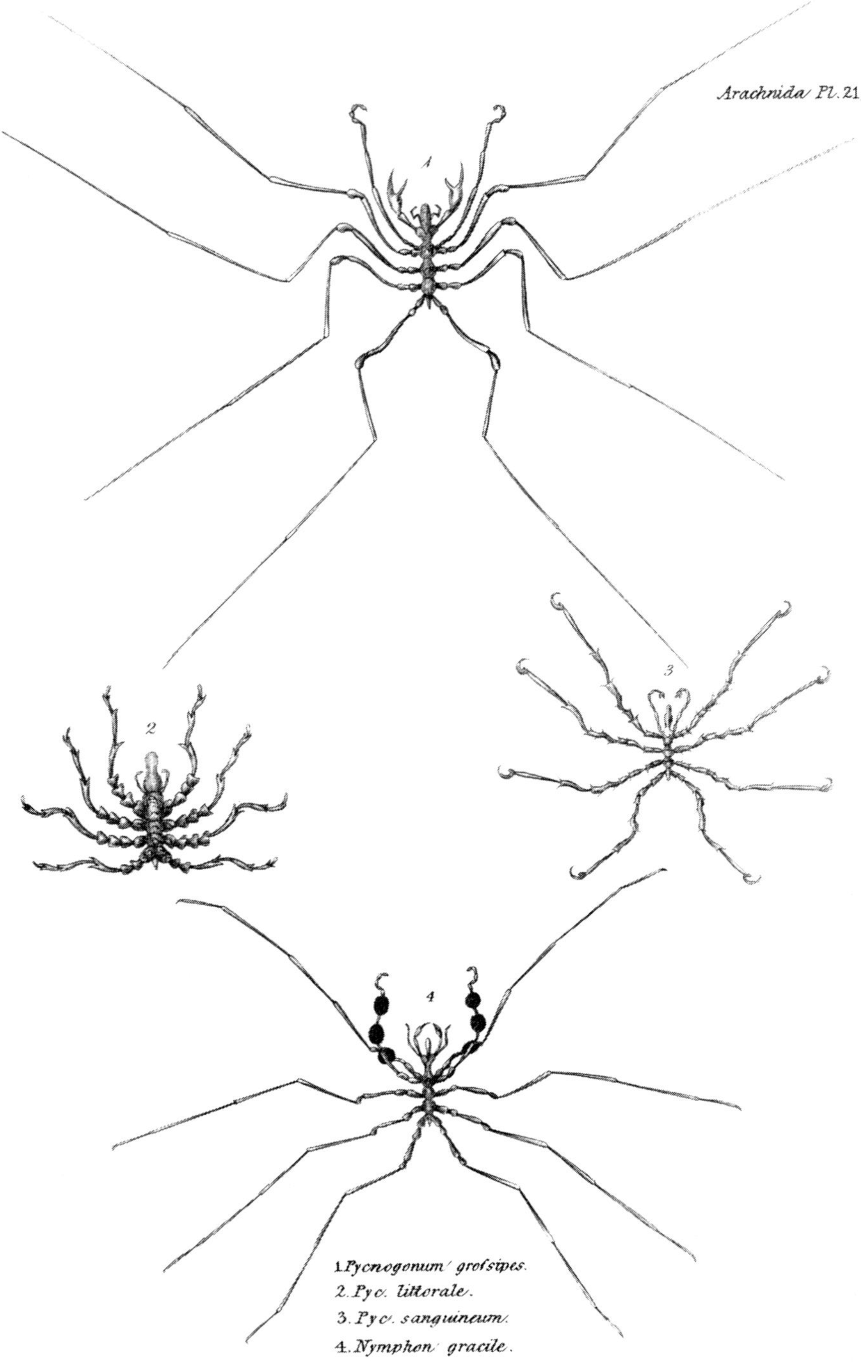

1. *Pycnogonum grossipes.*
2. *Pyc. littorale.*
3. *Pyc. sanguineum.*
4. *Nymphon gracile.*

London, Published by Whittaker & C°. Ave Maria Lane, 1833.

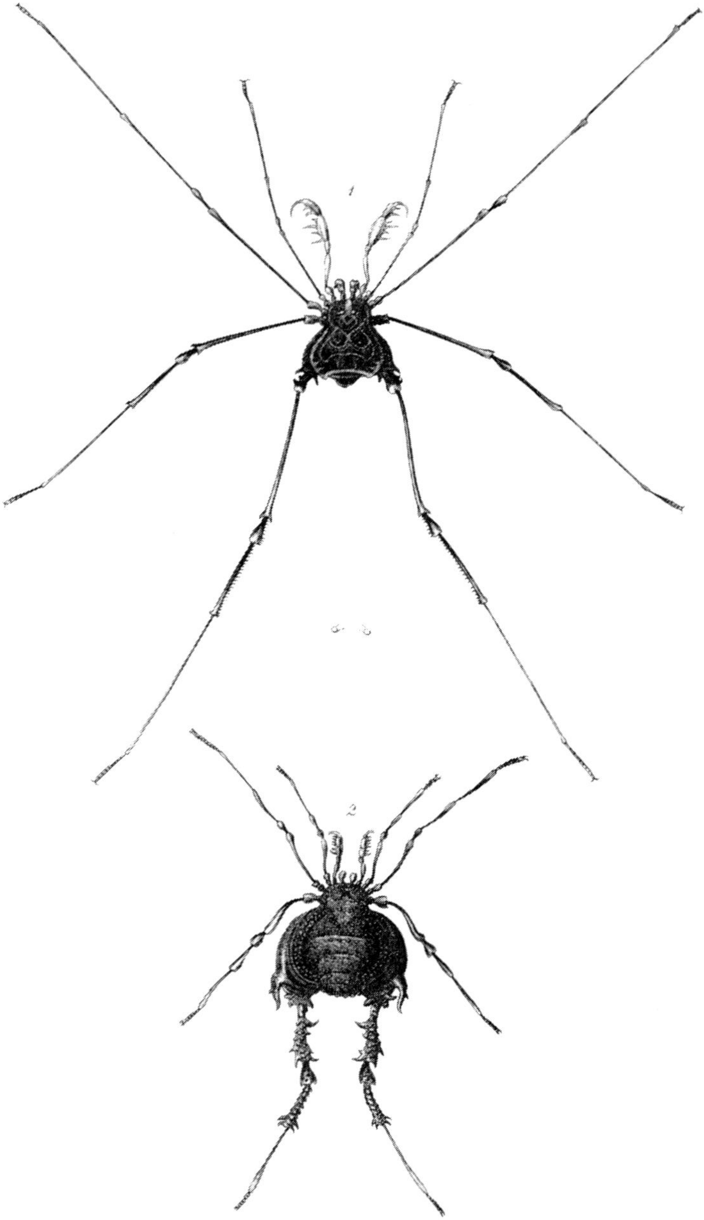

1. *Gonoleptes spinipes* _ G. R. Gray.

2. *Gonolephes Chilensis* _ G. R. Gray.

London, Published by Whittaker & C.º Ave Maria Lane, 1833

PHALANGIUM,

Which have the forceps projecting, much shorter than the body, and the eyes placed upon a common tubercle.

Their feet are very long, very slender, and when detached from the body, they exhibit, for a few instants, some signs of irritability. The generative organ of the male has the form of a dart, terminating in a semi-barb. The female has a membraneous oviduct, in the form of a thread, flexible, and annulated. The tracheæ are tubular.

Phalangium cornutum, Lin., male; *Opilio,* ejusd. female, Herbst., *Monog.* phal. i. 3. male, *ibid.* i. female. Body oval, reddish, or ash-colour above, white underneath, palpi long; two ranges of small spines on the tubercle which carries the eyes, and prickles on the thighs. Forceps horned in the male; a blackish band with its edges festooned in the female.

A celebrated English entomologist, Mr. Kirby, has formed, under the name of GONOLEPTES, a peculiar genus, or species, which have the palpi spiny, with the last two articulations almost of the same size, subovoid, and with a strong terminal claw, and the haunches of the two posterior feet very large, soldered together, and forming a plate under the body. These feet are remote from the others, and thrown backwards.

In *phalangium* proper, the palpi are filiform, without spines, terminated by an articulation, much longer than the preceding, and with a small hook at the end. All the feet are approximated, with haunches similar and contiguous at their origin. Such are all our indigenous species.

SIRO, *Latr.*,

With projecting forceps almost as long as the body; eyes apart, and each carried on an isolated tubercle, or without support.

MACROCHELES, *Lat.*,

Have also the forceps very projecting, and long; but the eyes are non-existent, or sessile. The two anterior feet are very long and antenniform. The upper part of the body forms a plate, or scale, without distinct rings.

I refer to this genus, the *Acarus marginatus*, and *testudinarius*, of Hermann the younger, (Mémoire, apterol. p. 76, pl. vi. fig. 6, and pag. 80, pl. ix. fig. 1.)

TROGULUS, *Latr.*

The anterior extremity of the body is advanced in the form of a hood, and receives, in a lower cavity, the forcep, and the other parts of the mouth.

Their body is very flatted, and covered with a very firm skin. Under stones.

The second tribe of HOLETRA, that of ACARIDES, sometimes has forceps, but simply composed of a single pincer, either didactylous, or talon-like, and concealed in a sternal labium; sometimes a sucker, formed of laminæ, like lancets, and united; or even but a mere cavity for a mouth, without any other apparent pieces.

This tribe is formed of the genus of the

MITES. ACARUS, *Lin.*

Most of these animals are very small, or almost microscopic. They are dispersed every where: some are erratic, and we find them under stones, leaves, the barks of trees, in the earth, the water, or in household provisions, such as flour, dried meat, old cheese, or animal substance in a state of putrefaction; others live, parasitically, on the skin, or in the flesh of various animals, and often greatly enfeeble them from their excessive multiplication. To some species the origin of certain maladies is even attributed, and especially

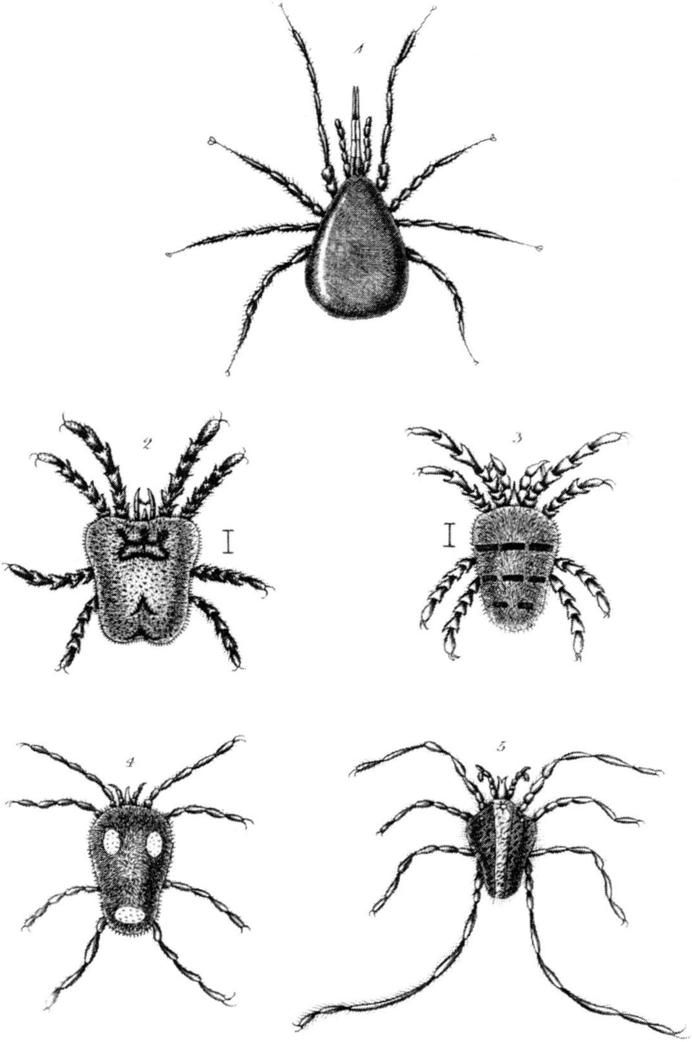

1 *Macrocheles testudinarius.*

2 *Trombidium holosericeum.*

3 *Tro. fuliginosum.*

4 *Tro. trimaculatum.*

5 *Erythræus phalangioides.*

London, Published by Whittaker & C° Ave Maria Lane, 1833.

that of the itch. It appears to result from the experiments of Dr. Galès, that the mites of the human itch, placed upon the body of a sound person, will inoculate him with the virus of this malady. We also find various sorts of mites on insects, and many coleoptera which live on cadaverous or excrementitious substances, are sometimes altogether covered by them. They have been observed even in the brain and the eyes of man.

The mites are oviparous, and multiply exceedingly. Many are born with only six feet, and the other two are developed a little time after. Their tarsi are terminated in various ways, and appropriate to their habits.

Some (ACARIDES proper) have eight feet, exclusively adapted for running, and forceps.

TROMBIDIUM, *Fab.*,

Which have the forceps talon-like, or terminated by a mobile hook; projecting palpi, pointed at the end, with a mobile appendage, or sort of claw, under their extremity; two eyes, situated each at the end of a small fixed pedicle, and the body divided into two parts, of which the first or anterior is very small, and carries, besides the eyes and mouth, the first two pair of feet.

T. holosericeum, Fab., Hermn., Mem. Apt. pl. i. 2. and ii. 1. Very common in spring in gardens; of a blood-red; the abdomen almost square, narrowed behind, with an emargination; the back charged with papillæ, hairy at their base, and globular at their extremity.

Another species is found in the East Indies, three or four times larger, and which yields a red tincture. (*T. tinctorium*, Fab.) Herm. Mem. Apt. I. i. 4.

ERYTHRÆUS, *Latr.*,

Which have the forceps and palpi of *Trombidium*, but the eyes are not carried on a pedicle, and the body is not divided.

GAMASUS, *Latr.*, *Fab.*

The forceps are didactylous, and the palpi projecting, or very distinct, and in the form of a thread.

Some have the upper part of the body clothed, altogether or in part, with a scaly skin; others have the body entirely soft. Some species of this division live on different birds and quadrupeds; some are known, such, especially, as the *acarus telarius* of Linnæus, which form on the leaves of several vegetables, particularly on those of the elm, very fine webs, and injure these plants very much. This species is reddish, with a blackish spot on each side of the abdomen.

CHEYLETUS, *Latr.*,

Which have also didactylous forceps, but the palpi are thick, in the form of an arm, and terminated like a scythe.

ORIBATA, *Lat.* NOTASPIS, *Hermn.*

The forceps are again didactylous, but the palpi are very short, or concealed. The body is covered with a firm coriaceous or scaly skin, in the form of a buckler or shield, and the feet are either long or of middle size.

The fore-part of the body is advanced like a muzzle. We often see an appearance of corslet. The end of the tarsus is terminated by a single hook in some, by two or three in the others, without vesicular cushion.

They are found on stones, trees, and moss, and walk slowly.

UROPODA, *Latr.*,

Which have, as analogy would lead us to presume, the forceps pincer-like. The palpi are not apparent or projecting. The body is entirely covered with a scaly skin, but the feet are very short, and there is a thread at the anus, by means of which they fix themselves on the body of some coleopterous insects, and suspend themselves in the air.

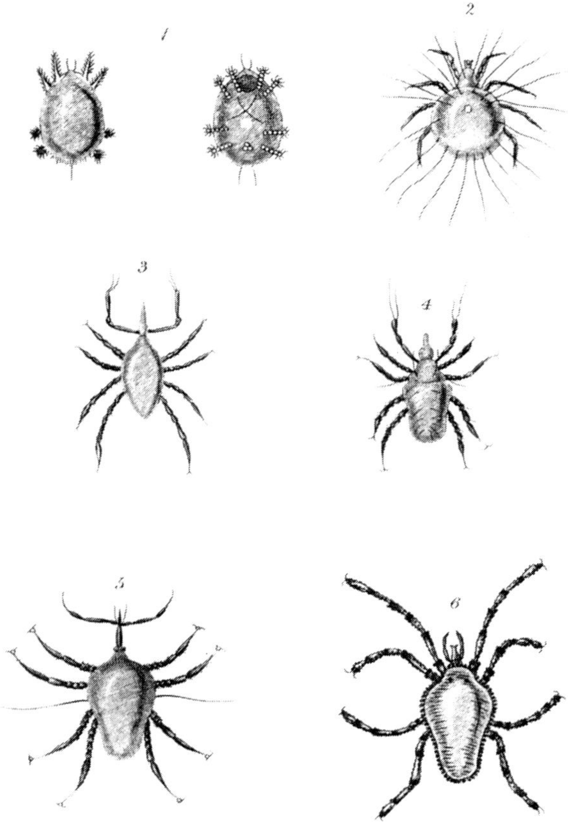

1 *Uropoda spinitarsis.*

2 *Acarus dimidiatus.*

3 *Bdella longirostris.*

4 *B. latirostris.*

5 *B. setirostris.*

6 *Smaridia papillosa*

London, Published by Whittaker & Co. Ave Maria Lane 1833.

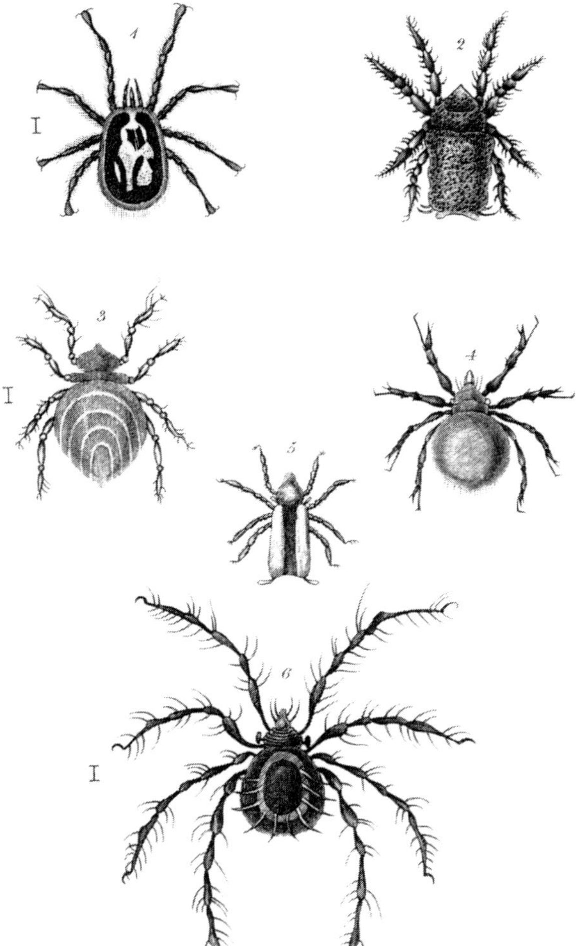

1 Gamasus hirundinis.

2 Oribata horrida.

3 O. thaleproctos.

4 O. castaneus.

5 O. segnis.

6 O. clavipes.

London, Published by Whittaker & C.º Ave Maria Lane, 1833.

ACARUS, *Fab.*, *Latr.* SARCOPTES, *Latr.*,

Having, as well as the preceding, two didactylous antennæ-pincers, palpi very short or concealed, but the body very soft, or without scaly crust.

The tarsi have at their extremity a vesicular pellet. Many species feed on our alimentary substances; others are found in the ulcers of the human itch, in those of the horse, dog, and cat.

Other *mites* (the TICS, *Riciniæ*, Lat.), have also eight feet exclusively proper for running, but are without antennæ-pincers or forceps properly so called. These organs are replaced by two laminæ, like lancets, forming, with the tongue, a sucker.

Sometimes they have distinct eyes, projecting palpi, filiform, and free; a sucker composed of membranaceous pieces, and without denticulations; and the body very soft. They are erratic.

BDELLA, *Latr.*, *Fab.* SCIRUS, *Hermn.*,

Which have elongated palpi, elbowed with threads or hairs at the end; four eyes; and the hinder feet the longest. Their sucker is advanced in the form of a conical bill, or like an awl. They are found under stones, the barks of trees, or in moss.

Acarus longicornis, Lin., *La Pince rouge*, Geoffr., *Scirus vulgaris*, Hermn. Mem. Apt. iii. 9; ix. Scarcely half a line in length, of a scarlet-red, with the feet paler; sucker in the form of an elongated and pointed bill; palpi, with four articulations, the first and last of which are the longest; the last is a little shorter than the first, and terminated by two threads. Common in the environs of Paris, under stones.

SMARIDIA

Are distinguished from Bdella by the palpi, which are scarcely longer than the sucker, straight, and without threads at the end—by their eyes, which are two in number, and by their two anterior feet being longer than the others.

Sometimes these mites with eight feet and without pincers have no perceptible eyes. Their palpi are either anterior and advanced, but in the form of valvules, widened or dilated towards the end, serving as a sheath to the sucker, or inferior. The pieces of the sucker are corneous, very hard, and denticulated. The body is clothed with a coriaceous skin, or has, at least in front, a scaly plate.

These tics are parasites, gorge themsclves with the blood of several vertebrated animals, and being at first very flatted, acquire by suction a very great volume and a vesicular form. They are round or oval.

IXODES, *Lat.*, *Fab.* CYNORŒSTHES, *Hermn.*,

Whose palpi sheath the sucker, and form with it an advanced, short, and truncated beak, a little dilated at the end.

The ixodes frequent thick woods, hook themselves to plants of no great elevation by the two anterior feet, and keep the others extended. They attach themselves to dogs, oxen, horses, and other quadrupeds, and even to tortoises, and engage their sucker so deeply in the flesh, that they cannot be detached from it but by force, and removing a portion of the flesh which adheres to the sucker. They lay a prodigious quantity of eggs, and through the mouth, according to M. Chabrier. Their multiplication on one ox or horse is sometimes so great, that these animals perish from exhaustion. Their tarsi are terminated by two hooks, inserted on a pallet, or united at their base by a common pedicle.

It would seem that the ancients designated these arachnides under the name of *ricinus*. The French huntsmen give the term *louvette* to the species which fixes on the dog, or the following,

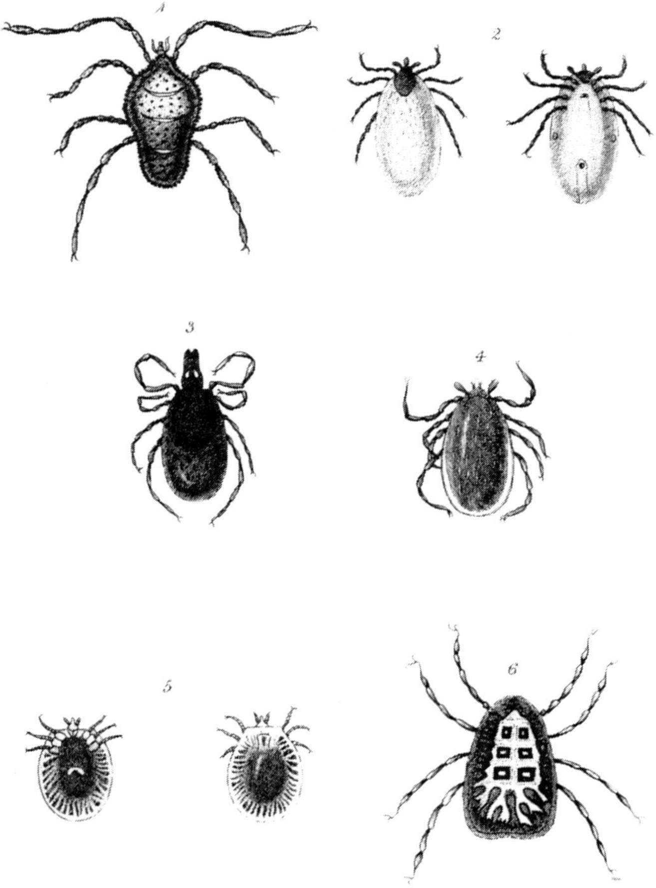

1 *Smaridia squamata.*

2 *Ixodes erianei.*

3 *Ix. trabeatus.*

4 *Ix. reduvius.*

5 *Argas pipistrella.*

6 *Argas columba.*

London, Published by Whittaker & C.o Ave Maria Lane, 1833.

Ixodes ricinus, Acarus Ricinus, Lin.; *Acarus reduvius,* D. G. Insect. VII. vi. 1, 2.: of a deep blood-red, with the anterior scaly plate deeper, sides of the body edged, a little hairy palpi sheathing the sucker.

Ixodes reticulatus, Latr., Fab.; *Acarus reduvius,* Schrank. Enum. Insect. Aust. No. 1043, iii. 1, 2; *Cynorhœstes pictus,* Hermn.: ash-coloured, with small spots, and small annular lines of a reddish brown; edges of the abdomen striated; palpi almost oval. It attaches itself to oxen, and is, when swelled up, five or six lines in length.

The study of the species of this genus has not been sufficiently pursued.

ARGAS, *Lat.* RHYNCOPRION, *Herm.*,

Differ from *Ixodes* by the inferior situation of the mouth, and by the palpi, which do not sheath the sucker; have a conical form, and are composed of four articulations, and not of three, as in the preceding genus.

Ixodes reflexus, Fab., Lat., Gen., Crust., et insect, &c. Of a pale yellowish, with lines of a deep blood-colour, or obscure and anastomosed. On pigeons, whose blood it sucks.

Another species, *Argas perricus,* described by travellers, under the name of the venomous bug of Miana, has been, as well as other ixodes, the object of a curious notice, by M. Fischer.

Other MITES, HYDRACHNELLÆ, have still eight feet, but ciliate, and adapted for swimming.

They form the genus HYDRACHNA of Muller, or *Athax* of Fabricius, and live exclusively in the water. Their body is generally oval, almost globular, and very soft. That of some males is narrowed posteriorly, in a cylindrical manner, or in the form of a tail. Their genital parts are placed at its extremity; the female has them under the belly. The number of eyes varies from two to four, and goes even as far as six, according to Muller.

The mouth of the species, which I have studied, presents the three following modifications, which have served as bases for three generic sections, but to which it is impossible to refer all the species of Muller's Hydrachnæ, this naturalist not having described them with sufficient detail.

EYLAIS, *Lat.*,

Which have antennæ-pincers, terminated by a mobile hook.

HYDRACHNA, *Lat.*,

Whose mouth is composed of laminæ, forming an advanced sucker, and whose palpi have under their extremity a mobile appendage.

LIMNOCHARES, *Lat.*,

Similar to hydrachna, in the sucker-formed mouth, but whose palpi are simple.

Other MITES, MICROPHTHIRA, *Lat.*, are remote from all the other aranchnides, by the number of feet, which is only six.

They are all parasite.

CARIS, *Lat.*,

Which have a sucker and apparent palpi; the body rounded very flat, and clothed with a scaly skin.

LEPTUS, *Lat.*,

Having also a sucker, and apparent palpi, but whose body is very soft and ovoid.

L. Autumnalis, Acarus Autumnalis, Shaw, Zool., Misc. t. II. pl. xlii., a species very common in autumn, on gramineous and other plants. It climbs and insinuates itself into the skin at the root of the hairs, and occasions itchings as insupportable as those produced by psora. In French it is called *Rouget*, and is in fact of a red colour, and is very small.

The other species are found on different insects, and enter into the division of *Trombidia hexapoda* of Hermann.

Aclysia, *Aud.*,

Whose body has the form of a bag-pipe, with a siphon, without distinct palpi, situated underneath its anterior extremity, which is narrowed, curved and obtuse; the feet are very small.

The aclysiæ live on the bodies of the dytisci. At first but one species was discovered, *A. dytisci,* that after which, M. Victor Audouin established this subgenus. But M. Le Count de Manheim, a Russian naturalist. who has already well deserved of science, by his Entomological Essays, and his zeal in seconding the efforts of those who devote their attention to it, has discovered, as it would seem, another species.

Atoma, *Latr.*,

Have neither sucker nor palpi visible. Their mouth consists only of a small aperture situated in the breast. Their body is oval, soft, and the feet are very short.

Ocypete, *Lat.*,

Of Dr. Leach; they belong to this tribe, from the number of their feet; but, according to him, possess mandibles.

SUPPLEMENT

ON THE

TRACHEAN ARACHNIDA.

THE genus GALEODES of the first family of this order was established by Olivier, in 1791, in the "*Encyclopedie Methodique*," under this denomination, as well as in almost all the works of the French naturalists. But Fabricius has preserved that of *Solpuga*, which was bestowed upon this genus a few years after by Lichtenstein. Respecting the general characters of these arachnida, it may be necessary to enlarge a little on the text. The body is oblong, and annulated; the anterior segment is much the larger, and supports two very strong projecting and compressed mandibles, terminating in a denticulated forceps, the lower branch of which is mobile. There are two simple eyes, which are dorsal, and approximated to each other on a common tubercle; and two large filiform palpi, without any terminal hook. The first feet are also filiform, imperfect, and in the form of palpi. The mouth is composed of two sciatic jaws, formed each by the union of the base of one of these palpi, and one of the anterior feet, and of a sternal subulate ligula, situated between the mandibles. There are six other filiform feet, terminated each by two species of long mobile digits, with a little hook at the end. The two posterior feet are larger, with a range of small pedicellated scales under the haunches.

The celebrated Pallas was the first who described with considerable detail a species of this genus, the *Galeodes araneoides*. In his *Spicilegia Zoologica*, and in Herbst.'s *Monograph* of the *Solpuga*, is a full development of the characters of these arachnids. Sonnini's Voyage into Greece also contains some valuable information on this subject, and more particularly some good figures, drawn by Marèchal, painter to the Museum of Natural History in Paris. Olivier has published a notice of those species which he observed in his travels in the Levant.

The Galeodes, as we have observed, have an oblong body, generally covered with a skin of a weak consistence, or slightly scaly, brown or yellowish, often bristling with long hairs, some of which, belonging to the mandibles, very distinctly appear to be tubular. The anterior part of the body presents two enormous mandibles, of a form pretty nearly conical, contiguous to each other all along their internal side, and terminating in a point. Each mandible is armed at its extremity with two talons, or scaly teeth, vertical, crossed one upon the other, denticulated internally, and finishing in a hooked point. In some species, or rather, perhaps, in the individuals of different sexes, a small scaly appendage is remarked, brown, and almost filiform, on the upper part of each mandible, and inclining against the posterior part. This appendage originates from the base of the interval between the hooks. Its use is unknown.

The palpi are very large in this genus: they exceed the feet in bulk, and are longer than the two or three anterior pair; they are advanced, filiform, of six articulations, the radical one of which is prolonged into a point at its internal and superior angle, and cemented with the corresponding articulation of the two following feet, to form a jaw. The second is very short, the three following very long, and the sixth very short, rounded at the end, appearing to be closed by a membrane, and without a hook. The two anterior feet, united with the palpi at their origin, and similarly annexed to the anterior segment of the body, which forms the major part of the thorax, resemble them in the form, the direction, and the manner in which they terminate; but they are shorter, and particularly much more slender. Their haunches present an additional articulation, and which appears to be formed by a division of the lower part of the third, or of the first of the elongated articulations, that which corresponds to the thigh. Analogy does not permit us to doubt that these organs represent the two anterior feet of the other arachnida; but from the nature of their functions we may conclude with M. Walckenaer, that the galeodes have four palpi, or rather six palpated feet; they approximate in this respect to Phryne and Thelyphone. The jaws are separated only by a linear cleft, and are even confounded together at their internal and superior angle. This angle is dilated in front, and forms, in the interval of the mandibles, at their origin, a small bifid ligula, and terminated by two silky appendages. It conceals a scaly piece, likewise ligulate, and having at the superior end a claw or tooth, arched towards its base, and in the form of a scythe. The two mandibles are very large, ovoid, compressed, applied one against the other by their internal faces, entirely uncovered, and advancing straight forward; they are terminated by two very strong teeth, scaly, vertical, crossed at their point, and denticulated at their internal side. One might suppose that these arachnida had four jaws. These mandibles, the first segment of the thorax, which is much more extended than the following, and its accessory parts, present, united, the appearance of an enormous head. This segment, or rather its dorsal plate, has the figure of a triangle, whose base is a little curved from the anterior edge of the corslet. We see near its middle, a small elevation, having on each side a small simple eye, placed a little obliquely. Another, but a very small plate, covers the second segment, that one which supports the second pair of feet. The rest of the body is soft, forms an oblong oval, and is composed of ten rings, which are simply folds of the skin; the third and fourth serve as an attachment for the four posterior feet.

These four feet, as well as the preceding two, are united in pairs, at their origin, by means of a common transverse articulation, divided in its middle by a depressed line, and corresponding to the maxillary articulation of the first two feet, and of the palpi. There is no sternum properly so called, the space which occupies its place being taken up by this series of radical articulations. These six feet, the lengths of which augment progressively, present, as usual, those parts which are distinguished by the names of haunch, thigh, leg, and tarsus. The haunch is composed of two very short articulations; the thigh and leg are long, and of a single piece; the tarsus, more slender, as well as the leg, is divided into three articulations, the first of which is very long, and the last very short; at the superior extremity of this last are implanted two filiform appendages, arched, hairy, similar to two fingers, and each terminated by a small scaly hook. All these divers articulations are cylindrical.

The feet are bristling with long hairs, and have here and there some small mobile spines. The haunches of the two last have, underneath, a range of small scales, very thin, transparent, and composed each of a pedicle, at the end of which is a mobile triangular piece, broad, folded in two, or forming an angle, and inclined in some species. These scales might be compared to one-half of a funnel, compressed, and cut in two, in the direction of its length. They represent, though but very remotely, the combs of the scorpions. The abdomen is oblong, soft, and more or less hairy, as well as the body, and without any appendage, at least any projecting one, at its extremity. It is composed of nine segments, of which that of the base, viewed underneath, represents a scaly plate. A stigma is perceptible on each side of the breast, near the second pair of feet.

The galeodes are peculiar to the warmer climates, especially to those of Southern Europe, of Asia, and Africa. MM. Dufour, and Dejean, have discovered one species in Spain; MM. de Humboldt and Bonpland, brought back another, but a very small one, from the equatorial countries of America. It appears from the travels of Pallas and Gmelin, that these arachnida are not rare in southern Russia, along the banks of the Volga and the Borysthenes, and that they are distinguished by the Calmucs, under the name of *bychorcho*.

They are greatly dreaded in all the countries where they are found. Not only is the history of these animals but little known, but the descriptions of their species are still considerably defective.

The *Monograph* of Herbst, which, like most of his other works on entomology, is little more than a mere compilation, furnishes us but with little matter in this point of view. Olivier, in his travels to Persia, has given us, on the species of the same genus which he has found, some useful notices, which we shall give in his own words. After having spoken of a prodigious number of locusts, by which himself and the caravan were infested, and which even came into the tents, he thus expresses himself:—

"In the evening, these small locusts were succeeded by another insect not less troublesome, and more disagreeable to behold. It belongs to that genus which I have established in the *Encyclopedie Methodique* under the name of *galeode*. The Arabs regard it as very venomous, and at first endeavoured to prevent us from touching it. When they saw, however, that we took sufficient precautions to avoid being bitten, they contented themselves by telling us an infinity of stories, one more appalling than the other. According to them, the bitten place swells very considerably, soon grows black, and is speedily followed by mortification and death. This opinion is equally established in Egypt and the south of Persia. M. Pallas relates several facts, of which he declares himself to have been a witness, which appear to prove that the poison of this insect is mortal, if a timely remedy be not applied. He considers oil and all unctuous substances as the fittest for this purpose. We confess, that in spite of the assertion of the Arabs, of the Egyptians, and of all the natives of those countries where the galeodes are found,—in spite of the assertion of M. Pallas himself,—we doubt that these insects are as venomous as is reported. Has not a similar reputation been given in Persia to the scorpion, in Italy to the tarantula, in almost all the east, and in the south of Europe, to the different species of geckos which live in houses or in ancient ruins? In Egypt and in Crete, are not the skinks equally considered as venomous?

"We have found the galeodes very common in Persia, in the desert of Mesopotamia, and in that of Arabia. Every evening it used to run over us, over our effects, our table, and our beds, with the utmost celerity, and without ever stopping. Nobody was bitten; and we have never been able to ascertain a well-authorized fact which could prove that this insect is as dangerous as they say. The bite of the galeodes must doubtless be very painful, if we may judge from the powerful forceps with which the mouth is armed; but is it perfectly certain that this bite is accompanied with an effusion of poison, as in the vipers? The inspection of the mouth of the animal does not appear to prove this. This insect conceals itself pretty generally during the day, and seldom comes forth except at night. It would seem that it is attracted by the light of a lamp or candle, for it was particularly in our tent, the only one which was lit, that the galeodes assembled. We saw less of them afterwards, because we had no longer any occasion for light.

"The species which ran with most celerity, and which was most commonly observed, seems referrible to that which Pallas observed to the north of the Caspian Sea, and which he has described under the name of *Phalangium araneoidès*. The feet are very long, and the whole body is hairy, of an ash-colour, a little reddish; the mandibles are entirely ciliated, and armed with strong teeth.

"We took a second species, *Galeodes Phalangium*, which less frequently presented itself, and which ran with considerably less rapidity. Its feet are not nearly half as long. The body is hairy, and of the same colour as that of the preceding; but the mandibles arė of a ferruginous red. They are less denticulated, and on the internal side of the upper piece may be remarked an arched, recurved, mobile hook, which is wanting in the *Galeodes araneoides*.

"We also saw, in the neighbourhood of our tent, two other galeodes, which differ but little one from the other, and which probably might be, perhaps, like the two preceding, not two species, but the two sexes of a single species. In the one, *Galeodes melanus*, the body is very black, the feet short and hairy, and there is an arched, recurved, mobile hook, at the internal part of the mandibles.

"The other, *Galeodes arabs*, which is evidently a female, has the feet very short, hairy, and the body of a velvety black. Its mandibles are denticulated, and without a lateral hook."

Pallas has made his description of the *Phalangium araneoides* from two individuals in the collection of Natural History of St. Petersburg, without informing us what country they belonged to. He considers the differences which he observed between these individuals, as merely sexual differences of the same species; but M. Latreille does not consider this opinion sufficiently authorized, and rather considers the individual taken by Pallas for a female, to be the *galeodes* of the Cape of Good Hope, represented by Petiver, in consequence of the hairs with which the breast, and the base of the feet are furnished. The species which Pallas gives as the male, M. Latreille would rather refer to the second, than to the first species of Olivier. Its palpi, which Pallas calls arms, and its feet, are much shorter than those of the *galeodes araneoides*, figured by Olivier. The mandibles have been represented in Olivier's figure in a position contrary to that which they naturally have, to render the forms of the forceps more sensible. This species is the *Solpuga arachnoides* of Herbst. Its body is nearly an inch and a half long. It is of a pale reddish yellow, with the extremity of the claws brown. It is bristling with hair, particularly on the palpi. The tubercle supporting the eye is blackish.

It is found in southern Russia and in the Levant.

Some passages of Pliny would lead us to conclude that the galeodes were known in his time. But the species which he might have observed, must have been the *araneoïdes* which is found in the Levant, and not a species of Bengal, as the citation made by Herbst would indicate.

The *galeodes setifera*, Oliv., is a little smaller than the preceding, of a brown red colour, and hairy. The abdomen has a lateral white stripe. The mandibles have each, at their upper part, an appendage in the form of a seta, and recurved. But it may be questioned whether this is a specific character. This species is from the Cape of Good Hope.

The *Galeodes dorsalis*, which MM. Dufour and Dejean found in Spain, is little more than half an inch long. Its body is whitish, and a little red underneath, of an ashen black above, with the forceps of the mandibles of a ferruginous colour, and the feet a very pale fulvous; the top of the palpi is more obscure. But in the southern part of Spain, a species has been found as large as the *araneoïdes.*

The *Solpuga fatalis* of Herbst, a native of Bengal, should, according to this author, be distinguished from its congeners by its forceps being horizontal. But the question is, whether this position be natural and constant, and not the result of some forced change.

The Greeks anciently employed in the composition of the medical substance called *theriaca,* a species of phalangium. It was the same *solpuga,* according to the opinion of the naturalist just quoted.

The genus CHELIFER is the *phalangium* of Linnæus, the *scorpio* of Fabricius, and the *obisium* of Illiger.

The most known species of this genus, the *Ch. cancroides,* or *scorpion-araignée* of Geoffroy, *faux scorpion d'Europe* of Degeer, was at first placed by Linnæus with *acarus.* He afterwards united it to *phalangium,* with which it has but a very trifling resemblance. Geoffroy, with reason, has made a genus proper of it, which he has named *chelifer,* in French *pince;* but he has placed in the same genus the *acarus longicornis,* an arachnid of quite another family. The *Ch. cancroides,* with Fabricius, is a species of *scorpion,* and these animals are in fact very nearly related. The cheliferi, however, differ from the scorpions by their body not being terminated by a tail, by having but two or four simple eyes, and by being destitute of those pectinated laminæ called *combs,* which are exhibited by these latter arachnida. The younger Hermann, in his excellent work on the apterous insects of Linnæus, has adopted the genus *chelifer* of Geoffroy, and has made known several species, which he has separated into two divisions, founded on new and acute observations. He has made of the *acarus longicornis* of Linnæus, and of some other analogous arachnida, a genus proper, *scirus,* but which had been previously established by M. Latreille.

Illiger, in a table, merely nomenclatory of the genera of the class *insecta*, which he has placed at the end of his work on the Coleoptera of Prussia, separates from the scorpions the species which Fabricius names *cancroïdes*, and *cimicoïdes*, to form a particular genus, which he calls *obisium*, and which corresponds exactly with the *chelifer* of M. Latreille in his Summary of the generic characters of insects, published anteriorly to the work of M. Illiger. Dr. Leach, adding some new observations to the preceding, has preserved the genus *obisium* of Illiger, but restrains it to the genera of cheliferi, which have four simple eyes, the body almost cylindrical, and the eight posterior feet composed of six articulations. The species in which the feet have but five articulations, whose body is depressed, and presents but two simple eyes alone, form the genus chelifer, of this gentleman. (*Zool. Miscell.* vol. iii. p. 48.) He places these two genera in the family of the *scorpionidea*.

Although it cannot be denied that these arachnida have, in their general structure, a great resemblance to the scorpions, they appear, nevertheless, to differ from them in some anatomical considerations. They present but two stigmata, situated, one on each side, above the origin of the two anterior feet. M. Latreille has, therefore, placed this genus, in his family of *Pseudo-scorpiones*, which immediately succeed that of the *scorpionides*. If it be true, that M. Fischer has observed in these animals, and the galeodes, respiratory organs, similar to those of the araneïdes, the family of false scorpions, composed of these two genera, should naturally terminate the order of pulmonary arachnida.

The cheliferi have the body ovoid and depressed, or oblong, and almost cylindrical, invested with a dermis somewhat coriaceous, almost smooth, or but slightly furnished with hairs. It is composed, 1st. of an anterior segment much the largest, almost square, or triangular, occupying the place of head or corslet, supporting two or four simple eyes, situated laterally, the organs of mastication, two palpated feet, large, in thê talon form, terminated by a didactylous forceps, and the first six feet; 2. of eleven other segments, but transverse, annuliform, and on the first of which, the fourth, and last pair of feet, appear to be inserted; the subsequent rings compose the abdomen. The mouth is formed, 1st. of two mandibles, situated at the anterior and superior extremity of the corslet, contiguous, advanced, and entirely uncovered in the *obisia*, corneous, of a single piece in the form of a forceps, ovoid, didactylous, and of which the external digit is mobile, denticulate, or ciliated; 2. of two large palpated feet, or claws, terminated also in a didactylous forceps, composed of six articulations, the mobile digit included; 3. of two jaws, formed by the internal prolongation of the radical articulation of the claws, valvular, a little gibbous, or convex in the middle, depressed, and with reflected edges, near the internal margins, joining along these margins, thus closing the mouth inferiorly, and terminated in a point; 4. of a sternal tongue, situated in the interior of the mouth, between the jaws and mandibles, cuspidated at its superior extremity, and presenting, at each side of this point, a small appendage, according to the observations of M. Savigny. The younger Hermann had already observed this piece; he says that it is a conical papilla, embraced by two sorts of valvules, (the jaws) and that it is doubtless the proboscis of these animals. The feet are divided into five (*chelifer*, properly so called) or six articulations, (the *obisia*) according as the tarsus is composed of one or two pieces. The extremity of the last articulation is always armed with hooked, elongated, and approximated teeth, under which is a small pellet. The articulation which corresponds to the thigh is in general broader, and elongated. The length of the feet goes on increasing, beginning from the second pair. They are shorter and thicker in *chelifer* (proper) than in *obisium*.

These arachnida are very small. The most common species, the *scorpion araignée* of Geoffroy, is found in humid places, under stones and garden flower-pots, in the unfrequented parts of houses, in dust, old books, and herbals. It lives on those little insects called *psocus pulsatorius* by Fabricius, on small acari, and it even attaches itself to flies. Goetze assures us that he has fed it with small aphides. Linnæus tells us that it sometimes introduces itself into the skin, and there produces a painful swelling of the size of a pea. He even relates, on the faith of Dr. Bergius, that a peasant, during the night, having had his thigh pierced by one of these arachnida, there formed a pustule, of the size of a nut, which occasioned most frightful torment. But these facts require authentication. When this arachnid is pursued, or when it meets in its way some object that it is desirous to avoid, it walks tolerably fast, both forwards, backwards, and sideways, like the scorpion and the crab. Rœsel has seen the female lay small eggs of a greenish white, and assemble them one beside the other. But he has not told us whether the young took a long time to come forth from these eggs.

The elder Hermann, according to the testimony of his son, has seen the same animal carrying its eggs, gathered up in a pellet under its belly, after the fashion of many araneïdes. He once found one of them enclosed in a silken follicle, covered with dust, and attached to a wall by one of its sides. It is M. Latreille's opinion that this cocoon was foreign to this animal, and that it merely made it its temporary domicile.

The PYCNOGONIDES appear to M. Savigny to form the passage from the crustacea to the arachnida. The PYCNOGONA differ from the other genera of the same family, not only by the absence of mandibles and palpi, but still further by the shorter proportions of the body and the feet. The feet appear to have an articulation less than in the other pycnogonides. The last but one appears to form, in the pycnogona, only a small inferior knot, and joining the last articulation of the tarsus with the preceding.

But a single species appears as yet to be known, the *P. balænarum*, figured by Brünniche, Müller, and some other naturalists. It is found under the stones on the shores of the European ocean; it is rare on the coasts of England and France.

The genus PHOXICHILUS M. Latreille at first thought to differ from the preceding, in having mandibles, and from the succeeding (*Nymphon*), by these organs being terminated by a single digit, and by the want of palpi as in the former genus. But having subsequently examined, with more attention, the mandibles of the species in which he had established this new generic section, he observed that they terminated in a didactylous forceps, in the same manner as those of the *nymphons*, and that the inferior digit, covered by the ordure which at first had prevented him from perceiving it, was only smaller than the superior one, or that which is mobile. But the Phoxichili are, nevertheless, strongly distinguished from the nymphons, by the absence of palpi. They are also removed from them in other points of view—1. The tube, or siphon, forming the sucker, and the aperture of which represents a trefoil, as is the case with that of the same part in other animals of this family, is bellied in the middle, a little narrowed afterwards, and is terminated by a rounded expansion furnished with hairs. Each of its sides, and the middle of its back, presents two impressed longitudinal lines, and which, uniting by their extremities, describe a sort of ellipsis. In considering these lines as sutures of united pieces, this tube would be composed of six valvules, or laminæ, cemented together, and which would represent the principal parts of the mouth of the insects. The siphon of the *nymphons* exhibits no division at its surface. 2. In this last genus, the first articulation of the body is narrowed a little in front of the insertion of the two anterior feet, and of the two oviferous feet; then it is prolonged and dilated insensibly, in the manner of a neck, or obconical pedicle, the anterior extremity of which serves as a support to the mandibles. This support is formed by the union of the radical articulation of each of these mandibles. These two articulations, intimately united, and dilated inferiorly, embrace the anterior extremity of the pedicle in the manner of a collar. The tubercle supporting the four eyes is situated at the posterior and dorsal extremity of the first segment behind the pedicle; but in the phoxichili, the first segment of the body is short and transverse, so that the two anterior feet, and the oviferous feet are inserted near the origin of the siphon. At the posterior and dorsal extremity of this part, are set back to back, and united longitudinally, the two radical articulations of the mandibles. The tubercle supporting the eyes, is situated between this common articulation and the following; the mandibles, exclusive of the support we have just mentioned, seem to take their origin from the anterior base of the oculiferous tubercle. 3. We see, on each side of the base of the siphon, in front of the insertion of the oviferous feet, a small articulation in the form of a rounded tubercle. 4. The feet, less elongated proportionably than those of the nymphons, have an articulation less than the latter, that is to say, eight, the lateral and projecting prolongations of the segments of the body included, instead of nine, the one which in the nymphons forms the last but one, or the eighth, being here almost insensible. The oviferous feet of the phoxichili resemble in their form and proportions, those of the nymphons. They are composed of ten articulations, as well as in this last genus; the mandibles are filiform, and terminate a little beyond the end of the siphon; but the upper digit is curved towards the internal side of these organs. The species which furnished these details is found in the seas of Australasia. *(Phalangioïdes)*. Its body is entirely of an obscure brown, and five lines in length. The feet are about three times as long, a little hairy, with the first two articulations, as well as the fifth and sixth, terminated by some salient angles, in the form of conical tubercles; these are visible at the superior extremity of the fifth.

The genus NYMPHON was at first confounded with that of *phalangium*, and subsequently with that of *pycnogonum*. Fabricius placed it among the diptera; but in one of his later works, in which he has more especially treated of the insects of that order (*Systema antliatorum*) he neither mentions this genus, nor that of pycnogonum. He neither says any thing about it in his System of Rhyngotes, which silence may be attributed either to negligence, or to an intention of this naturalist, of forming a particular order of these animals.

Olivier placed the nymphons in this third section of the order aptera, taking for antennæ the parts which are now considered as palpi. He also considered that the two feet which exclusively carry the eggs, were not less genuine feet than the others, and thus extended the total number of those organs of locomotion to ten; and then resting on some other approximations founded on habits, he was induced to believe, that these animals have more affinity with the crustacea than with the arachnida. M. Savigny appears to have adopted the same opinion, or at least to think that the nymphons form the passage from the cyami, a genus of crustacea, to the arachnida. It is evident, he says, that the nymphon has lost the antennæ, the composite eyes, and the masticatory organs of the cyamus; but it appears equally certain that it has preserved the fourteen feet. When we consider, he adds, the changes which take place externally, in the genera which conduct from the crabs to phalangium, we might believe, that nature, in subtracting from the crustacea their anterior organs, and replacing their tail by an abdomen, thus converted them into arachnida. But, admitting this hypothesis, it would still be necessary to pass from the nymphons to the araneïdes, or to the arachnida called pedipalpi, and it would not be very easy to explain the mode adopted by nature in operating this new transformation.

As the interior organization of the pycnogonides is unknown, it is not possible to determine the place which these singular animals should occupy in the natural series of created beings. Nevertheless, as they appear to have, in spite of some anomalies, great relations with *chelifer* and *phalangium*, affinities which had already been remarked by celebrated naturalists; —as the body of many trachean arachnida also presents an anterior articulation, supporting analogous mandibles and palpi;—as the tubular sucker of the pycnogonides may be nothing but an union of jaws and under lip prolonged and cemented;—as the absence of composite eyes, and the existence of a tubercle supporting the simple eyes confirm these relations;—as the feet of the pycnogonides are composed of nine articulations, a character for which we should seek in vain in the crustacea, but which we shall find in several of these arachnida;—as in giving to the pycnogonides feet as long, and a linear form adapted to their habits, Nature has been obliged to extend their thorax to the prejudice of their abdomen, which is here represented by a small articulation in the form of a tail;—M. Latreille was originally determined to place these animals between the pseudo-scorpions, and the phalangia. In his work called " General Considerations on the Natural Order of Crustacea, Arachnida, and Insects," the pycnogonides alone form an order, which unites the parasite insects, such as the pediculi and ricini, to the acerated, or arachnida.

The nymphons are distinguished from the pycnogona by

having mandibles and palpi; and from the phoxichili, because their mandibles are forceps-like or didactylous.

The nymphons are, of all the pycnogonides, those whose body and feet are most slender and most long. They also differ in the form of the first articulation of the body, which may be considered as the head: it is proportionally longer, and more narrowed in the middle. The sucker is cylindrical, which is likewise the case with the phoxichili, but not with the pycnogona, when this part has the form of an elongated cone, and truncated at the point. The two neighbouring feet in the females have two intermediate articulations, much longer than the others, and curved. The external and general organization of the nymphons, being otherwise similar to that of the other arachnida of the same family, it would be superfluous to dilate upon it here. Of their history nothing particular is known.

In the third family of Trachean Arachnida, that of HOLETRA, we have now to treat of the genus PHALANGIUM.

The phalangia are very remarkable for the length of their feet. The first naturalists who wrote upon these insects called them *long-legged spiders;* but they differ from the spiders, not only in their internal organization, but also with respect to the general form of the body, the number of the eyes, the parts of the mouth, and the mode of living. They are to be seen every where. In the country they may be found upon plants, and also are observed in houses, upon plastered walls, to which they are fond of hooking themselves.

Their body is ovoid or rounded, often depressed, and enclosed beneath a skin slightly coriaceous. Their corslet, the anterior of which is angular, and which is about one-third and a half of the length of the body, is separated from the abdomen only by a transverse line. This abdomen is covered with a skin of a single piece, forming several folds, which mark the rings. It has a stigma on each side, near the origin of the posterior feet; these stigmata are concealed by the haunches.

The feet, eight in number, are very long, very slender, cylindrical, composed of the haunch, of the thigh, of the leg formed of two articulations, and of the tarsus, whose length at least equals that of the leg and thigh taken together, and which is composed of a great number of articulations, the first of which is very long, and the last provided with a small hook, which appears simple, and arched.

The naturalists who have treated of the phalangia, with the exception of Lister and the younger Hermann, whose observations, however, were not published until after those of M. Latreille, were not acquainted with the sexual organs of these insects. These organs have a singular form, especially those of the males, and in the two sexes their position is curious: the male organ is a sort of elongated dart, composed of two pieces, the first of which, forming the base, is short, thick, and of a soft consistence; it serves as a case to the second, which is a little longer, more narrow, almost scaly, terminated in *Ph. cornutum* by a triangular membranaceous piece, hooked at the internal side, with a small setaceous point, black, and arched, which proceeds from the superior angle of this piece. In a state of inaction, this part is concealed in a sheath, situated immediately under the mouth. The sexual part of the female is placed similarly to that of the male. We discover there a membranaceous tube, compressed, and very flexible, which serves as an oviduct. By pressing a small eminence called *lip*, which is found between the last two pairs of feet, at the base of the abdomen, these parts may be made to protrude in both sexes.

These arachnida do not spin, as some authors have pretended. Many species have a strong odour of the walnut-tree, and all are carnivorous. They feed on little insects, which they seize with their mandibles: they pierce them with the crooks with which they are armed, and suck them. They also engage together in mortal combat, and constantly devour their consimilars.

The long legs with which nature has provided them not only serve the purpose of enabling them to walk with very great facility, but also to escape from the pursuit of their enemies, and to advertise them of their presence. In a state of repose, placed upon a wall, or on the trunk of a tree, the phalangium extends its feet circularly around its body. As they occupy a considerable space, if an animal touches at one of its parts, the phalangium places itself immediately on its feet, which form so many arcades, under which the animal passes, if it be small. Should this stratagem not succeed, the arachnid leaps to the ground, and soon escapes. It frequently escapes in the same manner from the hands of the observer, but seldom without leaving between the fingers which has seized it one or more of its feet, which will continue to move for hours together, folding and unfolding alternately. This phenomenon takes place because each foot is a hollow tube, which contains in the whole length of its cavity a sort of very fine tendinous thread, on which the air acts, when the foot is detached from the trunk. Geoffroy, having found a phalangium with the third foot much shorter than the others, presumes that this foot had replaced one that was lost, as it happens to crabs and lobsters in a similar predicament. But the shortness of life in the phalangium militates against this conjecture.

In spring only small phalangia are found, which proceed from the eggs deposited the preceding autumn. It is hardly until towards the end of summer that they have acquired their full growth, and then they couple. Sometimes this does not occur without a combat on the part of the males, and some resistance on that of the female. After this, the female deposits in the earth, at a certain distance from its surface, some eggs, of the size of a grain of sand, of a whitish colour, heaped one upon the other.

Although these animals are akin to the spiders, nevertheless they do not live, like them, for many years. Almost all of them perish at the end of autumn. One of their enemies, which fixes on their bodies to suck them, is a species of mite of the subgenus *Leptus*. This insect sometimes holds to the phalangium only by its bill; the rest of its body appears suspended in the air. A *gordius*, similar to that which is often found in the interior of locusts, being found in the abdomen of the *Phalangium cornutum*, would lead us to believe that these arachnida are subject to be infested by these worms. That which was observed was very smooth, a little transparent, and filled with a milky matter. It was about seven inches four lines in length, and two-tenths of a line in breadth.

We are now come to the ACARIDES, or last family of the Arachnida, and our general observations upon them will be comprized in what we have to say respecting the genus ACARUS, or the MITES.

The name of *Acarus* in the method of Linnæus designates a genus of apterous insects very numerous in species. The *acarus* (proper) of M. Latreille comprehends the species of this tribe which have eight feet: simply ambulatory mandibles like forceps, palpi very short or concealed, and the body very soft, with the tarsi usually terminated by a vesicular pellet.

The *Acarus domesticus*, or *common mite*, is of all the species the best known. It is found in great abundance upon old cheese, on dry or smoked meat, on birds and insects, in collections of natural history, on old bread, and dried up confectionary, which have been kept too long. It is for this reason that Degeer has named this species *domestic.* He also observed some of these mites in the flower-pots which he had in his chamber. This insect is almost invisible to the naked eye; its colour is a dirty white, bordering a little on the brown, with two brown spots produced by the internal parts, which appear through the skin, which is transparent. The body is bristling with hairs, thick, oval, a little narrowed in the middle; its anterior part is terminated in a cone, or a sort of muzzle, containing the organs of manducation; the mandibles have been distinguished; the palpi are very short and setaceous; the skin is smooth and tense; the eight feet are rather long, always curved towards the plane of position, terminated by an oval piece, transparent, and swelled like a small bladder with a long neck, having in front a sort of small cleft or separation. The insect can impart to it all kinds of inflexions, swell and contract it. It dilates it when walking, and contracts it, so as to make it disappear, when the foot does not touch the plane of position, and is raised. The vesicle can be folded in two in its length, by reason of the cleft which we have just mentioned. Each moiety is furnished with a small hook, which enables the mite to fix itself on the object upon which it walks. The feet are of equal length, but the two anterior pair are much thicker than the two last.

The females are larger than the males, and have at the hinder part a small cylindrical tube, perhaps an oviduct, and a small eminence underneath.

The numerous hairs with which the body is bristled are barbed on both sides, and what is singular is, that the insect can move them on one side and the other. Each hair, says Degeer, must necessarily be attached to, or have communication with, a muscle, which gives it motion. What marvellous mechanism in so small an object! These sorts of prickles are placed upon the body in regular order: two are observed on the upper part of its anterior extremity, which represent, as it were, two small antennæ. There are some on the feet which are finer, and on which Degeer has observed no barbs.

The sexual intercourse of these mites takes place in the same manner as that of other insects. The female lays some oval eggs, very white, and which appear to be reticulated or spotted with brown.

Leuwenhoëk, who has particularly observed this species, saw but six feet upon the little ones just disclosed; and the same has been remarked in relation to the *mite of the itch.* Propagation goes on even in winter, at least in our houses, the temperature which reigns there being favourable to the activity of these little animals.

Acarus farinæ, Deg. (the mite in flour), is elongated, white, with the anterior part reddish, and advanced in the form of a thick and conical muzzle. The eight feet are thick, and tolerably long, especially those of the first two pair. The body also has hairs, and those of the hinder part are very long. Degeer could not perceive at the extremity of the feet the vesicle which we have mentioned belonging to the feet of the preceding species. These animals walk tolerably fast.

Linnæus at first distinguished separately the *mite of the itch*, but he subsequently confounded it with the two preceding species.

"The mites," says Degeer, "which I had occasion to extract from the ulcers of the itch, were very small, and not larger than grains of ordinary sand. The colour of the body is white, and transparent; but the head and feet have a slight tinge of red or yellowish brown. The body is of a rounded or almost circular figure, and its surface is rugged, having inequalities, and here and there some hairs, but in small quantities. The head is in the form of a muzzle, short, cylindrical, rounded at the end, and furnished with some hairs; but the smallness of the insect prevented me from detecting its parts, and ascertaining their true construction. Not having been able to remark upon the back two curved brown lines of which Linnæus speaks, I have reason to believe that the mite of which

I present the description here, is of a species different from the one observed by that naturalist. It appears to me more conformable to the species which the same author designates by the name of *Acarus exulcerans*, judging merely by the phrase which he applies to it, as he has given no other description."

Its eight feet are rather short, especially the first two pairs. These last are thick, conical, have some hairs, of which a few are tolerably long. These feet are terminated by a slender, straight, cylindrical part, having at the end a little ball, in the form of a vesicle, which the animal rests on the level on which it walks, and guides in various directions. The four posterior feet are equally terminated by a slender and brown part; but Degeer could not perceive the vesicle of the preceding. These posterior feet have a very long hair, and are placed at a certain distance from the first two pairs. When removed from the epidermis, this little animal remains at first inactive, but it moves its feet by degrees, and commences to walk, though but slowly.

Whether this mite be not that of the common human itch, or whether Degeer has made his observations on individuals of different forms, or did not study the subject with sufficient attention, the figure which he gives of it does not accord with that of the itch-mite published by Dr. Galès in a dissertation of great research on this subject, and whose observations have been confirmed by several celebrated naturalists. He had the courage to inoculate himself with the itch by means of this acarus. Other researches on this subject, made on divers animals, have proved that the mites, taken from their wounds when they were afflicted with this malady, differed from that of the human itch.

M. Latreille saw a quadruped of New Holland, (the *phascolomys*) brought alive to Paris, where it died a few days after, as it would seem, of the itch. The surface of its spoil was covered, from the effect of the preparation it underwent to preserve it, with an innumerable quantity of mites, almost invisible, being scarcely the twentieth part of a line in their greatest diameter. Examined with the microscope, some of these mites, the smallest, appear to have great relations with the itch-mite, and others, with that which infects sparrows, *Acarus passerinus*, Lin.

The persons who prepared the animal, had their arms very speedily covered with small irritating pustules, occasioned by the introduction of the mite into the skin.

The study of these little animals is of the greatest interest, not only to the naturalist, but also to the physician. It appears by the best observations that the ulcers of the itch, both in man, horse, dog, and cat, almost always exhibit mites; and that these animals, impregnated with the morbific virus, can communicate it. But this circumstance excepted, it remains an undecided question, whether or no they are the primary causes of this malady. That they may establish themselves, and propagate in sores favourable to their development; that they may aggravate the malady in proportion to their multiplication; and that they may spread over other parts of the body, we may very naturally and easily conceive; but to draw any further conclusions, appears to be somewhat precipitate. Through an injudicious rage for generalization, the origin of the dysentery has been attributed to a species of the same genus; and Olivier was even of opinion that a similar cause might have given rise to the plague.

ALPHABETICAL LIST

OF

SPECIES OF ANNELIDA, CRUSTACEA, AND ARACHNIDA.

ANNELIDA.

CRUSTACEA.

ARACHNIDA.

THE END.

GILBERT & RIVINGTON, PRINTERS, ST. JOHN'S SQUARE, LONDON.